UCLA Symposia on Molecular and Cellular Biology, New Series

Series Editor, C. Fred Fox

UCLA Symposia Board

C. Fred Fox, Ph.D., Director
Professor of Microbiology, University of California, Los Angeles

Molecular Biology
of Plant Growth Control

Molecular Biology of Plant Growth Control

Proceedings of the ARCO Plant Cell Research Institute–UCLA
Symposium Held in Lake Tahoe, California, February 22–28, 1986

Editors

J. Eugene Fox
ARCO Plant Cell Research Institute
Dublin, California

Mark Jacobs
Department of Biology
Swarthmore College
Swarthmore, Pennsylvania

Alan R. Liss, Inc. • New York

Address all Inquiries to the Publisher
Alan R. Liss, Inc., 41 East 11th Street, New York, NY 10003

Copyright © 1987 Alan R. Liss, Inc.

Printed in the United States of America

Library of Congress Cataloging-in-Publication Data

Molecular biology of plant growth control.

(UCLA symposia on molecular and cellular biology ; new ser., v. 44)

Proceedings of the ARCO Plant Cell Research Institute-UCLA Symposium on Molecular Biology of Plant Growth Control, organized through the Molecular Biology Institute of UCLA.

Includes index.

1. Growth (Plants)—Congresses. 2. Plant hormones—Congresses. 3. Plant molecular biology—Congresses. 4. Plant regulators—Congresses. 5. Plants, Effect of light on—Congresses. I. Fox, J. Eugene, 1934– II. Jacobs, Mark, 1950– . III. ARCO Plant Cell Research Institute. IV. ARCO Plant Cell Research Institute–UCLA Symposium on Molecular Biology of Plant Growth Control (1986 : Tahoe City, Calif.) V. University of California, Los Angeles. Molecular Biology Institute. VI. Series.

QK731.M633   1987   581.3'1   86-34274
ISBN 0-8451-2643-1

Contents

viii   Contents

Contributors

Klaus Apel, Botanisches Institut der Christian-Albrechts-Universität, 2300 Kiel, Federal Republic of Germany **[413]**

Z. Ariffin, Division of Plant Industry, CSIRO, Canberra, ACT 2601, Australia **[23]**

D.A. Armitage, School of Life Sciences, Leicester Polytechnic, Scraptoft Campus, Leicester LE7 9SU, United Kingdom **[245]**

Frederick Ausubel, Department of Molecular Biology, Massachusetts General Hospital, Boston, MA 02114 **[145]**

Helen M. Bailey, School of Life Sciences, Leicester Polytechnic, Scraptoft Campus, Leicester LE7 9SU, United Kingdom **[245]**

Jean C. Baker, Department of Biochemistry, University of Georgia, Athens, GA 30602 **[51]**

R.D.J. Barker, School of Life Sciences, Leicester Polytechnic, Scraptoft Campus, Leicester LE7 9SU, United Kingdom **[245]**

Richard F. Barker, Agrigenetics Advanced Research Laboratory, Madison, WI 53716; present address: Plant Breeding Institute, Cambridge CB2 2LQ, United Kingdom **[425]**

Alfred Batschauer, Botanisches Institut der Christian-Albrechts-Universität, 2300 Kiel, Federal Republic of Germany **[413]**

Roger N. Beachy, Department of Biology, Washington University, St. Louis, MO 63130 **[97]**

David L. Brandon, USDA Western Regional Research Center, Albany, CA 94710 **[209]**

Elizabeth A. Bray, Department of Biology, Washington University, St. Louis, MO 63130; present address: Department of Botany and Plant Sciences, University of California, Riverside, CA 92521 **[97]**

Winslow R. Briggs, Department of Plant Biology, Carnegie Institution of Washington, Stanford, CA 94305 **[413]**

Mark R. Brodl, Department of Biology, Washington University, St. Louis, MO 63130 **[35]**

Peter H. Brown, Department of Biology, Washington University, St. Louis, MO 63130 **[35]**

Ann M. Callahan, Department of Plant and Soil Sciences, West Virginia University, Kearneysville, WV 25430 **[157]**

The numbers in brackets are the opening page numbers of the contributors' articles.

Peter M. Chandler, Division of Plant Industry, CSIRO, Canberra, ACT 2601, Australia **[23]**

Joanne Chory, Department of Molecular Biology, Massachusetts General Hospital, Boston, MA 02114 **[145]**

James T. Colbert, Department of Botany, University of Wisconsin, Madison, WI 53706; present address: Department of Botany, Colorado State University, Fort Collins, CO 80523 **[425]**

Joseph W. Corse, USDA Western Regional Research Center, Albany, CA 94710 **[209]**

Martha L. Crouch, Department of Biology, Indiana University, Bloomington, IN 47405 **[73]**

Susan M. Daniels, Department of Botany, University of Wisconsin, Madison, WI 53706; present address: Department of Plant Pathology, University of Wisconsin, Madison, WI 53706 **[425]**

Ronald W. Davis, Department of Biochemistry, Stanford University School of Medicine, Stanford, CA 94305 **[133]**

Alice J. DeLisle, Department of Biology, Indiana University, Bloomington, IN 47405 **[73]**

Hans Depta, Pflanzenphysiologisches Institut, Universität Göttingen, Cytologische Abteilung, D-3400 Göttingen, Federal Republic of Germany **[267]**

Leon Dure III, Department of Biochemistry, University of Georgia, Athens, GA 30602 **[51]**

Joseph R. Ecker, Department of Biochemistry, Stanford University School of Medicine, Stanford, CA 94305 **[133]**

Malcolm C. Elliott, School of Life Sciences, Leicester Polytechnic, Scraptoft Campus, Leicester LE7 9SU, United Kingdom **[245]**

Michael L. Evans, Department of Botany, Ohio State University, Columbus, OH 43210 **[361]**

Donna E. Fernandez, Department of Molecular, Cellular and Developmental Biology, University of Colorado, Boulder, CO 80309-0347 **[323]**

Ruth R. Finkelstein, Department of Biology, Indiana University, Bloomington, IN 47405; present address: MSU-DOE Plant Research Laboratory, Michigan State University, East Lansing, MI 48824 **[73]**

Susan Flores, Department of Biology, University of California, Los Angeles, CA 90024 **[123,401]**

Anders Follin, Laboratorium voor Genetica, Rijksuniversiteit Gent, B-9000 Gent, Belgium **[181]**

Glenn A. Galau, Department of Botany, University of Georgia, Athens, GA 30602 **[197]**

Wilhelm Gruissem, Department of Botany, University of California, Berkeley, CA 94720 **[167]**

Tom Guilfoyle, Department of Botany, University of Minnesota, St. Paul, MN 55108 **[85]**

Gretchen Hagen, Department of Botany, University of Minnesota, St. Paul, MN 55108 **[85]**

Karin Hajek, Institut für Genetik, Universität Bonn, D 5300 Bonn 1, Federal Republic of Germany **[257]**

J.F. Hall, School of Life Sciences, Leicester Polytechnic, Scraptoft Campus, Leicester LE7 9SU, United Kingdom [245]

Michael A. Hall, Department of Botany and Microbiology, University College of Wales, Aberystwyth, Dyfed SY23 3DA, Wales, United Kingdom [335]

Karl-Heinz Hasenstein, Department of Botany, Ohio State University, Columbus, OH 43210 [361]

Howard P. Hershey, Department of Botany, University of Wisconsin, Madison, WI 53706; present address: Department of Central Research and Development, E.I. du Pont de Nemours and Co., Wilmington, DE 19898 [425]

Thomas Hesse, Botanical Institute, University of Bonn, D-5300 Bonn 1, Federal Republic of Germany [279]

Tuan-hua David Ho, Department of Biology, Washington University, St. Louis, MO 63130 [35]

Catherine J. Howarth, Department of Botany and Microbiology, University College of Wales, Aberystwyth, Dyfed SY23 3DA, Wales, United Kingdom [335]

Jim Hu, Department of Cellular and Developmental Biology, Harvard University, Cambridge, MA 02138 [145]

D. Wayne Hughes, Department of Botany, University of Georgia, Athens, GA 30602 [197]

L. Huiet, Division of Plant Industry, CSIRO, Canberra, ACT 2601, Australia [23]

Kenneth B. Idler, Agrigenetics Advanced Research Laboratory, Madison, WI 53716 [425]

Dirk Inzé, Laboratorium voor Genetica, Rijksuniversiteit Gent, B-9000 Gent, Belgium [181]

Mark Jacobs, Department of Biology, Swarthmore College, Swarthmore, PA 19081 [177]

Hans-Jörg Jacobsen, Institut für Genetik, Universität Bonn, D 5300 Bonn 1, Federal Republic of Germany [257]

J.V. Jacobsen, Division of Plant Industry, CSIRO, Canberra, ACT 2601, Australia [23]

Alan M. Jones, Department of Botany, University of Wisconsin, Madison, WI 53706 [425]

George A. Karlin-Neumann, Department of Biology, University of California, Los Angeles, CA 90024 [401]

Brian Keith, Plant Growth Laboratory/ Department of Vegetable Crops, University of California at Davis, Davis, CA 95616 [289, 299]

Joe L. Key, Botany Department, University of Georgia, Athens, GA 30602 [1]

Dieter Klämbt, Botanical Institute, University of Bonn, D-5300 Bonn 1, Federal Republic of Germany [279]

Tsune Kosuge, Department of Plant Pathology, University of California, Davis, CA 95616 [371]

Coralie C. Lashbrook, Plant Growth Laboratory/Department of Vegetable Crops, University of California at Davis, Davis, CA 95616 [299]

George G. Laties, Department of Biology, University of California at Los Angeles, Los Angeles, CA 90024 [107]

Jörg Lehmann, Institute of Plant Biochemistry, Academy of Sciences of the German Democratic Republic, 4020 Halle (S.), German Democratic Republic **[391]**

Silva Lerbs, Institute of Plant Biochemistry, Academy of Sciences of the German Democratic Republic, 4020 Halle (S.), German Democratic Republic **[391]**

Werner Lerbs, Institute of Plant Biochemistry, Academy of Sciences of the German Democratic Republic, 4020 Halle (S.), German Democratic Republic **[391]**

Leslie S. Leutwiler, Department of Biology, University of California, Los Angeles, CA 90024 **[401]**

Jane A. Lewis, School of Life Sciences, Leicester Polytechnic, Scraptoft Campus, Leicester LE7 9SU, United Kingdom; present address: Biotechnology Department, Glaxo Group Research Ltd., Greenford, Middlesex UB6 OHE, United Kingdom **[245]**

K.R. Libbenga, Department of Plant Molecular Biology, University of Leiden, NL-2311 VJ Leiden, The Netherlands **[229,245]**

Liang-shiou Lin, Department of Biology, Washington University, St. Louis, MO 63130 **[35]**

James L. Lissemore, Department of Botany, University of Wisconsin, Madison, WI 53706 **[425]**

Zin-Huang Liu, Department of Biological Sciences, Simon Fraser University, Burnaby, British Columbia, Canada V5A 1S6 **[315]**

Marian Löbler, Botanical Institute, University of Bonn, D-5300 Bonn 1, Federal Republic of Germany **[279]**

Jane Longland, Department of Botany, Edinburgh University, Edinburgh EH9 3JH, Scotland, United Kingdom **[345]**

Arieh Maoz, ARO Volcani Center, Bet Dagan 50250, Israel **[209]**

James J. McFadden, Department of Botany, Ohio State University, Columbus, OH 43210 **[361]**

A.M. Mennes, Department of Plant Molecular Biology, University of Leiden, NL-2311 VJ Leiden, The Netherlands **[229,245]**

Peter H. Morgens, Department of Plant and Soil Sciences, West Virginia University, Kearneysville, WV 25430 **[157]**

Egon Mösinger, Biologisches Institut II der Albert-Ludwigs-Universität, D-7800 Freiburg i. Br., Federal Republic of Germany **[413]**

Randall C. Nolan, Department of Biology, Washington University, St. Louis, MO 63130 **[35]**

Paul W. Oeller, Department of Biological Chemistry, Washington University School of Medicine, St. Louis, MO 63110 **[63]**

Neil Olszewski, Department of Molecular Biology, Massachusetts General Hospital, Boston, MA 02114 **[145]**

Angela M. O'Sullivan, School of Life Sciences, Leicester Polytechnic, Scraptoft Campus, Leicester LE7 9SU, United Kingdom **[245]**

Benno Parthier, Institute of Plant Biochemistry, Academy of Sciences of the German Democratic Republic, 4020 Halle (S.), German Democratic Republic **[391]**

Birgit Piechulla, Department of Botany, University of California, Berkeley, CA 94720 **[167]**

Jana B. Pyle, Department of Plant and Soil Sciences, West Virginia University, Kearneysville, WV 25430 **[157]**

Peter H. Quail, Department of Botany and Genetics, University of Wisconsin, Madison, WI 53706 **[425]**

Natasha V. Raikhel, Department of Botany, University of Georgia, Athens, GA 30602 **[197]**

Lawrence Rappaport, Plant Growth Laboratory/Department of Vegetable Crops, University of California at Davis, Davis, CA 95616 **[289, 299]**

Francisco Roberto, Department of Plant Pathology, University of California, Davis, CA 95616 **[371]**

Duncan Robertson, Department of Botany and Microbiology, University College of Wales, Aberystwyth, Dyfed SY23 3DA, Wales, United Kingdom **[335]**

Gillian M. Robinson, School of Life Sciences, Leicester Polytechnic, Scraptoft Campus, Leicester, LE7 9SU, United Kingdom; present address: Wyeth Laboratories, Taplow, Berks, United Kingdom **[245]**

Patrick Rüdelsheim, Departement voor Biologie, Universitaire Instelling Antwerpen, B-2610 Wilrijk, Belgium **[181]**

Ian O. Sanders, Department of Botany and Microbiology, University College of Wales, Aberystwyth, Dyfed SY23 3DA, Wales, United Kingdom **[335]**

Eberhard Schäfer, Biologisches Institut II der Albert-Ludwigs-Universität, D-7800 Freiburg i. Br. Federal Republic of Germany **[413]**

Jeff Schell, Laboratorium voor Genetica, Rijksuniversiteit Gent, B-9000 Gent, Belgium, and Max-Planck-Institut für Züchtungsforschung, D-5000 Köln 30, Federal Republic of Germany **[181]**

Elias A. Shahin, ARCO Plant Cell Research Institute, Dublin, CA 94568 **[381]**

Jane Silverthorne, Department of Biology, University of California, Los Angeles, CA 90024 **[401]**

Anne E. Simon, Department of Biology, Indiana University, Bloomington, IN 47405; present address: Department of Biology, University of California at San Diego, La Jolla, CA 92093 **[73]**

Karoline Simon, Botanical Institute, University of Bonn, D-5300 Bonn 1, Federal Republic of Germany **[279]**

Robert B. Simpson, ARCO Plant Cell Research Institute, Dublin, CA 94568 **[381]**

Cheryl Smart, Department of Botany, Edinburgh University, Edinburgh EH9 3JH, Scotland, United Kingdom; present address: Department of Botany, Cambridge University, Cambridge CB2 3EA, United Kingdom **[345]**

Aileen R. Smith, Department of Botany and Microbiology, University College of Wales, Aberystwyth, Dyfed SY23 3DA, Wales, United Kingdom **[335]**

Peter G. Smith, Department of Botany and Microbiology, University College of Wales, Aberystwyth, Dyfed SY23 3DA, Wales, United Kingdom **[335]**

Lalit M. Srivastava, Department of Biological Sciences, Simon Fraser University, Burnaby, British Columbia, Canada V5A 1S6 **[309, 315]**

L. Andrew Staehelin, Department of Molecular, Cellular and Developmental Biology, University of Colorado, Boulder, CO 80309-0347 **[323]**

Robert J. Starling, Department of Botany and Microbiology, University College of Wales, Aberystwyth, Dyfed SY23 3DA, Wales, United Kingdom **[335]**

Charles L. Stinemetz, Department of Botany, Ohio State University, Columbus, OH 43210 **[361]**

Kitisri Sukhapinda, ARCO Plant Cell Research Institute, Dublin, CA 94568 **[381]**

Zhao-Da Tang, Department of Biology, Nankai University, Tianjin, People's Republic of China **[335]**

Athanasios Theologis, Department of Biological Chemistry, Washington University School of Medicine, St. Louis, MO 63110 **[63]**

Christopher J.R. Thomas, Department of Botany and Microbiology, University College of Wales, Aberystwyth, Dyfed SY23 3DA, Wales, United Kingdom **[335]**

Elaine M. Tobin, Department of Biology, University of California, Los Angeles, CA 90024 **[123,401]**

Anthony Trewavas, Department of Botany, Edinburgh University, Edinburgh EH9 3JH, Scotland, United Kingdom **[345]**

Anna J. Trulson, ARCO Plant Cell Research Institute, Dublin, CA 94568 **[381]**

P.C.G. van der Linde, Department of Plant Molecular Biology, University of Leiden, 2311 VJ-Leiden, The Netherlands; present address: Bulb Research Centre, 2160 AB-Lisse, The Netherlands **[229]**

E.J. van der Zaal, Department of Plant Molecular Biology, University of Leiden, 2311 VJ-Leiden, The Netherlands **[229]**

Marc Van Montagu, Laboratorium voor Genetica, Rijksuniversiteit Gent, B-9000 Gent, Belgium **[181]**

Harry Van Onckelen, Departement voor Biologie, Universitaire Instelling Antwerpen, B-2610 Wilrijk, Belgium **[181]**

H.J. van Telgen, Department of Plant Molecular Biology, University of Leiden, 2311 VJ-Leiden, The Netherlands **[229]**

J.E. Varner, Department of Biology, Washington University, St. Louis, MO 63130 **[441]**

Michael A. Venis, East Malling Research Station, Maidstone, Kent ME19 6BJ, United Kingdom **[219]**

Daniel Voytas, Department of Molecular Biology, Massachusetts General Hospital, Boston, MA 02114 **[145]**

Reinhard A. Weidhase, Institute of Plant Biochemistry, Academy of Sciences of the German Democratic Republic, 4020 Halle (S.), German Democratic Republic **[391]**

Richard A.N. Williams, Department of Botany and Microbiology, University College of Wales, Aberystwyth, Dyfed SY23 3DA, Wales, United Kingdom **[335]**

Reinhold Wollgiehn, Institute of Plant Biochemistry, Academy of Sciences of the German Democratic Republic, 4020 Halle (S.), German Democratic Republic **[391]**

Nasser Yalpani, Department of Biological Sciences, Simon Fraser University, Burnaby, British Columbia, Canada V5A 1S6 **[309]**

J.A. Zwar, Division of Plant Industry, CSIRO, Canberra, ACT 2601, Australia **[23]**

Preface

The ARCO Plant Cell Research Institute-UCLA Symposium on Molecular Biology of Plant Growth Control, organized through the Molecular Biology Institute of UCLA, was held at Lake Tahoe, California, February 22 to 28, 1986. The year preceding the meeting was characterized by interesting and exciting new research in plant hormone-regulated gene expression, second messengers operating in plant hormone action, plant pathogen regulation of plant hormone synthesis and metabolism, and molecular aspects of light's regulation of plant growth. The field of plant growth control also saw a rapid increase in the use of techniques employing mutants and monoclonal antibodies. These topics and a thorough and enlightening discussion of plant hormone receptors and binding moieties by many of the world's most active research groups provided the substance for the conference and for these proceedings.

The success of a meeting is largely dependent on those who speak. We sincerely thank all those who accepted our invitations, both those included in this volume and those who are not. We included in this volume contributions from speakers in plenary sessions, those who took part in workshops discussions, and participants who gave poster presentations. We also are thankful to the session chairmen for their help and counsel. For this we particularly thank J. Key, R. Quatrano, G. Laties, M. Venis, M. Elliott, S. Roux, T. Kosuge, and W. Briggs. We also are grateful to J. Varner for his keynote address.

We thank ARCO Plant Cell Research Institute for its generous sponsorship of this meeting. We also gratefully acknowledge gifts from Agrigenetics Corporation, and Campbell Institute for Research and Technology, Campbell Soup Company. Additional support was provided by gifts from contributors to the Director's Fund, Biosciences Laboratory, Corporate Technology, Allied-Signal Corporation; AMOCO Corporation; Allelix, Inc.; AMGen, Inc.; Boehringer Ingelheim Pharmaceuticals, Inc.; Hoeschst-Roussel Pharmaceuticals, Inc.; New England Biolabs Foundation; Ayerst Laboratories Research, Inc.; Bristol-Myers Company; Celanese Research Company; Wyeth Laboratories and Codon Corporation.

We wish to extend our gratitude to Bill Parson, Manager of Granlibakken Resort, and the Chef, Ron for dealing so well with the elements. Finally, we thank W. Coty, B. Handy, R. Yeaton and R. Harwood and the entire Symposia staff for their hard work in putting the conference and this volume together.

J. Eugene Fox
Mark Jacobs

Molecular Biology of Plant Growth Control, pages 1-21
© 1987 Alan R. Liss, Inc.

AUXIN-REGULATED GENE EXPRESSION: A HISTORICAL
PERSPECTIVE AND CURRENT STATUS

Joe L. Key

Botany Department, University of Georgia
Athens, GA 30602

INTRODUCTION

The work of Skoog and colleagues in the early 1950s
demonstrated that the concentration and ratio of auxins to
cytokinin in the growth medium markedly altered not only the
growth pattern but also the relative levels of DNA and RNA in
callus tissue. Skoog (1954) suggested that the mechanism of
action of plant hormones might involve changes in nucleic
acid metabolism. A number of studies over the next 15 years
showed that auxin treatment, including herbicidal levels of
the synthetic auxin 2,4-D (see Hanson and Slife, 1969),
markedly affected the RNA content of seedlings and tissues
excised from them and incubated with auxin in shake culture
(see reviews by Trewavas, 1968; and Key, 1969).

In discussing auxin-regulated developmental processes,
auxins should be recognized as a class of hormones required
both for cell division and cell enlargement. Thus, auxin-
regulated gene expression in mature hypocotyl that has the
potential to undergo cell division in response to auxin and
excised elongating hypocotyl tissue that requires auxin to
maintain rates of elongation that are comparable to that of
intact tissue will be discussed. I will review primarily
features of auxin-regulated gene expression from studies done
in my laboratory using the soybean system including some
historical perspective since other manuscripts in this volume
(especially those of Guilfoyle and Hagen and of Theologis)
will cover related current work. Details of much of the
early work are covered in reviews by Trewavas (1968) and Key
(1969) and the more recent work in a review by Guilfoyle
(1986).

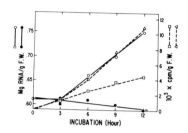

FIGURE 1. Time Course of Auxin-Induced RNA Synthesis in Excised Mature Soybean Hypocotyl. Control, ●——●, □---□; Auxin, o——o, Δ---Δ, (From Key and Ingle, 1968).

The Influence of Auxin on RNA Metabolism in Mature Tissue of Soybean Hypocotyl

When intact 3 to 4 day old etiolated soybean seedlings are sprayed with high concentrations of auxin, the typical response is the inhibition of cell division and cell elongation in the apical regions of the hypocotyl and the induction of cell division in the mature region. Cell division is initiated at the base of the hypocotyl within 12 hr after auxin treatment and extends up the hypocotyl to within about 0.5 cm of the cotyledons after about 24 hr (Key et al. 1966). Although auxin treatment causes radial enlargement of cells, due presumably to enhanced ethylene production, the responses in altered RNA metabolism discussed below are a consequence of auxin and are not ethylene responses.

Auxin treatment rapidly and dramatically increases synthesis of RNA in the mature zone of excised or intact hypocotyls. In excised tissue, there is a significant increase in radioactive nucleotide incorporation into RNA (Fig. 1). This auxin-enhanced RNA synthesis occurs with a lag of 1 to 2 hr, is linear for several hr, and precedes DNA synthesis by 8 to 12 hr (Key et al., 1966). The enhancement of RNA synthesis by auxin is concentration-dependent, reaching an optimum at about 5 x 10⁻⁵ M then decreasing as the concentration increases (Key and Shannon, 1964). Although synthesis of all classes of RNA is increased, rRNA, as a component of ribosomes, is the major accumulating class of RNA. A much higher proportion of the ribosomes are present as polyribosomes in the auxin-treated tissue relative to untreated control samples (Travis et al., 1973). The auxin-induced accumulation of rRNA is at least partly a consequence of increased synthesis which can be demonstrated directly in nuclei, nucleoli, or chromatin isolated from auxin-treated tissue (see Chen et al., 1975; Guilfoyle et al., 1975; Lin et al., 1976 and inclusive references). The nuclei of auxin-treated tissue are enlarged, primarily as a result of increased nucleolar volume. The nucleoli have a much higher RNA and protein content per unit of DNA, primarily in the form of pre-ribosomal particles. Nucleolar chromatin from auxin-treated tissue has a higher protein content per unit DNA than comparable chromatin from control tissue and altered patterns of proteins as noted following gel fractionation (Chen et al., 1983).

Enhanced rRNA synthesis results from an increase both in the number of RNA polymerase I molecules and in the in situ specific activity (i.e. enhanced chain elongation rates) of those molecules (Guilfoyle et al., 1975; Lin et al., 1976; Olszewski and Guilfoyle, 1980). The basis of the enhanced specific activity of the RNA polymerase I molecules has not yet been defined, but a change in level and/or activity of other nucleolar chromatin components seems likely. Such a change could involve the activity of trans-acting factors which interact with enhancers and/or the promoter region(s) of rDNA (see Reeder, 1984) to effect enhanced RNA polymerase I initiation; equally plausible would be a change in chromatin structure or factor activity that increase rates of rRNA chain elongation. Changes in RNA polymerase II activity by auxin treatment of tissue, as measured in isolated nuclei or when assayed after solubilization on naked DNA templates, are small when compared to RNA polymerase I. Thus, the activity of RNA polymerase I is much more subject to modulation during development and auxin treatment than is RNA polymerase II (discussed in more detail by Guilfoyle, 1986).

Coupled with the increase in RNA polymerase I activity, the earlier work showed that ribosomes accumulate in mature hypocotyl tissue in response to auxin treatment, suggestive of enhanced accumulation of ribosomal proteins (r-proteins). In a study on the regulation of r-protein synthesis, Gantt and Key (1983) showed that the level of translatable mRNA for r-proteins increased an average of about 8-fold following auxin treatment. A cDNA library was then constructed and used to identify cDNAs to r-protein mRNAs. Nine recombinant plasmids containing r-protein sequences were identified by hybrid selection-translation followed by gel electrophoresis (Gantt and Key, 1985). The individual plasmids were used to determine the amount of r-protein mRNA in the mature hypocotyl at various times after auxin treatment. After a lag of 1 hr, the level of r-protein mRNAs increased up to

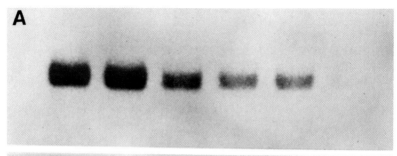

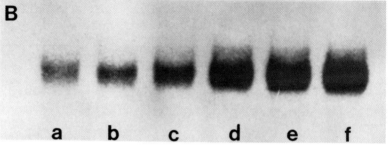

FIGURE 2. Effect of Auxin on the Concentration of Two Soybean mRNAs. Northern analysis of poly(A)RNA isolated from auxin-treated 4 day-old etiolated mature hypocotyl. The blots were probed with cDNA inserts labeled by nick-translation. A, probed with cDNA p11 (Baulcombe and Key, 1980); B, probed with cDNA encoding cytoplasmic ribosomal protein L13 (Gantt and Key, 1985); a-f, 0, 2, 4, 8, 12 and 24 hr of auxin treatment, respectively. (Data of Kroner and Key.)

3- to 8-fold within 18 hr (see Fig. 2). The kinetics of increase in r-protein mRNAs closely parallels the increase in rRNA synthesis noted above. While the mechanism of auxin-enhanced accumulation of r-protein mRNA is not understood, nuclear run-off experiments indicate that increased rates of transcription account at least in part for this accumulation (unpublished observations of Kroner and Gantt).

The complexity of total poly(A)RNA is not significantly affected by auxin treatment of mature hypocotyl tissue (Baulcombe et al., 1980; 1981) as assessed by kinetic hybridization analyses of cDNA to poly(A)RNA in excess or by saturation hybridization of single copy DNA to poly(A)RNA in excess. The kinetic hybridization analyses demonstrate changes in relative concentration of some sequences, however; for example, an especially abundant component of poly(A)RNA in control tissue decreases markedly following auxin treatment. In vitro translation of poly(A)RNAs isolated from mature tissue in a wheat germ system followed by resolution of the translation products on O'Farrell 2-D gels indicate that several of the more abundant poly(A)RNAs change in concentration following auxin treatment, with about equal numbers (20 or so) showing an increase and a decrease. Zurfluh and Guilfoyle (1982b) have found that auxin alters the level of translatable mRNAs similarly in excised and intact mature soybean hypocotyl.

Following up on the observation from hybridization analyses that auxin treatment resulted in depletion of an abundant component of poly(A)RNA in mature soybean hypocotyl, cDNAs made to control poly(A)RNA were cloned and recombinant colonies screened with radiolabeled cDNAs made to poly(A)RNAs from control and auxin-treated tissue (Baulcombe and Key, 1980). Several cDNA clones representative of three different poly(A)RNAs which responded rapidly to auxin treatment were isolated. These clones represent three very abundant poly(A)RNA sequences of control tissue (about 2.1% of the total). The concentration of each of these sequences decreases markedly and rapidly, representing only about 0.01% to 0.05% each of the total poly(A)RNA after 24 hr of auxin treatment. Most of this decrease occurs over the initial 8 hr following auxin treatment, prior to the onset of auxin-induced cell division (Fig. 2). These studies were the first direct demonstration that auxin regulates the levels of specific mRNAs. We do not yet understand the mechanism by which auxin regulates the concentration of these poly(A)RNAs.

Gene Expression during Auxin-Enhanced Cell Elongation

Cell elongation is the area of auxin biology that engenders the most interest of "auxinologists". Although the physiology/biochemistry of auxin-regulated cell elongation may be more amenable to experimentation than auxin-induced cell division, we may be a long way from gaining real insight into the mechanism(s) of auxin control of cell elongation (see reviews by Vanderhoef, 1980; Hanson and Trewavas, 1982; Evans, 1985; Cross, 1985; Guilfoyle, 1986).

Auxin enhances RNA synthesis in excised elongating hypocotyl tissue, but in contrast to elongation in intact hypocotyl and the response in auxin-treated mature hypocotyl,

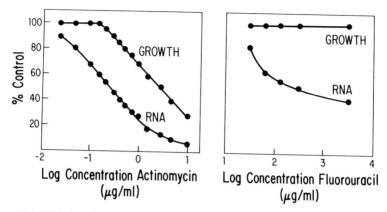

FIGURE 3. Influence of Actinomycin D and 5-Fluorouracil on Auxin-Induced Cell Elongation and RNA Synthesis in Excised Elongating Sections of Soybean Hypocotyl. Tissue was incubated for 8 hr in presence of 50 μM 2,4-D and indicated concentrations of actinomycin D and 5-fluorouracil (From Key and Ingle, 1964).

there is no net accumulation of RNA in response to auxin; in fact total RNA content declines during excised elongation and auxin only slows the process (e.g. Key and Hanson, 1961; Key and Shannon, 1964).

As a follow-up to studies which showed that auxin enhances RNA synthesis in excised elongating tissues, a number of inhibitors of RNA and protein synthesis were used to assess whether RNA and protein synthesis were required for auxin enhancement of the rate of cell elongation. The results from several studies (e.g. Key, 1964; Key and Ingle, 1964; Key et al; 1967) demonstrated clearly that cell elongation requires continued RNA and protein synthesis. Actinomycin D, an effective inhibitor of RNA synthesis in soybean hypocotyl, is a potent inhibitor of cell elongation. At low concentrations of the drug, rRNA synthesis is selectively inhibited while cell elongation is not affected (Fig. 3). Likewise, the base analogue 5-fluorouracil dramatically inhibits RNA synthesis (i.e. labeled nucleotide incorporation into RNA) up to a maximum of about 60% without affecting cell elongation. Other data show, however, that 5-fluorouracil selectively and effectively inhibits rRNA synthesis and/or accumulation without affecting the synthesis of "DNA-like" RNA (heterodisperse non-polyadenylated RNA similar in nucleotide composition to DNA often referred to as

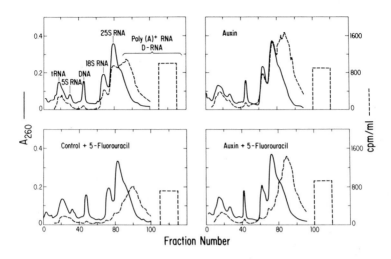

FIGURE 4. Influence of Auxin on RNA Synthesis in Control and 5-Fluorouracil-Treated Excised Elongating Soybean Hypocotyl. Profiles represent nucleic acid absorbance and ^{32}P-labeled newly synthesized RNA during a 4-hr incubation following elution from methylated albumin Kieselghur columns by salt gradient fractionation. (From Key and Ingle, 1968).

hnRNA) and poly(A)RNA (see Fig. 4 and Key and Ingle; 1964, 1968). The data of Fig. 4 demonstrate that auxin increases the synthesis of all classes of RNA analyzed including rRNA, poly(A)RNA and the non-polyadenylated DNA-like RNA. Taken together these results suggest that mRNA is the only RNA species whose synthesis is required for cell elongation.

When the actinomycin D inhibition of cell elongation was assessed in the presence of 5-fluorouracil, there was a parallel inhibition of auxin-induced cell elongation and of 5-fluorouracil-insensitive RNA synthesis by increasing concentrations of actinomycin D (Fig. 5). The 5-fluorouracil inhibition of RNA synthesis had no effect on protein synthesis. Actinomycin D inhibited protein synthesis up to a maximum of about 60 to 70 percent over an 8 hr period. However, for each increment of protein synthesis inhibited by actinomycin D, there was a corresponding inhibition of auxin-induced cell elongation (e.g. Key et al., 1967). This

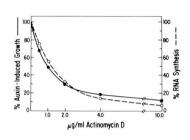

FIGURE 5. Parallel Inhibition by Actinomycin D of Auxin-Induced Cell Elongation and 5-Fluorouracil-Insensitive RNA Synthesis in Excised Soybean Hypocotyl. Tissue was incubated for 4 hr in the indicated concentrations of actinomycin D and 2.5×10^{-3} M 5-fluorouracil. ^{14}C-Adenosine incorporation and growth were measured over the next 4 hr in the presence of 10 µg/ml 2,4-D. (From Key et al., 1967).

partial and progressive inhibition of protein synthesis is consistent with the fact that the mRNAs of soybean have different half-lives, with some being less than 1 hr and others ranging up to many hours for an average half-life of about 5 hr (see Silflow and Key, 1979).

The protein synthesis inhibitor, cyclohexamide, was also a potent inhibitor of cell elongation. The extent of inhibition of auxin-induced growth at various cyclohexamide concentrations was proportional to the amount of inhibition of protein synthesis (Fig. 6). Thus, the inhibition of protein synthesis either directly by cyclohexamide or indirectly by the inhibition of mRNA synthesis by actinomycin D results in the marked inhibition of cell elongation. These results are consistent with the general enhancement of poly(A)RNA (or mRNA) synthesis (e.g. Key, 1964 and Key et al., 1967) and the maintenance of a higher level of polyribosomes in auxin treated tissue (e.g. Travis et al., 1973). Even though a somewhat higher level of protein is maintained in auxin-treated than in control tissue, the protein content of both tissues declines during excised incubation. (It should be noted that cell elongation in intact hypocotyl tissue is associated with a large net increase in protein in contrast to the decline which occurs during excised growth; see Key and Hanson, 1961). The interpretation of the experiments described above using inhibitors of RNA and protein synthesis to establish an obligatory link of RNA and protein synthesis to cell elongation are somewhat tenuous due to their presumed pleiotropic effects. However, several lines of evidence

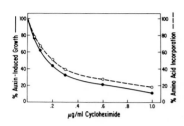

FIGURE 6. Parallel Inhibition by Cyclohexamide of Auxin-Induced Cell Elongation and Protein Synthesis in Excised Soybean Hypocotyl. Hypocotyl sections were pre-incubated for 4 hr in the indicated concentrations of cyclohexamide prior to the addition of auxin (10 μg/ml 2,4-D) and ^{14}C-leucine. Auxin-induced growth and protein synthesis were measured over the next 4 hr. (From Key et al., 1967).

demonstrate definitively that auxin regulates the levels of several specific mRNAs and proteins in elongating sections of soybean hypocotyl. Zurfluh and Guilfoyle (1982a) demonstrated by in vitro translation of poly(A)RNA in wheat germ and reticulocyte systems and 2-D O'Farrell gel fractionation of the translation products that auxin enhanced selectively the level of some translatable mRNAs of excised soybean hypocotyl. Similar results were obtained for the effects of auxin on translatable poly(A)RNA levels of excised pea epicotyl (Theologis and Ray, 1982). Changes in levels of translatable poly(A)RNAs following auxin treatment of excised soybean hypocotyl elongating sections also were observed in our studies (Fig. 7 and unpublished data of Walker, Legocka and Key).

A series of experiments were designed to isolate cDNA clones to poly(A)RNAs that increased in concentration following auxin treatment of excised elongating hypocotyl tissue (Walker and Key, 1982). The rationale behind the cloning strategy was to clone cDNA made to poly(A)RNA of the elongating region of the soybean hypocotyl prior to excised incubation, since this tissue contains all mRNAs necessary for maximum rates of elongation. The library was screened with cDNAs made to 1) poly(A)RNA from excised tissues depleted of endogenous auxin in which the rate of elongation had decreased dramatically compared to intact tissue (i.e. 4 hr of incubation in the absence of added auxin) and 2) poly(A)RNA from tissues treated in an identical manner to that above but induced to elongate at rates approaching that of intact tissue by the addition of auxin to the incubation

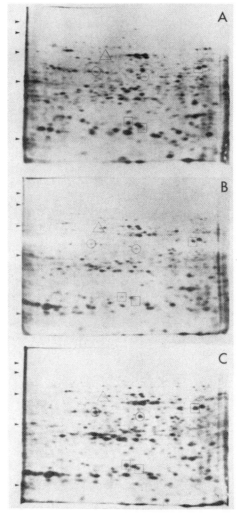

FIGURE   7.   Auxin-Induced   Changes   in   Translatable
Poly(A)RNAs   in   Excised   Elongating   Soybean   Hypocotyl.   The
poly(A)RNAs   were   translated   in   the   wheat   germ   system,   and   the
translation   products   were   fractionated   in   the   2D   O'Farrell
Gel   System.   Poly(A)RNA   was   isolated   from: A,   unincubated
elongating   hypocotyl;   B,   excised   elongating   hypocotyl
incubated   4   hr   in   the   absence   of   auxin;   C,   tissue   incubated
as   in   B   followed   by   2   hr   in 10^{-5}M   2,4-D.   O,Δ,□   used   to
indicate   changes   in   level   of   just   a   few   of   the   translation
products   affected   by   auxin.

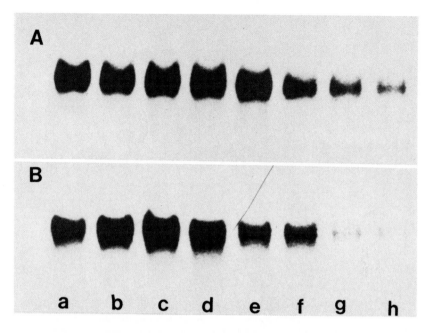

FIGURE 8. Depletion of JCW1 and JCW2 RNAs during Excised Incubation of Elongating Soybean Hypocotyl in the Absence of an Exogenous Auxin Supply as Assesed by Northern Blot Hybridization Analyses. Each lane contains 0.25 µg total poly(A) RNA (A, pJCW1) or 0.5 µg poly(A)RNA (B, pJCW2). Poly(A)RNA was isolated from unincubated tissue (lanes a and e), tissue incubated with auxin for 2(b), 4(c), or 6(d) hr, and tissue incubated without auxin for 2 (f), 4(g) or 6(h) hr. (From Walker and Key, 1982).

medium. Two cDNA clones, designated pJCW1 and pJCW2, were isolated (Walker and Key, 1982) which responded in a manner expected of mRNAs involved in the rate-limiting process(es) of cell elongation.

The early results suggested that the "rate-limiting" RNAs and encoded proteins for cell elongation had half-lives of about 1 hr (or even "used up" in the case of possible structural proteins) (e.g. Key and Ingle 1964). The levels of the poly(A)RNAs homologous to pJCW1 (JCW1 RNA) and pJCW2 (JCW2 RNA) are higher in the elongating tissue than in dividing or mature tissues, and JCW1 RNA is present at much higher levels than JCW2 RNA (Walker and Key, 1982). Both

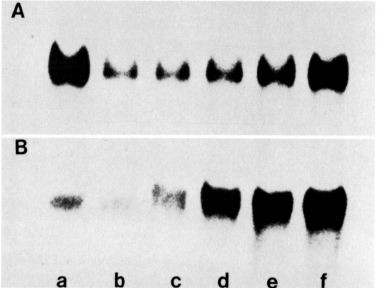

FIGURE 9. Kinetics of Auxin-Induced Accumulation of JCW1 and JCW2 RNAs in Excised Elongating Soybean Hypocotyl Using Northern Blot Hybridization Analysis. A, 0.25 µg poly(A)RNA probed with pJCW1; B, 0.5 µg poly(A)RNA probed with pJCW2. Poly(A)RNA was prepared from unincubated elongating hypocotyl (lane a), 4-hr incubated tissue without auxin (b), or 4-hr incubated tissue as in b followed by addition of auxin for 15, 30, 60 or 90 min (lanes c to f, respectively). (From Walker and Key, 1982).

sequences deplete rapidly during incubation of excised sections in the absence of auxin (Fig. 8); JCW1 RNA is maintained at a near constant level when auxin is present throughout the incubation of the elongating sections while the amount of JCW2 RNA increases above the level in intact tissue. The addition of auxin to tissues that were depleted in JCW1 and JCW2 sequences by a 4- to 6-hr incubation without an exogenous auxin supply resulted in a rapid accumulation of both poly(A)RNAs (Fig. 9). The JCW1 sequence increased significantly within 30 min and continued to increase up to 90 min of incubation to levels 3- to 5-fold above the auxin-depleted level and to about the levels in intact tissue. The JCW2 RNA responded very rapidly to auxin treatment, with increases being detectable within 15 min (the shortest time measured); JCW2 RNA increases 5- to 10-fold

within 30 to 60 min to levels considerably higher than in
intact elongating tissue.

Several lines of evidence suggest strongly that the
change in relative abundance of the JCW1 and JCW2 RNAs was in
fact an auxin-dependent response (Walker et al., 1985).
First, the sequences increase similarly in response to
several active auxins, but not to inactive related compounds.
Second, the change in concentration of these poly(A)RNAs is
not simply a response to changes in the rate of cell
elongation of the tissue: a) fusicoccin increases the rate
of cell elongation in tissue incubated without auxin to
rates above the maximum induced by auxin, but the JCW1 and
JCW2 RNAs remain at levels similar to that of the auxin-free
medium where cell elongation is minimal and b) cytokinin
significantly inhibits auxin-induced growth in excised
soybean hypocotyl but does not affect the auxin-induced
accumulation of JCW1 and JCW2 RNAs (Fig. 10).
Fusicoccin-induced cell elongation is not dependent on RNA
and protein synthesis (see Hanson and Trewavas 1982). It has
been speculated that fusicoccin somehow by-passes the
processes that require RNA and protein synthesis for
auxin-stimulated cell elongation.

In contrast to auxin-induced cell elongation,
fusicoccin-induced cell elongation is not inhibited by
cytokinin (Walker et al., 1985). If it is assumed, based in
part on data presented above and that covered in cited
reviews, that the auxin-enhanced rate of cell elongation is
the result of auxin-enhanced accumulation (synthesis) of
specific "rate-limiting" mRNAs and their translation
products, the action of cytokinin in inhibiting auxin-induced
cell elongation (but not fusicoccin-induced cell elongation)
may well relate to the inhibition of synthesis of the
"rate-limiting" protein(s) or the inhibition of the
function(s) of such a protein(s). Since cytokinin does not
affect the accumulation of these RNAs, either their protein
products are not a part of the rate limiting step(s) of cell
elongation or cytokinin inhibition of auxin-induced
elongation occurs at the level of protein synthesis or
function. Cytokinin does in fact inhibit protein synthesis
in the excised soybean hypocotyl tissue (e.g. Zurfluh and
Guilfoyle, 1980).

Since auxin treatment of tissue may induce ethylene
production, an apparent auxin response can be either a direct
response to auxin or an indirect response mediated by
ethylene. Walker et al., (1985) showed that ethylene has no
effect on JCW1 and JCW2 RNAs alone or in combination with

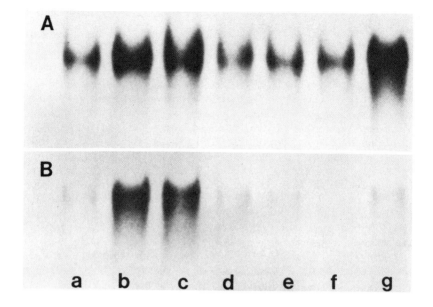

FIGURE 10. Northern Blot Hybridization Analysis of pJCW1 (A) and pJCW2 (B) cDNA Clones to Poly(A)RNAs Isolated from Growth   Regulator-Treated   Excised   Elongating   Soybean Hypocotyl.   (a) elongating tissue incubated 4 hr in absence of auxin; (b) as in a followed by a 2-hr incubation in auxin, (c)   auxin   plus   isopentenyladenosine   (50 µm),   (d) isopentenyladenosine, (e) fusicoccin (10 mm), (f) fusicoccin plus isopentenyladenosine, and (g) RNA from unincubated control tissue.   (From Walker et al., 1985).

auxin. A precursor of ethylene, ACC, and an inhibitor of ethylene biosynthesis, AVG, have no effect on the expression of JCW1 and JCW2 RNAs in the presence or absence of auxin. Thus, there is no effect of either exogenously applied or endogenously produced ethylene on the concentration of these two auxin-responsive mRNAs, and ethylene does not alter the influence of auxin on their accumulation.

     In hybrid-selection translation experiments (Walker and Key, unpublished data), pJCW1 hybrid-selects poly(A)RNAs which translate into polypeptides of 31 and 30 kD with pIs of about 6.1 and 6.7, respectively. Clone pJCW2 hybrid-selects mRNA encoding a major polypeptide of 22 kD on 1D SDS gels and a minor polypeptide of 19 kD. The 22 kD polypeptide resolves in O'Farrell 2-D gels into 5 polypeptides with pIs ranging from 5.9 to 7.2; the 19 kD polypeptide(s) does not focus on the O'Farrell system.

Hagen et al. (1984) have recently isolated four different cDNA clones to poly(A)RNAs that increase in soybean hypocotyl in response to auxin; these changes are marked and rapid and appear to result from increased rates of transcription as assessed in nuclei run-off transcription experiments using nuclei isolated from control and auxin-treated soybean tissues (Hagen and Guilfoyle, 1985). None of these clones show homology to pJCW1 and pJCW2 based on hybridization analyses (unpublished data of Hagen and Guilfoyle). Additional clones to IAA-induced pea epicotyl mRNAs were recently isolated by Theologis et al. (1985). It is unknown whether the proteins encoded by the mRNAs homologous to these three sets of cDNAs have any common functions. Any relationships (homologies) among these different cloned sequences will be defined from DNA sequence analysis of the cDNAs, corresponding genomic clones, and/or the derived amino acid sequence of the encoded proteins.

Isolation of Genomic Clones to Auxin-Regulated Poly(A)RNAs

Genomic clones have been isolated that are homologous to pJCW1 and pJCW2 (Walker, 1985), to each of the three sequences isolated by Baulcombe and Key (1980) which are down-regulated by auxin and for several of the r-proteins mRNAs (Gantt and Key, unpublished data). Sequence analysis and mapping of the 5' and 3' ends of their corresponding mRNA have been completed on some of these genes as a first step to determine the mechanism by which auxin regulates their expression and the function of the encoded proteins.

Genes homologous to pJCW1 and pJCW2 have five and three introns, respectively. The 5' end of the mRNA corresponding to pJCW2 maps about 33 base pairs downstream of a TATA box sequence. Rather typical of plant genes sequenced to date, neither gene contains a canonical polyadenylation signal, AATAAA, found in most other higher eukaryotic genes. Each gene is a member of a small multigene family.

The sequence analyses revealed several interesting features of the JCW1 and JCW2 proteins (Walker and Key, unpublished data). Both are very hydrophilic proteins with very similar hydropathic profiles (Kyte and Doolittle, 1982). Both proteins have potential glycosylation sites, but the absence of a signal peptide in either protein would indicate that these may not be functional sites for glycosylation. A search of the Protein Identification Resource protein sequence data bank has revealed no significant homologies with the derived amino acid sequence of these clones and the sequences of other published proteins. While there are no continuous nucleotide homologies of more than 11 nucleotides in the cDNAs (or in the genomic sequences), there are five highly conserved (80 to 100%) regions of amino acid sequence homology consisting of 7, 8, 15, 17 and 33 amino acids in the proteins encoded by JCW1 and JCW2 RNAs.

SUMMARY AND CONCLUSIONS

The intent of this article was 1) to provide some historical perspective on the study of auxin-regulated gene expression, 2) to summarize the current state of knowledge on auxin-regulated gene expression from the perspective of research accomplished in this laboratory, and 3) to demonstrate how advancing technology permits greater understanding of complex biological systems. The understanding of auxin-regulated gene expression has advanced from the earliest studies which only provided quantitative estimates of changes in total RNA to recombinant DNA technologies which permit the detailed analysis of specific auxin-regulated genes and their protein products. The early studies provided the conceptual framework for more recent studies which analyze the expression of specific genes.

Definitive evidence is provided for a rapid, dramatic, and specific effect of auxin on the expression of a few genes in tissues undergoing both cell division and cell elongation. These observations, coupled with the earlier work which showed an absolute requirement for RNA and protein synthesis for cell elongation in response to auxin, provide strong supportive evidence for the concept that an auxin "response(s)" is tightly coupled to auxin-regulated gene expression in controlling cell division, cell elongation, and probably cell differentiation. Independent of the mechanism(s) of auxin action, the auxin-regulated expression of a few genes and their encoded proteins must be of major importance in the physiology of auxin-regulated growth processes.

Nuclear run-off transcription studies (Hagen and Guilfoyle, 1985) indicate that auxin treatment results in a very rapid induction of synthesis of those RNA sequences that are responsive to the hormone. Moreover, several auxin-regulated mRNAs have been shown to be induced within 15 to 30 min in tissue incubated with auxin. It is tempting to speculate that some of these rapid responses may represent primary responses to auxin. These changes certainly are induced in the same time frame that steroid hormone-induced synthesis of various mRNAs occurs in responsive tissues (see reviews by Shapiro, 1982; Chambon et al., 1984; Rousseau, 1984; and Yamamoto, 1985). The steroid hormone-induction or enhancement of transcription of responsive genes generally results from steroid hormone-receptor interaction with gene sequences (and/or chromatin structures) 5' to the transcription start sites of those genes. Even in these elegant studies that are the culmination of work done over many decades and with the enormous progress made over the last several years in understanding steroid hormone-regulated gene expression, there remain many unknowns about the details of these systems (see Chambon et al., 1984); as they pointed out, progress with in vitro transcription systems is required to elucidate the major unsolved problems. They further point out that "'elegant' molecular genetics will have to give way to 'trivial' biochemistry!" in order to make further progress. While the results on steroid hormone-regulated gene expression may serve as one of several models for the mechanism underlying auxin-regulated gene expression, to date there is no information available to indicate whether such a mechanism will apply to the auxin response.

Several aspects of auxin-regulated gene expression must be researched and defined before a causal relationship between an auxin-regulated physiological event (e.g. enhanced rate of cell elongation) and gene expression can be proven. The function of the protein products of auxin-regulated genes must be understood. Auxin "receptors", if they are directly involved in regulating gene expression, and other factors involved in this regulation must be purified, characterized, and available in sufficient quantity to permit an analysis of their interactions with an auxin-regulated gene. Does an auxin-receptor complex interact with a cis-acting element of the gene, does auxin or an auxin-receptor interact with some trans-acting transcription factor to enhance specific transcription by RNA polymerase II of auxin-regulated genes, or does auxin treatment result in rapid modification of factors which in the modified state alter the rate of transcription of the gene(s) in question? These only a few questions whose answers are crucial to gaining an understanding of the mechanism(s) of auxin-regulated gene expression. RNA polymerase II transcription systems which function accurately when reconstituted in vitro would be a valuable tool to approach these questions. With a number of auxin-regulated gene sequences in hand, the problem may also be approached by looking for protection of regions of the genes by auxin-receptor complexes or other factors in DNAseI foot-print analyses or affinity purification of proteins which bind specifically to auxin-regulated genes. Currently, these kinds of experiments are much easier to design than to bring to fruition especially in view of a lack of an in vitro RNA polymerase II transcription system from plants and lack of known auxin receptors which interact with DNA (see Lobler and Klambt, 1985a and b; Cross, 1985; and Venis, 1985 for information on auxin receptors).

Additional technical advances adapted to plant systems will be needed, however, to permit an analysis of how auxin regulates the expression of only a few genes out of many thousands normally expressed in plants. The future should be exciting in bringing this area of auxin biology to a higher level of understanding.

ACKNOWLEDGMENTS

The support of research related to "auxin-regulated gene expression" by NIH since 1961 is gratefully acknowledged; current research is supported by grant GM30317 from the NIH. Research contributions by many colleagues is most appreciated. Thanks go to Drs. Ron Nagao and Michael Ainley and Michael Mansfield for review and thoughtful comments on the manuscript. The assistance of Joyce Kochert and Ginger Vickery in manuscript preparation is acknowledged.

REFERENCES

1. Baulcombe D, Giorgini J, Key JL (1980). The effect of auxin on the polyadenylated RNA of soybean hypocotyls. In Leaver CJ (ed) "Genome Organization and Expression in Plants," New York: Plenum Press, p 175.
2. Baulcombe DC, Key JL (1980). Polyadenylated RNA sequences which are reduced in concentration following auxin treatment of soybean hypocotyls. J Biol Chem 255:8907.
3. Baulcombe DC, Kroner PA, Key JL (1981). Auxin and gene regulation. In Subtelny S, Abbott UK (eds) "Levels of Genetic Control in Development," New York: Alan R. Liss, p. 83.
4. Chambon P, Dierich A, Gaub M-P, Jakowley S, Jongstra J, Krust A, Le Pennec J-P, Oudet P, Reudelhuber T (1984). Promoter elements of genes coding for proteins and

modulation of transcription by estrogens and progesterone. Recent Prog Hormone Res 40: 1.

5. Chen Y-M, Huang D-H, Lin S-F, Lin C-Y, Key JL (1983). Fractionation of nucleoli from auxin-treated soybean hypocotyl into nucleolar chromatin and preribosomal particles. Plant Physiol 73:746.

6. Chen Y-M, Lin C-Y, Chang H, Guilfoyle TJ, Key JL (1975). Isolation and properties of nuclei from control and auxin-treated soybean hypocotyl. Plant Physiol 56:78.

7. Cross JW (1985). Auxin action: the search for the receptor. Plant Cell and Environ 8:351.

8. Evans ML (1985). The action of auxin on plant cell elongation. In CRC Crit Rev Plant Sci 2:317.

9. Gantt JS, Key JL (1983). Auxin-induced changes in the level of translatable ribosomal protein messenger ribonucleic acids in soybean hypocotyl. Biochemistry 22:4131.

10. Gantt JS, Key JL (1985). Coordinate expression of ribosomal protein mRNAs following auxin treatment of soybean hypocotyls. J Biol Chem 260:6175.

11. Guilfoyle TJ (1986). Auxin regulated gene expression in higher plants. CRC Crit Rev Plant Sci (In Press).

12. Guilfoyle TJ, Lin C-Y, Chen Y-M, Nagao RT, Key JL (1975). Enhancement of soybean RNA polymerase I by auxin. Proc Natl Acad Sci USA 72:69.

13. Hagen G, Guilfoyle TJ (1985). Rapid induction of selective transcription by auxins. Mol and Cell Biol 5:1197.

14. Hagen G, Kleinschmidt A, Guilfoyle T (1984). Auxin-regulated gene expression in intact soybean hypocotyl and excised hypocotyl sections. Planta 162:147.

15. Hanson JB, Slife FW (1969). Role of RNA metabolism in the action of auxin herbicides. Residue Reviews 25:59.

16. Hanson JB, Trewavas AJ (1982). Regulation of plant cell growth: the changing perspective. New Phytol 90:1.

17. Key JL, Hanson JB (1961). Some effects of 2, 4-dichlorophenoxy-acetic acid on soluble nucleotides and nucleic acid of soybean seedlings. Plant Physiol 36:145.

18. Key JL, Shannon JC (1964). Enhancement by auxin of ribonucleic acid synthesis in excised soybean hypocotyl tissue. Plant Physiol 39:360.

19. Key JL, Ingle J (1964). Requirement for the synthesis of DNA-like RNA for growth of excised plant tissue. Proc Natl Acad Sci USA 52:1382.

20. Key JL (1964). Ribonucleic acid and protein synthesis as essential processes for cell elongation. Plant Physiol 39:365.

21. Key JL, Lin C-Y, Gifford EM Jr, Dengler R (1966). Relation of 2,4-D-induced growth aberrations to changes in nucleic acid metabolism in soybean seedlings. Bot Gaz 127:87.

22. Key JL, Barnett NM, Lin C-Y (1967). RNA and protein biosynthesis and the regulation of cell elongation by auxin. In Ann New York Acad Sci 144: 49.

23. Key JL, Ingle J (1968). RNA metabolism in response to auxin. In Wightman F, Setterfield G (eds) "Biochemistry and Physiology of Plant Growth Substances," Ottawa: The Runge Press Ltd, p 711.

24. Key JL (1969) Hormones and nucleic acid metabolism. Ann Rev Plant Physiol 20: 449.

25. Kyte J, Doolittle RF (1982). A simple method for displaying the hydropathic character of a protein. J Mol Biol 157:105.

26. Lin C-Y, Chen Y-M, Guilfoyle TJ, Key JL (1976). Selective modulation of RNA polymerase I activity during growth transitions in the soybean seedling. Plant Physiol 58:614.

27. Lobler M, Klambt D (1985a). Auxin-binding protein from coleoptile membranes of corn (Zea mays L) I. Purification by immunological methods and characterization. J Biol Chem 260:9848.

28. Lobler M, Klambt D (1985b). Auxin-binding protein from coleoptile membranes of corn (Zea mays L) II. Localization of a putative auxin receptor. J Biol Chem 260:9854.

29. Olszewski N, Guilfoyle TJ (1980). A new method for determining the number of RNA polymerases active in chromatin transcription. Biochem Biophys Res Comm 94:553.

30. Reeder RH (1984). Enhancers and ribosomal gene spacers. Cell 38:349.

31. Rousseau GG (1984). Control of gene expression by glucocorticoid hormones. Biochem J 224:1.

32. Shapiro DJ (1982). Steroid hormone regulation of vitellogenin gene expression. In CRC Crit Rev in Biochem. 12:187.

33. Silflow CD, Key JL (1979). Stability of polysome-associated polyadenylated RNA from soybean suspension culture cells. Biochemistry 18:1013.

34. Skoog F (1954). Substances involved in normal growth and differentiation of plants. Brookhaven Symposium in Biology 6:1.

35. Theologis A, Ray PM (1982). Early auxin-regulated polyadenylated mRNA sequences in pea stem tissue. Proc Natl Acad Sci USA 79:418.
36. Theologis A, Huynh TV, Davis RW (1985). Rapid induction of specific mRNAs by auxin in pea epicotyl tissue. J Mol Biol 183:53.
37. Travis RL, Anderson JM, Key JL (1973). Influence of auxin and incubation on the relative level of polyribosomes in excised soybean hypocotyl. Plant Physiol 52:608.
38. Trewavas A (1968). Relationship between plant growth hormones and nucleic acid metabolism. Progress Phytochem. 1:113.
39. Vanderhoef LN (1980). Auxin-regulated cell enlargement: Is there action at the level of gene expression? In Leaver CJ (ed): "Genome Organization and Expression in Plants," New York: Plenum Press, p 159.
40. Venis M (1985). "Hormone Binding Sites in Plants." New York: Longman Inc.
41. Walker JC, Key JL (1982). Isolation of cloned cDNAs to auxin-responsive poly(A)$^+$RNAs of elongating soybean hypocotyl. Proc Natl Acad Sci USA 79:7185.
42. Walker, JC (1985). Auxin regulation of gene expression. Ph.D. Dissertation, University of Georgia.
43. Walker JC, Legocka J, Edelman L, Key JL (1985). An analysis of growth regulator interactions and gene expression during auxin-induced cell elongation using cloned complementary DNAs to auxin-responsive messenger RNAs. Plant Physiol 77:847.
44. Yamamoto KR (1985). Steroid receptor regulated transcription of specific genes and gene networks. Ann Rev Genet 19:209.
45. Zurfluh LL, Guilfoyle TJ (1980). Auxin-induced changes in the patterns of protein synthesis in soybean hypocotyl. Proc Natl Acad Sci USA 77:357.
46. Zurfluh LL, Guilfoyle TJ (1982a). Auxin-induced changes in the population of translatable messenger RNA in elongating sections of soybean hypocotyl. Plant Physiol 69:332.
47. Zurfluh LL, Guilfoyle TJ (1982b). Auxin- and ethylene-induced changes in the population of translatable messenger RNA in basal sections and intact soybean hypocotyl. Plant Physiol 69:338.

Molecular Biology of Plant Growth Control, pages 23–33
© **1987 Alan R. Liss, Inc.**

CONTROL OF α-AMYLASE EXPRESSION BY GIBBERELLIN
AND ABSCISIC ACID IN BARLEY ALEURONE

Peter M. Chandler, J.V. Jacobsen,
J.A. Zwar, Z. Ariffin and L. Huiet

CSIRO Division of Plant Industry, P.O. Box 1600,
Canberra, Australia

ABSTRACT   The expression of α-amylase genes in isolated
aleurone tissue of barley is stimulated by gibberellin
(GA) and inhibited by abscisic acid (ABA).   Changes
seen   in   levels   of   α-amylase   mRNA   in   aleurone
protoplasts treated with either GA or ABA + GA result,
in part at least, from the influence of these hormones
on transcription of α-amylase genes.   Based on this it
is likely that expression of α-amylase genes following
germination   of   an   intact   grain   involves   GA-mediated
transcriptional activation.
       Primer   extension   analysis   indicates   that   there   is
differential   expression   of   mRNAs   for   the   low   pI   and
high   pI   α-amylase   mRNA   families   following   grain
germination.   At   present   it   is   not   known   whether   this
difference   results   from   transcriptional   effects   or
other aspects of GA action.
       Situations   are   described   where   ABA   may   influence
expression of α-amylase genes in vivo.   We demonstrate
that   in   dehydrating   seedlings   there   is   induction   in
aleurone   tissue   of   mRNAs   corresponding   to   the   set   of
mRNAs   induced   by   ABA   treatment   of   mature   isolated
aleurone.

INTRODUCTION

The   aleurone   layer   from   barley   (Hordeum   vulgare   L.)
grains has been studied for many years as a model system for
the   action   of   phytohormones   on   gene   expression   in
differentiated   plant   tissues.   This   specialized   layer   of
cells is a major source of secreted hydrolytic enzymes (e.g.

α-amylase and protease) following germination of the seed; such enzymes are responsible for the mobilization of stored endosperm reserves which ultimately provides sucrose and amino acids (in addition to other compounds) for utilization by the developing seedling.

Gibberellin (GA) is the signal compound which elicits production of hydrolases by the aleurone layer. In a germinating barley grain GA_1 is probably the major GA species produced by the embryo following imbibition of water (1,2). Highest levels of this hormone are seen approximately 2-3 days after imbibition, which slightly precedes the peak in levels of aleurone α-amylase mRNA (3-4 days after imbibition, see below). Generally there is little, if any, production of α-amylase by embryo-less half-grains or isolated aleurone layers incubated in the absence of GA. In some harvests of grain however there is production of α-amylase by endosperm halves in the absence of added GA (5): this may reflect high endogenous levels of GA in the endosperm of the mature grain associated with particular environmental conditions during grain ripening.

Our understanding of hormone action on aleurone gene expression has mostly come from studies in which GA_3 and/or ABA is applied to aleurone layers isolated from mature seed, or to protoplasts derived from such layers. In general these two systems appear to respond similarly to aleurone from whole seedlings, although there are some differences in response. In assessing the physiological significance of any observed response by an isolated tissue to hormone application, it is desirable to be able to compare this response with that of the tissue in the whole plant to endogenous levels of the particular hormone.

RESULTS AND DISCUSSION

Effect of Hormones on α-Amylase mRNA Levels in Isolated Aleurone Layers

In the last few years hybridization studies using α-amylase cDNA clones (6,7,8) have provided clear evidence for hormonal regulation of α-amylase mRNA levels, supporting an earlier conclusion derived from cell-free translation experiments (9). The results of one such study are shown in Fig 1, and demonstrate an approximately 50-fold increase in α-amylase mRNA levels in aleurone layers treated with 1 μM GA_3 for 24 h relative to control layers. The presence of

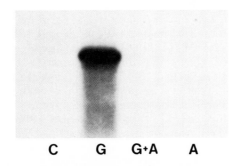

C   G   G+A   A

Figure 1. Hybridization of an α-amylase cDNA clone to size-fractionated RNA from aleurone layers incubated with or without hormones.
Aleurone layers were incubated 24 h without hormone (C) or in the presence of 1 μM GA₃ (G), 1 μM GA₃ + 25 μM ABA (G + A), or 25 μM ABA alone (A). RNA was extracted and equal amounts fractionated by size, blotted to a filter, hybridized with ^{32}P-labelled pHV19 and processed as described (8).

ABA (25 μM) by itself reduces α-amylase mRNA levels to about half of control levels, and in the case where both ABA and GA₃ are present a marked inhibition of the GA response is seen. These results imply that the well-known effect of GA on α-amylase synthesis by aleurone tissue (10) probably occurs by changes in mRNA level. ABA is known to inhibit the production of α-amylase by aleurone in response to GA₃ (11), and the data in Fig 1 suggests that this too may be a consequence of effects on mRNA levels.

Whereas the action of GA₃ in eliciting production of α-amylase by isolated aleurone is believed to mimic events which follow germination of intact grains, at present there is no clearly identified physiological counterpart for explaining the negative effects of ABA on α-amylase production, since ABA levels in the endosperm of germinated barley grains are very low. Two obvious possibilities can be considered. In the first instance high endogenous levels of ABA known to occur in cereal grains during maturation (12) may function in inhibiting expression of genes prior to grain ripeness which are not normally expressed until after germination. Secondly, it is possible that under some circumstances ABA levels may rise in a germinated grain or a young seedling; for instance if it begins to dry out it may be advantageous to switch off continued production of hydrolytic enzymes (and consequent reserve mobilization) as growth is being limited by water availability. Such "pre-germinated" grains remain viable for significant periods of time, and recommence growth when water is again available. When barley seedlings are dehydrated they accumulate in their aleurone a set of mRNAs characterized as being ABA-inducible in mature isolated aleurone layers. A set of cDNA clones has been constructed from polyA RNA of ABA-treated isolated aleurone layers and these clones define several unrelated RNAs which increase greatly in abundance following ABA treatment (PMC and ZA, unpublished). These same mRNAs increase in abundance in the aleurone of dehydrating seedlings, but not of well-watered controls (Fig 2). This observation provides indirect evidence that there may be higher levels of ABA in the aleurone of dehydrating seedlings than in control seedlings. We are currently attempting to measure ABA levels in such tissues, and eventually it may be possible to demonstrate an in vivo down-regulation by ABA of α-amylase expression to match the well known in vitro effect.

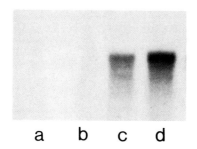

a   b   c   d

Figure 2.   Levels of pHVA39 mRNA in aleurone of dehydrating seedlings of barley.
Seedlings (2.5 d after imbibition) were allowed to continue development on moist filter paper, or placed in a desiccator over 25% glycerol (relative humidity 85-90%). RNA was extracted from the endosperm of seedlings after two additional days development on moist filter paper (lane a), or after 1, 2 or 3 d dehydration (lanes b-d respectively) and processed as described in the legend to Fig 1, except that the hybridization probe was pHVA39 (see text).

A Transcriptional Role for GA and ABA in Aleurone

The levels of mRNAs measured in hybridizations such as Fig 1 depend principally upon the rate of production of α-amylase transcripts in aleurone nuclei, and on the stability of the transcripts (pre-mRNA and mRNA) once formed: observed changes in the level of a particular mRNA can not therefore be immediately ascribed to either transcriptional or stability changes. It has long been suspected that GA might stimulate transcription of α-amylase genes in aleurone but until recently direct experimental support for this concept was lacking. The isolation of GA-responsive aleurone protoplasts, first in wild oat (Avena fatua L., 13) then in barley (14) provided the means for isolating transcriptionally active nuclei, as this approach had been generally unsuccessful starting from intact aleurone layers.

Aleurone protoplasts of barley respond to GA_3 and ABA in a similar manner to intact aleurone layers in terms of α-amylase production: GA stimulates production, and the simultaneous presence of GA and excess ABA prevents this stimulation (14). If nuclei are isolated from protoplasts after 24 h incubation with or without hormone, it is possible to assay the relative rate of transcription of the α-amylase genes by incorporating ^{32}P into the total population of elongating transcripts, and then assay the fraction of incorporated radioactivity which is α-amylase specific (by hybridization to immobilized α-amylase cDNA). The results of such experiments in both barley (15) and wild oat (16) indicate selective transcriptional stimulation of α-amylase genes (and other GA-inducible genes) relative to genes for ribosomal RNA (Fig 3). ABA present in the protoplast incubation medium prevents the stimulatory effect of GA, and restores the transcriptional profile of the nuclei to a pattern very similar to the "no hormone" control. Based on such results the effects of GA and ABA on α-amylase mRNA levels in aleurone may be accountable for solely by hormonal effects on the rate of α-amylase gene transcription; at this stage however it is still possible that in addition to the transcriptional effects there are also changes in transcript stability.

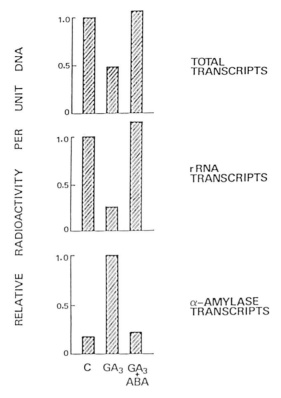

Figure 3. Levels of transcripts made by nuclei isolated from aleurone protoplasts incubated without hormones (C) or in the presence of GA$_3$ or GA$_3$ + ABA for 24 h (modified from ref. 15).

Assaying mRNAs for Particular α-Amylase Isozymes Following Grain Germination

α-Amylase from Himalaya barley consists of four principal isozymes, two each in the low pI (4.4-5.2) and high pI (5.9-6.6) groups (17). The genetic loci for these groups are located on different chromosomes (18,19), and the isozymes within a group exhibit closer relationship to each other than they do to isozymes of the other group (17). Based upon NH$_2$-terminal amino acid sequences, and amino acid composition of high pI and low pI isozymes it was suggested (8) that the α-amylase cDNA clone "E" (7) corresponded to a "low pI" mRNA whereas pHV19 (8) corresponded to a "high pI" mRNA. Subsequent amino acid sequencing of the high pI and low pI isozyme groups supports this assignment (20).

Differences have been noticed between the two groups of isozymes in terms of their regulation in isolated aleurone layers in response to applied GA - for instance the low pI group is induced by lower concentrations of GA_3 than is the high pI group (17). Many of the hybridization studies measuring α-amylase mRNA levels were probably measuring some composite of both high pI and low pI mRNAs, since under normal hybridization conditions cross-hybridization between low pI and high pI clones is observed. However Huang et al (21) have recently studied the expression of high pI and low pI mRNAs in response to GA_3 under conditions which minimize cross-hybridization. Their results support the finding that the low pI isozymes are induced by lower GA_3 concentrations, since maximal stimulation of low pI mRNA levels was seen at 10^{-9} M GA_3, whereas maximal levels of the high pI mRNA were not seen until the GA_3 concentration reached 10^{-6} M. These results add further support to the concept that it is mRNA levels which determine rates of α-amylase production.

We have approached the study of differential regulation of expression of high pI and low pI α-amylase mRNAs by primer extension analysis. This technique has several advantages over the differential hybridization described above in that (i) it allows a direct estimate of any cross-hybridization which might occur, and (ii) it allows regulation of the individual mRNAs within a group to be studied. Briefly, oligonucleotides have been synthesized which are complementary to regions surrounding the ATG initiator methionine codon of either the low pI mRNA (7) or the high pI mRNA (8). When such molecules are 5'-labelled with ^{32}P, hybridized to total RNA, and extended by synthesis with reverse transcriptase in the presence of non-radioactive dNTPs, the proportional representation of particular mRNA species in the starting RNA can be deduced from the lengths and relative radioactivities of extended primer bands revealed by denaturing gel electrophoresis and autoradiography.

An example of the products of this type of reaction is shown in Fig 4. The two major extended low pI products correspond to those seen previously: their nucleotide sequences are different but closely match those of Rogers

and Milliman (22). Two major extended products of the high
pI primer share the same sequence, and that sequence exactly
matches one determined for a high pI genomic clone (L.
Huiet, unpublished). The relationship of this sequence to
that of the third product is unknown at present. These
products may be derived from the mRNA of a single gene, in
which case there may be "nibbling" of nucleotides off the 5'
end of the mRNA, or different transcription initiation
sites. An alternative explanation is that the mRNAs are
derived from genes which differ in length in their 5'
untranslated region, but which have a conserved sequence.

A B

Figure 4. Extended primers synthesized off RNA from
aleurone protoplasts incubated 48 h with 1 µM GA$_3$.
Synthetic primers specific for either the low pI α-amylase
mRNA sequence (7) or the high pI α-amylase mRNA sequence (8)
were synthesized and 5'-labelled with ^{32}P. They were then
hybridized in separate reactions with 10 µg each of RNA from
GA-treated protoplasts. After 1 h at 57°C in 10 µl of 120
mM KCl, 75 mM Tris pH 8.3, 5 µl of a mixture containing 40
mM MgCl$_2$, 40 mM DTT, 500 µM each dNTP, and 2 units AMV
reverse transcriptase was added to each tube which was then
incubated at 41°C for 1 h. The samples were frozen after
the addition of 10 µl dye-formamide, and boiled 3 min prior
to electrophoresis on 12% polyacrylamide-urea sequencing
gels and autoradiography. Lane A = low pI, lane B = high
pI.

The lack of cross-hybridization in such experiments is shown by the failure to see bands which are common to both lanes, as both primers are complementary to the same region of their corresponding mRNAs.

We have investigated the relative levels of high pI and low pI extended products in RNA obtained from the aleurone layer of developing seedlings. Primer extensions were carried out on aleurone RNA from seedlings 2,3,4,5 and 6 days after imbibition. Regions of the gel corresponding to the major extended products were excised and radioactivity determined. The results (Table 1) show that the level of high pI-specific extended products reaches a maximal value at 3 days, and then declines. In contrast the level of low pI-specific products does not peak until 4 days after imbibition, and then declines more slowly. Furthermore, it is valid to make a direct comparison between the levels of the low pI and high pI extended products, since the primers have been labelled to the same specific activity. This means that the peak levels of high pI and low pI mRNAs are different in aleurone of developing seedlings (there is an excess of the high pI species), and they are temporally separated.

TABLE 1

RADIOACTIVITY IN EXTENDED PRIMERS USING
ALEURONE RNA FROM DEVELOPING SEEDLINGS[a]

| | Radioactivity | |
Days after imbibition	Low pI	High pI
2	475	3850
3	3316	7751
4	3441	2776
5	2078	946
6	611	237

[a]Primer extensions were carried out on RNA extracted from aleurone of seedlings of the indicated stage of development. The region of the gel including the extended primer bands was cut out and radioactivity determined by liquid scintillation counting.

The analysis of low pI and high pI isozyme and mRNA levels has established that differential regulation of expression of these two gene families occurs both in vitro and in vivo. It will be of great interest to determine whether this differential regulation reflects effects on transcription of the gene families, or on some other aspect of hormone regulation.

REFERENCES

1. Yamada K (1982) Determination of endogenous gibberellins in germinating barley by combined gas chromatography-mass spectrometry. J. Am. Soc. Brew. Chem. 40 18-25

2. Gaskin P, Gilmour SJ, Lenton JR, McMillan J and Sponsel VM (1984) Endogenous gibberellins and kaurenoids identified from developing and germinating barley grains. J. Plant Growth Regul. 2 229-242

3. Yomo H (1960) Studies on the α-amylase activating substances IV. On the amylase activating action of gibberellin. Hakko Kyokaichi 18 600-602

4. Paleg LG (1960) Physiological effects of gibberellic acid. II On starch hydroyzing enzymes of barley endosperm. Plant Physiol. 35 902-906

5. Nicholls PB (1982) Influence of temperature during grain growth and ripening of barley on the subsequent response to exogenous gibberellic acid. Aust. J. Plant Physiol. 9 373-383

6. Muthukrishnan S, Chandra GR and Maxwell ES (1983) Hormonal control of α-amylase gene expression in barley: studies using a cloned cDNA probe. J. Biol. Chem. 258 2370-2375

7. Rogers JC and Milliman C (1983) Isolation and sequence analysis of a barley α-amylase cDNA clone. J. Biol. Chem. 258 8169-8174

8. Chandler PM, Zwar JA, Jacobsen JV, Higgins TJV and Inglis AS (1984) The effects of gibberellic acid and abscisic acid on α-amylase mRNA levels in barley aleurone layers: studies using an α-amylase cDNA clone. Plant Molec. Biol. 3 407-418

9. Higgins TJV, Zwar JA and Jacobsen JV (1976) Gibberellic acid enhances the level of translatable mRNA for α-amylase in barley aleurone layers. Nature 260 166-169

10. Varner JE (1964) Gibberellic acid controlled synthesis of α-amylase in barley endosperm. Plant Physiol. 39 413-415

11. Chrispeels MJ and Varner JE (1966) Inhibition of gibberellic acid induced formation of α-amylase by abscisin II. Nature 212 1066-1067

12. King RW (1982) Abscisic acid in seed development. in Khan AA (ed): "The Physiology and Biochemistry of Seed Development, Dormancy and Germination" Elsevier Biomedical Press pp 157-181

13. Hooley R (1982) Protoplasts isolated from aleurone layers of wild oat (Avena fatua L.) exhibit the classic response to gibberellic acid. Planta 154 29-40

14. Jacobsen JV, Zwar JA and Chandler PM (1985) Gibberellic-acid-responsive protoplasts from mature aleurone of Himalaya barley. Planta 163 430-438

15. Jacobsen JV and Beach L (1985) Control of transcription of α-amylase and rRNA genes in barley aleurone protoplasts by gibberellin and abscisic acid. Nature 316 275-277

16. Zwar JA and Hooley R (1986) Hormonal regulation of α-amylase gene transcription in wild oat (Avena fatua L.) aleurone protoplasts. Plant Physiol. in press

17. Jacobsen JV and Higgins TJV (1982) Characterization of the α-amylases synthesized by aleurone layers of Himalaya barley in response to gibberellic acid. Plant Physiol. 70 1647-1653

18. Brown AHD and Jacobsen JV (1982) Genetic basis and natural variation of α-amylase isozymes in barley. Genet. Res. Camb. 40 315-324

19. Muthukrishnan S, Gill BS, Swegle M and Ram Chandra G (1984) Structural genes for α-amylases are located on barley chromosomes 1 and 6. J. Biol. Chem. 259 13637-13639

20. Svensson B, Mundy J, Gibson RM and Svendsen IB (1985) Partial amino acid sequences of α-amylase isozymes from barley malt. Carlsberg Res. Commun. 50 15-22

21. Huang J-K, Swegle M, Dandekar AM and Muthukrishnan S (1984) Expression and regulation of α-amylase gene family in barley aleurones. J. Molec. and Appl. Genet. 2 579-588

22. Rogers JC and Milliman C (1984) Coordinate increase in major transcripts from the high pI α-amylase multigene family in barley aleurone cells stimulated with gibberellic acid. J. Biol. Chem. 259 12234-12240

Molecular Biology of Plant Growth Control, pages 35–49
© 1987 Alan R. Liss, Inc.

REGULATION OF GENE EXPRESSION IN BARLEY ALEURONE LAYERS[1]

Tuan-hua David Ho, Randall C. Nolan, Liang-shiou Lin, Mark R. Brodl, and Peter H. Brown

Department of Biology, Washington University
St. Louis, Missouri 63130

ABSTRACT   The aleurone layers of barley seeds offer many convenient features for a detailed investigation of the action of two plant hormones, gibberellic acid (GA_3) and abscisic acid (ABA), which regulate the synthesis and secretion of α-amylase and several other hydrolytic enzymes in this tissue. The hormone-regulated synthesis of α-amylase isozymes correlates with the levels of their specifc mRNAs indicating a potential transcriptional control. However, regulation at the level of mRNA stability also exists. α-Amylase mRNAs are very stable in GA_3 treated tissue probably due to their association with the GA_3-induced endoplasmic reticulum (ER). When the cells are under heat stress there is a destruction of ER with a concomitant decline in the stability of mRNAs encoding secretory proteins such as α-amylases and protease. Midcourse addition of ABA also reduces the stability of α-amylase mRNAs. This effect of ABA is dependent on the expression of other genes because it is prevented by transcription and translation inhibitors. In addition to the inhibition of α-amylase synthesis, ABA also induces a set of new proteins. Some of these ABA induced proteins are probably related to the conversion of ABA to phaseic acid which has been suggested to be an active component in ABA action.

INTRODUCTION

The aleurone layers of barley and wheat seeds play a crucial role in the mobilization of endosperm nutrients to support post-germination seedling growth. After the onset of germination, aleurone layers respond to the

[1]This work was supported by NSF DCB 8316319.

hormone, gibberellin (GA) from the embryo by synthesizing and secreting several hydrolytic enzymes, including α-amylase (1), protease (2), β-1,3-1,4 glucanase (3) and nuclease (4), into the endosperm where these enzymes hydrolyze the stored starch, proteins, cell wall polysaccharides and remnant nucleic acid. Another hormone, abscisic acid (ABA) which causes seed dormancy, prevents all of the known GA effects in this tissue. The aleurone layers have been considered as a convenient system for studies of the action of GA and ABA because of the following advantages: First, this tissue consists of a homogeneous cell population which responds to the two hormones. Second, at least for GA, the source (embryo) and the target tissue (aleurone layers) of the hormone can be physically separated, thus the target tissue can be treated with known concentrations of hormones. Third, many enzymes and proteins are available which can serve as biochemical markers for studies of hormone action (Table 1). Fourth, protoplasts of aleurone cells that still respond to GA can be prepared (5). The protoplasts are ideal for studies of hormonal regulation on secretion. Furthermore, organelles such as nuclei can be more easily isolated from protoplasts than from the intact cells.

COMPLEXITY AND EXPRESSION OF α-AMYLASE ISOZYMES

Both GA and ABA alter the expression of several sets of genes in barley aleurone layers. As summarized in Table 1, gibberellic acid (one of the GAs; GA_3) not only induces α-amylase isozymes, thio-protease, nuclease and a few other enzymes, but also suppresses the expression of some other genes. In contrast, ABA represses the expression of the aforementioned hydrolytic enzymes, yet induces a set of new proteins (see detailed discussion later). Because of the abundance of α-amylase, the effect of hormones on its synthesis has been most extensively studied. α-Amylase is composed of two sets of isozymes (high and low pI species) encoded by two sets of structural genes each located on different chromosomes (5,10-12). The expression of high pI α-amylase is not detectable before the addition of GA_3 to aleurone layers, yet the low pI α-amylase is expressed at a low level in the same tissue. After hormone treatment, the expression of both groups of α-amylase is enhanced within a couple of hr. The GA_3-enhanced expression of high pI α-amylase reaches a maximum around 20 hr and then declines afterwards (Fig 1). Very little of this isozyme group is synthesized beyond 30 hr of GA_3 treatment. In contrast, the synthesis of low pI α-amylase continues through 40 hr of GA_3 treatment (Fig 1). This differential expression of α-amylase isozymes in GA_3-treated barley aleurone layers can be observed at the protein level by analyzing newly

TABLE 1
HORMONAL REGULATION OF GENE EXPRESSION
IN BARLEY ALEURONE LAYERS

Gene	None	GA	ABA	GA+ABA	Ref
		Hormone Treatment			
α-Amylase-high pI	-	+++++	-	-	(6)
α-Amylase-low pI	+	+++++	+	+	(6)
Thio-protease	-	+++	-	-	(8)
Nuclease (RNase+DNase +3'-nucleotidase)	+	+++	-	ND[a]	(4)
β-1,3-1,4 glucanase	-	+++	-	-	(3)
Actin	++	++	++	++	b
Non-differential[c]	++	++	++	++	b
Alcohol dehydrogenase	+	-	+	+	b
GA suppressed[c]	+++	-	+++	+	b
ABA induced p29	+	-	+++	+	(9)
ABA-induced p36 (glcNH$_2$-lectin-?)	-	-	++	+	(9)

a, ND: not determined. b, Nolan RC, Brodl MR, Ho THD, unpublished.   c, unidentified cDNA clones.

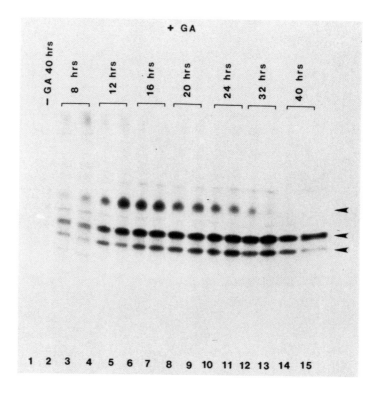

Figure 1. Time course of the GA₃ induction of α-amylase isozymes in barley aleurone layers. Isolated aleurone layers were incubated with 10⁻⁶ M GA₃ for the lengths of time indicated on top of the gel. The tissues were labeled with [³⁵S]methionine during the last hr of incubation. Duplicate samples of newly synthesized proteins were analyzed by non-denaturing gel electrophoresis and the fluorogram of the gel is shown. The top arrow on the right indicates the position of the high pI isozymes (the two major high pI isozymes run to the same position on this gel) and the two lower arrows indicate the low pI isozymes.

synthesized proteins with native gel electrophoresis (Fig 1). Similar results have been obtained at the RNA level by dot blot analysis probed with cDNAs specific for the two groups of isozymes (Fig 2). Since the results of protein analysis are similar to those of RNA analysis, the GA_3-mediated induction of α-amylase isozymes is mainly at the transcriptional level. Results similar to what has been described here have also been reported by Huang *et al* .(13). More recently, employing the *in vitro* nuclear run-on transcription technique, Jacobsen and Beach (14) have shown that GA_3 enhances the rate of transcription of α-amyalse genes.

Regulation of α-amylase synthesis at levels other than transcription also exists in barley aleurone layers. For example, α-amylase mRNAs appear to be quite stable. It has been shown that transcription inhibitors, such as cordycepin (3'-deoxyadenosine), are very effective in inhibiting GA_3 induced α-amylase synthesis (15). However, cordycepin fails to prevent the continued synthesis of α-amylases when it is added 12 hr or more after GA_3 indicating that the α-amylase mRNAs are synthesized during the first 12 hr of hormone treatment and that the turnover rate of this mRNA is very slow (15). The half-life of these mRNAs has been estimated to be longer than 100 hr (16). What causes the stability of these particular mRNAs? Investigating the effect of heat shock in barley aleurone layers, we have recently revealed a potential mechanism underlying mRNA stability. A heat shock treatment (40° C) of barley aleurone layers not only induces the heat shock proteins (hsp) but also effectively diminishes the synthesis of GA_3- induced α-amylases (17). In the analysis of α-amylase mRNA sequences by Northern gels and dot blots probed with specific cDNAs, it is observed that the normally stable α-amylase mRNAs in heat shocked barley aleurone layers are actively degraded (Fig 3). This is contrary to the case of *Drosophila* cells where normal mRNAs are sequestered and preserved, and only hsp mRNAs are translated during the heat shock treatment (18). The timing of the heat shock induced α-amylase mRNA destruction is closely correlated with another heat stress induced phenomenon, a fast delamination of endoplasmic reticulum (ER) lamellae (Fig 4). Both of these events start within 30 min of heat shock and reach the maximum within 3-4 hr. When the tissue is recovered from heat shock treatment, the ER lamellae are reformed and the synthesis of α-amylases is resumed (17, Fig 4). The resumption of α-amylase synthesis, of course, is dependent on the new synthesis of their mRNAs (17). We have found that the heat shock treatment not only affects the synthesis of α-amylases but also of all of the other secretory proteins. In contrast, the mRNA encoding actin, a protein synthesized on free polyribosomes, is unaffected by the heat shock treatment (Fig 3). Since ER is the site for the synthesis and processing of secretory proteins, we suggest that the association of mRNAs encoding

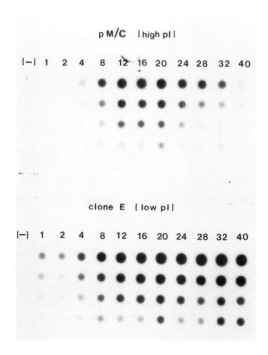

Figure 2. Dot blot analysis of the levels of mRNA sequences for α-amylase isozymes in GA$_3$ treated barley aleurone layers. Isolated aleurone layers were incubated with 10^{-6} M GA$_3$ for the number of hr shown on top of the blots. Total RNA was isolated and applied to "Gene Screen Plus" membrane (New England Nuclear, Boston, MA) after a series of two-fold dilutions (the highest concentration being the topmost). The membrane was then probed with two nick-translated cDNA probes, pM/C for the high pI isozyme mRNA sequences and clone E for the low pI isozyme sequences. The fluorogram of the blots is shown. "-", aleurone layers incubated without GA$_3$ for 40 hr.

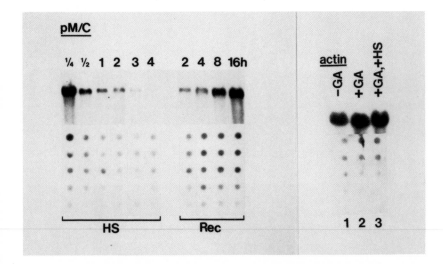

Figure 3. Effect of heat shock on the levels of mRNA in barley aleurone layers. The aleurone layers were incubated with 10^{-6} M GA_3 for 16 hr before they were heat shocked at 40° C for the lengths of time in hr as indicated on top of the blots. Some of the heat shocked layers were allowed to recover at 25° C for up to 16 hr as indicated. Total RNA was isolated and analyzed by Northern gel (top) and dot blot (bottom) hybridizations. The probes used were pM/C, a cDNA clone for the high pI α-amylase isozymes, and a soybean actin cDNA clone which cross-hybridizes with barley RNA.

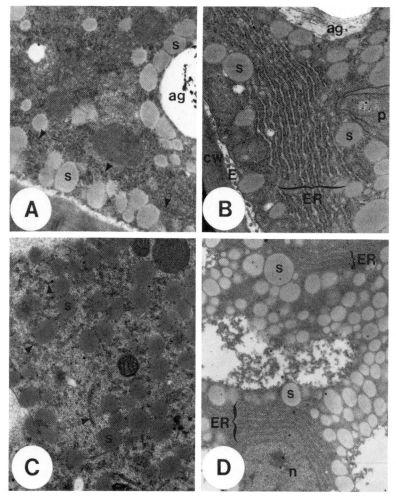

Figure 4. Effect of heat shock on the ultrastructure of barley aleurone layers. The aleurone layers were incubated with 10^{-6} M GA_3 for 16 hr before they were heat shocked at 40° C for 3 hr. The tissues were then processed for observation by transmission electron microscopy. Panel A, tissue treated with no GA_3 and also not heat shocked. Panel B, tissue treated with GA_3 but not heat shocked. Note the massive proliferation of ER lamellae that are induced by GA_3. Panel C, GA_3-treated tissue that has been heat shocked. Note that the GA_3-induced ER lamellae are disrupted. Panel D, same as in Panel C except the tissue was recovered from heat shock at 25° C for 8 hr. Note that ER lamellae have reappeared.

secretory proteins with the ER leads to the stabilization of these mRNAs. Once the ER is destroyed by heat shock, the mRNAs normally associated with the ER become unstable and are rapidly degraded. It is well documented that GA_3 also induces the formation of ER in barley aleurone layers (19). Thus, besides regulating α-amylase synthesis at the transcription level, this hormone may also indirectly stabilize α-amylase mRNA.

EFFECT OF ABA ON THE LEVELS OF α-AMYLASE mRNA

Abscisic acid is quite effective in inhibiting the synthesis of both high and low pI α-amylase isozymes when it is added at the same time as GA_3 (Fig 5). However, it has a much more noticeable effect on the synthesis of high pI isozymes than that of the low pI isozymes when it is added 20 hr or longer after GA_3 (Fig 5). This differential effect of ABA on α-amylase synthesis can be easily observed by non-denaturing gel electrophoresis, yet not by SDS gel electrophoresis because the isozymes are not resolved by the latter technique. The physiological significance of this differential inhibition of the synthesis of α-amylase isozymes is still unclear. Probing mRNA sequences by Northern gels, we have observed that the ABA inhibition of α-amylase synthesis correlates with the decrease in mRNA sequences for the individual groups of isozymes (Fig 6). Thus, the ABA inhibition on α-amylase synthesis appears to be regulated at the level of mRNA rather than at the translational level as previously suggested (20,21). Since α-amylase mRNAs appear to be quite stable after more than 12 hr of GA_3 treatment and the mid-course addition of ABA alters the levels of α-amylase mRNA, it appears that ABA affects the stability of these mRNA. However, it should be noted that affecting mRNA stability is by no means the only mode of action of ABA in this tissue. Jacobsen and Beach (14) have shown that ABA decreases the transcription rate of α-amylase genes when it is added at the same time as GA_3.

In the course of investigating the action of ABA on α-amylase mRNA stability, we have observed an interesting phenomenon, i.e. the action of ABA itself is dependent on the continuous synthesis of some RNA and proteins. As shown in Fig 6, inhibition of RNA or protein synthesis by cordycepin or cycloheximide blocks the effects of ABA on the decrease of α-amylase mRNA sequences. One cannot argue that the effects of these metabolic inhibitors are the consequence of induced cellular toxicity because the α-amylase mRNA sequences are even more abundant in the presence of these compounds. Thus, the action of ABA probably relies on the expression of another gene(s).

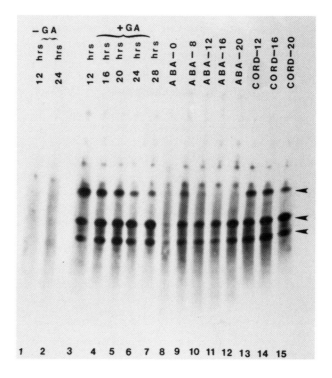

Figure 5. Effect of ABA on the synthesis of α-amylase isozymes in barley aleurone layers. Aleurone layers were incubated with various hormones and inhibitors as indicated and labeled with [^{35}S]methionine during the last hr of incubation. The newly synthesized proteins were analyzed by non-denaturing gel electrophoresis and the fluorogram of the gel is shown. The numbers for GA$_3$ treated samples indicate the numbers of hr of incubation. The numbers for the ABA and cordycepin (CORD) treated samples indicate the time when ABA or cordycepin was added (in hr after GA$_3$ addition). GA$_3$ was present continuously for these samples and the total incubation time was 25 hr. [GA$_3$]=10^{-6} M, [ABA]=2 x 10^{-5} M, and [CORD]=10^{-4} M. The top arrow on the right indicates the position of high pI α-amylase isozymes and the two lower arrows indicate the low pI isozymes.

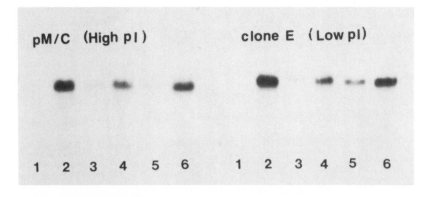

Figure 6. Effect of ABA on the levels of mRNA for α-amylase isozymes. Aleurone layers were incubated for 25 hr in the absence of GA_3 (lane 1), 25 hr in the presence of 10^{-6} M of GA_3 (lane 2), 25 hr in the presence of GA_3 and 2×10^{-5} M ABA with a second addition of ABA at 12 hr (lane 3), 25 hr with ABA added 10 hr after GA_3 (lane 4), 25 hr with ABA added 20 hr after GA_3 (lane 5), or 25 hr with ABA and 10^{-4} M cordycepin both added 20 hr after GA_3 (lane 6). Total RNA from these samples was analyzed by Northern gels probed with pM/C for high pI isozymes and clone E for low pI isozymes. Note that cordycepin prevents the effect of ABA (compare lanes 5 & 6).

ABA INDUCED PROTEINS AND THEIR POTENTIAL ROLES

In order to study the gene(s) whose expression is crucial to the action of ABA in the inhibition of α-amylase synthesis, we have analyzed newly synthesized proteins in ABA treated aleurone layers. As shown in Fig 7A, there are nine ABA induced protein bands on one-dimensional gels, and the number increases to 16 when the samples are analyzed by two-dimensional gels (9). Treatment of aleurone layers with ABA also increases the levels of mRNAs encoding the ABA-induced proteins as determined by *in vitro* protein synthesis (Fig 7B). ABA concentrations as low as 10^{-8} M are able to induce some of these proteins. The identities of these ABA-induced proteins are still not clear. However, some interesting features of these proteins have been revealed. An ABA-induced protein with a size of 36 kD can be precipitated with antiserum against a barley lectin which is specific for glucosamine, galactosamine and mannosamine (Fig. 8) (22). The most

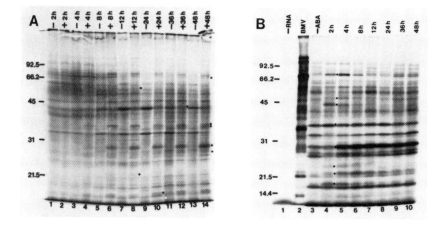

Figure 7. Time course of ABA induction of new proteins in barley aleurone layers. (A) *In vivo* studies. Aleurone layers were incubated with (+) or without (-) 2 x 10⁻⁵ M ABA for the indicated time and labeled with [³⁵S]methionine during the last hr of incubation. (B) *In vitro* studies. Total RNA was isolated from tissues treated the same way as described under (A). The RNA samples were translated in rabbit reticulocyte lysates with [³⁵S]methionine as the label. No RNA (lane 1) and Brome Mosaic Virus (BMV) mRNA controls were included. All the samples were analyzed by SDS gel electrophoresis and the fluorograms of the gels are shown. The dots indicate protein bands that are induced by ABA. Molecular weight markers in kD are indicated by bars at left.

abundant among all of the ABA induced proteins is 29 kD in size. This protein can also be induced by osmotic stress and salt stress, thus it can be classified as a stress-induced protein. A few of the ABA induced proteins can also be induced by ABA in developing barley seeds. It is not known whether they are part of barley seed storage proteins.

Some of the ABA induced proteins may be related to the metabolism of ABA to phaseic acid (PA), the first stable ABA metabolite which has been suggested to be an active component in ABA action (23). Phaseic acid is

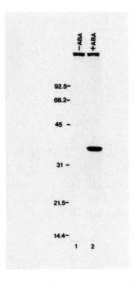

Figure 8. Immunoprecipitation of barley aleurone proteins with antiserum against a barley lectin specific for glucosamine, galactosamine and mannosamine. Aleurone layers were incubated with or without 2×10^{-5} M ABA for 24 hr and labeled with [^{35}S]methionine for the last hr. Proteins were extracted and incubated with antiserum against the lectin and precipitated with *Staphylococcus aureus* Cowan strain 1. The pellets were washed, resuspended and analyzed by SDS gel electrophoresis. The fluorogram of the gel is shown. Molecular weight markers are indicated by bars at left.

very effective in the inhibition of α-amylase synthesis, yet it does not induce any of the ABA inducible proteins (23). It has been shown by Uknes and Ho (24) that the *in vivo* conversion of ABA to PA in barley aleurone layers is enhanced by pretreating the tissue with ABA. Thus, it seems that ABA is capable of enhancing its own metabolism to PA similar to the case of substrate induction of nitrate reductase by nitrate that has been well documented in many plant tissues. The conversion of ABA to PA is dependent on RNA and protein synthesis, and the enzyme involved in this process is thought to be a cytochrome P450-linked monooxygenase (25). Work is underway in our laboratory to test whether the known inhibitors of this enzyme system can block the action of ABA on the inhibition of α-amylase synthesis.

REFERENCES

1. Chrispeels MJ, Varner JE (1967). Gibberellic acid-enhanced synthesis and release of α-amylase and ribonuclease by isolated barley aleurone layers. Plant Physiol 42:398.
2. Jacobsen JV, Varner JE (1967). Gibberellic acid-induced synthesis of protease by isolated aleurone layers. Plant Physiol 42:1596.
3. Mundy J, Fincher GB (1986). Effect of gibberellic acid and abscisic acid on levels of translatable mRNA (1-3, 1-4) β-D-glucanase in barley aleurone. FEBS Let 198:349.
4. Brown PH, Ho THD (1986). Barley aleurone layers secrete a nuclease in response to gibberellic acid. Purification and partial characterization of the associated ribonuclease, deoxyribonuclease, and 3'-nucleotidase activities. Plant Physiol , in press.
5. Jacobsen JV, Zwar JA, Chandler PM (1985). Gibberellic-acid-responsive protoplasts from mature aleurone of Himalaya barley. Planta 163:430.
6. Higgins TJV, Jacobsen JV, Zwar JA (1982). Gibberellic acid and abscisic acid modulate protein synthesis and mRNA levels in barley aleurone layers. Plant Mol Biol 1:191.
7. Jacobsen JV, Higgins TJV (1982). Characterization of the α-amylase synthesized by aleurone layers of Himalaya barley in response to GA3. Plant Physiol 70:1647.
8. Hammerton RW, Ho THD (1986). Hormonal regulation of the development of protease and carboxypeptidase activities in barley aleurone layers. Plant Physiol 80:692.
9. Lin LS, Ho THD (1986). Mode of action of abscisic acid in barley aleurone layers: induction of new proteins by ABA. Plant Physiol, in press.
10. Callis J, Ho THD (1983). Multiple molecular forms of the gibberellin-induced α-amylase from the aleurone layers of barley. Arch Biochem Biophys 224:224.
11. Muthukrishnan S, Gill BS, Swegle M, Chandra R (1984). Structural genes for α-amylase are located on barley chromosomes 1 and 6. J Biol Chem 259:13637.
12. Brown AHD, Jacobsen JV (1982). Genetic basis and natural variation of α-amylase isozymes in barley. Genet Res 40:315.
13. Huang, JK, Swegle, M, Dandekar, AM, Muthukrishnan (1984). Expression and regulation of α-amylase gene family in barley aleurones. J Mol App Genet 2:579.
14. Jacobsen JV, Beach LR (1985). Control of transcription of α-amylase and rRNA genes in barley aleurone layer protoplasts by

gibberellic acid and abscisic acid. Nature 316:275.
15. Ho THD, Varner JE (1974). Hormonal control of messenger ribonucleic acid metabolism in barley aleurone layers. Proc Natl Acad Sci 71:4783.
16. Ho THD (1976). "On the Mechanism of Hormone Controlled Enzyme Formation in Barley Aleurone Layers." Ph D dissertation Michigan State University, East Lansing, MI.
17. Belanger FC, Brodl MR, Ho THD (1986). Heat shock causes destabilization of specific mRNAs and destruction of endoplasmic reticulum in barley aleurone layers. Proc Natl Acad Sci 83:1354.
18. Storti RV, Scott MP, Rich A, Pardue, ML (1980). Translational control of protein synthesis in response to heat shock in *D. melanogaster* cells. Cell 22:825.
19. Jones RL, Jacobsen JV (1982). The role of the endoplasmic reticulum in the synthesis and transport of α-amylase in barley aleurone layers. Planta 156:421.
20. Ho THD, Varner JE (1976). Response of barley aleurone layers to abscisic acid. Plant Physiol 57:175.
21. Mozer, TJ (1980). Control of protein synthesis in barley aleurone layers by the plant hormones gibberellic acid and abscisic acid. Cell 20:479.
22. Patridge J, Shannon L, Gumpf D (1976). A barley lectin that binds free amino sugars. I. Purification and characterization. Biochim Biophys Acta 451:470.
23. Ho THD, Nolan RN, Uknes SJ (1985). On the mode of action of abscisic acid in barley aleurone layers. Current Topics in Plant Biochem Physiol 4:118.
24. Uknes SJ, Ho THD (1984). Mode of action of abscisic acid in barley aleurone layers: abscisic acid induces its own conversion to phaseic acid. Plant Physiol 75:1126.
25. Gillard DF, Walton, DC (1976). Abscisic acid metabolism by a cell free preparation from *Echinocystis lobata* liquid endosperm. Plant Physiol 58:795.

Molecular Biology of Plant Growth Control, pages 51–62
© 1987 Alan R. Liss, Inc.

DNA AND AMINO ACID SEQUENCES OF SOME ABA RESPONSIVE GENES EXPRESSED IN LATE EMBRYOGENESIS IN COTTON COTYLEDONS

Jean C. Baker and Leon Dure III

Department of Biochemistry, University of Georgia
Athens, Georgia USA 30602

INTRODUCTION

During the embryogenesis of the cotton seed at least 7 different sets of genes are active to the extent of providing the embryo with abundant proteins (1,2,3,). Two of these sets are very active in late embryogenesis and produce the most abundant mRNAs of the mature seed. The high expression of these genes is not continued in germination and these residual mRNA disappear during the first 24 hours of germination (1,2,3,4). The proteins coded by these genes we call LEA (late embryogenesis abundant) proteins; thus LEA mRNAs, LEA genes. These two sets differ from one another in that the up-modulation of expression in embryogenesis of one set occurs developmentally a few days before the other.

These two sets of genes are of interest because their up-modulation in expression may be linked to the exposure of the seed tissue to elevated levels of abscisic acid. Young embryos that have yet to reach the stage in embryogenesis when the LEA mRNAs and proteins become easily demonstrable can be induced to express the LEA genes at a high level prematurely by exposing them to ABA (10^{-6}M) upon excision. (1,4)

We have made cDNA libraries to the mRNA of the mature seed, cloned these libraries by conventional means, and identified and isolated cDNA representing some of the mRNAs produced from the LEA gene sets. Several of these cDNAs have been sequenced as have several of the genes themselves, isolated from our cotton genomic library. From these sequences we have deduced the amino acid structure of several proteins and determined features of the genes.

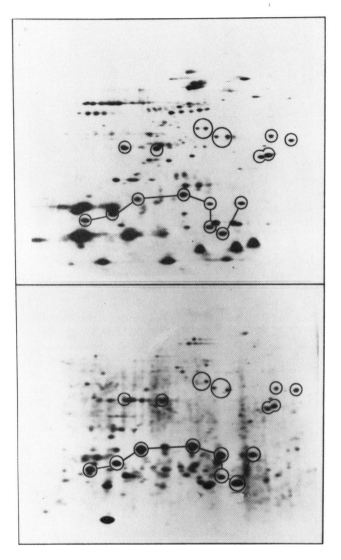

FIGURE 1.   LEA proteins are abundant in late
embryogenesis. The top panel is a 2D gel of extant soluble
protein from the mature seed shown by staining. The bottom
panel is of translation products obtained in vitro with
mature seed mRNA. LEA proteins are circled.

Our goal in this is ultimately to:
1) identify the cis-acting DNA sequences flanking the genes of the two sets that function in their up-modulated expression in late embryogenesis.
2) isolate and characterize the trans-acting proteins that, via binding to the cis-acting sequences, effect the up-modulation in expression.
3) test the possibility that the trans-acting proteins are induced by ABA or perhaps utilize ABA as a co-effector.
4) determine the function of the proteins specified by the LEA genes.
We report here progress made in sequencing LEA cDNAs and genes and some characteristics of the LEA proteins.

RESULTS

LEA Proteins are Abundant in Cotton Cotyledons in Late Embryogenesis.

Figure 1 (bottom) shows the 2D profile of proteins synthesized in the wheat germ translation system using mRNA extracted from mature cotton cotyledons. The encircled proteins are those previously identified as LEA proteins via 2D electrophoresis of in vivo and in vitro synthesized proteins during embryogenesis and early germination.(1,3) Clearly these proteins are the dominant products in this translation implying that their mRNAs are the most abundant mRNAs at this point in ontogeny. The top panel is a stained gel of the readily soluble proteins of mature cotyledons (storage proteins removed) and the LEA proteins are again encircled. Although there are other very abundant proteins that have accumulated in this tissue during embryogenesis, it is apparent that the LEA proteins are members of the abundant class by seed maturity. Notice also that the LEA proteins migrate to identical positions on both gels indicating that they are not processed in anyway after their synthesis.
Figure 2 shows some of the LEA proteins whose in vitro translation can be arrested by single cDNA clones (e.g. proteins included in a single circle or circled proteins connected by a line). In most cases a single clone was found to hybridize with at least 2 different mRNAs. This suggets that most of the LEA proteins belong to small multigene families, which is not surprising in view of the

fact that commercial cotton used in these experiments is a
natural tetraploid (amphidiploid).

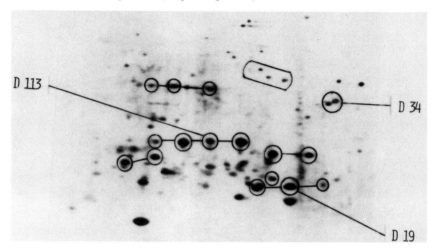

FIGURE 2. Multigene families among LEA genes. In
vitro translation products obtained with RNA from mature
cotyledons separated on a 2D gel. Those proteins whose
translation is arrested by a single cDNA species are
enclosed in a common circle or connected by a common line.

As mentioned. one group of LEA genes is up-modulated
in expression before the other. The former group we have
labelled "gradual" and the latter "abrupt", since their
appearance on stained developmental 2D gels is rather sudden
in late embryogenesis. To date we have cDNA and genomic DNA
sequence data on 3 members of the abrupt group, trivally
named D-19, D-11 and D-34, and a single member of the
gradual group, D-113. The proteins encoded by the gene
families represented by these cDNA are shown on Figure 2.
D-11 is not shown on the figure since we have not observed
the D-11 cDNA to arrest the translation of any protein found
on the 2D gels. This may indicate that the mRNA for the
protein is not abundance enough to give use to scorable
protein on the gels, or to the fact that the protein pI is
not within the pH span of the gels (pH 4.2 7.5). The
latter seems more likely in view of large number of
potential phosporylation sites in the amino acid sequence of
the defined protein (see below).

Developmental Time Course of Expression of Some LEA
Proteins.

Figure 3 shows dot blot hybridization of mRNA prepared
from different developmental time points with several cDNA
probes including those that we have sequenced. The top 3
lanes show hybridizations with probes that represent the
gradual group of LEA genes and the next 3 lanes with abrupt
probes. The difference in time of up-modulation of
expression is apparent. By 48 hours of germination, none of
these mRNAs are abundant indicating a down-modulation or
even a total shut down in the expression of these genes in
germination. The bottom three lanes show hybridization with
probes representing genes under totally different
regulation. (1,5).

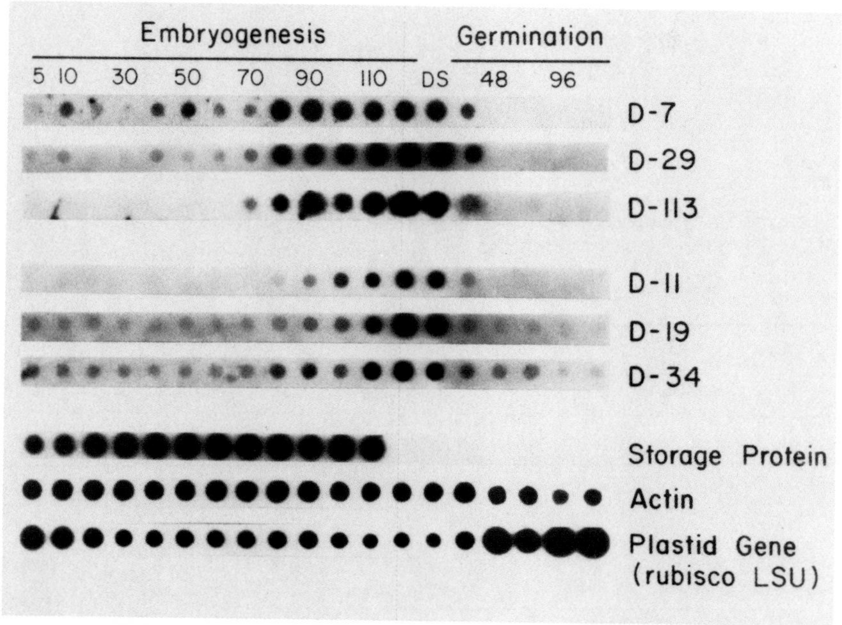

FIGURE 3. Developmental dot blot of some LEA mRNAs.
Numbers across the top refer to embryo weight
(embryogenesis) and days germinated (germination).
Identical amounts of total RNA was applied to each dot. The
top three lanes were probed with gradual cDNAs, the middle
three lanes with abrupt cDNAs and the bottom three with
cDNAs for genes under different regulation.

Characteristics of cDNAs/Genes.

Figure 4 is a diagram of the cDNAs and genomic
fragments sequenced to date. Three of the 4 cDNAs sequenced
are essentially full length copies of their mRNAs in that
they contain the ATG translation start site and extend
through the poly (A) 3' tail. One clone, C-34 contains only
the 3' half of the mRNA. Some general features emerge from
these cDNA sequences. Only 1 of the 4 has the perfect
consensus poly (A) signal AATAAA. The others have single
base variants. The distance between these signals and the
beginning of the poly (A) tail varies from a low of 18 to
over 50 bases. Two clones (D-19, D-113) contain a curious
cloning artifact. The 3' terminal 100 nucleotides are

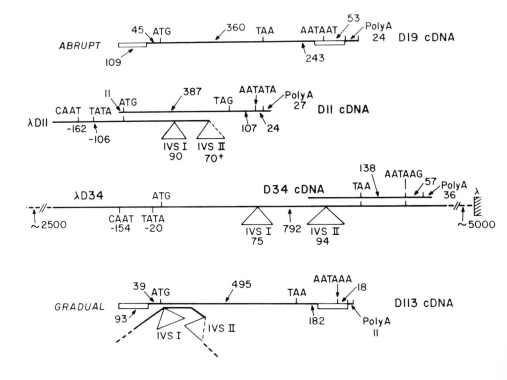

FIGURE 4. Schematic of cDNAs/genomic fragments
sequenced to date.

found at the 5' end of these clones in the inverse
complement orientation (boxed sequences in Figure 4). We
have observed this same phenomenon in some cDNA clones for
other groups of genes. We have posited a possible mechanism
for how this artifact is generated (6) in which we propose
that at the end of cDNA second strand synthesis, a loop is
formed and a "third" cDNA strand is synthesized to
completion. This would artifically introduce 3' terminal
nucleotides at the 5' terminus of the cDNA in an inverse
complement arrangement.

Genomic DNA hybridizing with D-11, D-34 and D-113
cDNAs has been sequenced. In the case of D-11 the sequence
extends from 200 nucleotides upstream from the ATG
translation start site through about 2/3 of the mRNA. Two
small introns have been encountered. The TATA and CAAT
sequences required for correct expression of the gene are
unusually far upstream from the ATG; -106 and -162
respectively. In the case of C-34 we have the sequence from
-1000 through the coding region and the poly (A) site to
about 800 nucleotides further down stream. Two small
introns characterize this gene also. In both cases the
genomic sequences match precisely the cDNA sequences
indicating that the mRNAs originate from these genes and not
from other genes of the multigene families.

This is not the case with genomic DNA sequences
isolated by hybridization with cDNA D-113. Two different
genomic fragments have been obtained. One of these show
about 5% mismatch with the DNA indicating that is not the
family member that gave rise to the mRNA. The other appears
to have no homology with the cDNA except in a region of
about 60 nucleotides that codes for a highly charged region
of 20 amino acids this sequence is diagrammed in Figure 4.
Even in this region there are several mismatches. We do not
know if this latter gene is a member of the multigene the
family represented by D-113 or if it is a gene for a protein
that shares with it only a common functional amino acid
domain.

Computer Analysis of DNA Sequences

Quite obviously we do not have enough sequence data to
search for 5' consensus sequences unique to each set of LEA
genes or even sequences consensus to all embryo-abundant
genes. The identification of possible cis-acting regulatory
sequences, either with general functions in transcription

(phasing of nucleosomes, binding of topoisomerases, etc.) or with specific functions (up-modulation of expression at specific developmental time points) requires much more sequence data for the LEA genes and for other gene sets having different coordination in embryogenesis.

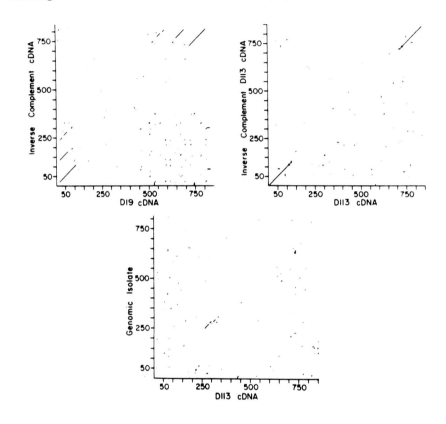

FIGURE 5.   Dot matrix analyses of DNA sequences.   Top two panels are of cDNAs for D-19 and D-113 plotted against the inverse complement of their sequence.   The bottom panel is of cDNA D-113 plotted against a genomic DNA D-113 plotted against a genomic DNA fragment obtained by hybridization with D-113.

However a computer analysis comparing sequence relationships allows for other phenomena to be visualized. Figure 5 shows examples of this. The top panels of this figure show a dot matrix analysis of the cDNAs D-19 and D-113 vs their inverse complement sequence. (The span of nucleotides is 8 of which 7 must be identical to generate a dot.) The presence of the 3' terminal stretch at nucleotides of the 5' terminus of the cDNA in the inverse complement orientation mentioned above is clearly demonstrable by this means. The series of lines seen in the D-19 plot shows that a sequence in the 3' terminal stretch is repeated 3 times.

The bottom panel compares directly the sequence of the cDNA D-113 with the genomic fragment mentioned above that has no homology other than in the 60 nucleotide stretch. This stretch is easily scored by this means and the lack of homology elsewhere in the sequences is obvious.

Characteristics of Derived Proteins

These proteins are of interest in their own right in that becoming abundant in late embryogenesis suggests that they may play a role in equipping the seed for survival, i.e. they may constitute "defense proteins". Furthermore, if these proteins are indeed induced into the abundant class by ABA, they would constitute part of the arsenal of proteins regulated by this interesting depressant hormone. Unfortunately, screening their amino acid sequences against the Dayhoff Protein Data Base has not revealed any homologies to date.

Since many seeds accumulate large amount of trypsin and chymotrypsin inhibitors, presumably to discourage insect or microorganism predation (plant proteases are generally not serine-histidine proteases), we asked C.A. Ryan (Washington State University) to screen the soluble protein fraction of the cottons seeds for such inhibitors by affinity chromatographic techniques. The inhibitors he found were in low abundance, and, by 2D electrophoresis, we identified them as minor proteins present throughout embryogenesis at constant levels, i.e. they are not LEA proteins.

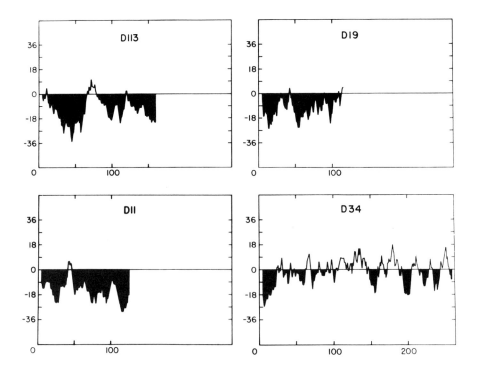

FIGURE 6.   Hydropathy analyses of the amino acid
sequences of 4 LEA proteins.

There are some interesting features of the amino acid
sequences of the 4 proteins as shown in Table 1.   None have
cys or trp.   Several are very high in gly which would seem
to allow for high flexibility.   D-11 is also very high in
the hydroxylated amino acids, some of which are clustered as
shown in the table.   If any of these ser or thr residues are
phosphorylated, the pI of the D-11 protein could be lower
than 4 (see above).   Hydropathy analysis, (7) shown in
Figure 6, of the 4 known amino acid sequences show that the
3 rather small proteins have similar profiles and are
extremely hydrophilic.   The D-34 protein gives a profile
more typical of a globular protein.

TABLE 1

AMINO ACID COMPOSITIONS

	Gradual D-113 (17,486 kD)	Abrupt D-11 (13,978 kD)	D-19 (13,186 kD)	D-34 (26,934 kD)
gly	17.6%	14.7%	13%	
ala				18.7
val				10.5
ser		10.1		
thr		10.8		
cys	0	0	0	0
trp	0	0	0	0

D-11

...ser ser ser ser ser ser ser ser ser...
...thr ser thr thr thr...

REFERENCES

1. Dure L III, Greenway S, Galau GA (1981). Developmental biochemistry of cottonseed embryogenesis and germination XIV. Changing mRNA population as shown by in vivo and in vitro protein synthesis. Biochemistry 20: 4162-4168.

2. Galau GA, Dure L III (1981). Ibid XV Changing mRNA population as shown by nucleic acid hybridization. Biochemistry 20:4196-4178.

3. Dure L III, Chlan CA Galau, GA (1983) Developmentally regulated gene sets in cotton embryogenesis. In Goldberg RB (ed) "Plant Molecular Biology" New York, Alan R. Liss Inc. pp 331-342.

4. Galau GA, Hughes W, Dure L III (submitted). Identification and characterization of cDNA clones complementary to mRNAs coding for LEA (Late Embryogenesis Abundant) proteins. Plant Molecular Biology.

5. Dure L III Pyle JB, Chlan CA, Baker JD, Galau GA. Ibid
    XVII (1983). Expression during development of cottonseed
    storage protein genes. Plant Molecular Biology 2:199-
    206.

6. Chlan. CA. Pyle JB. Legocki AB and Dure L III
    (submitted). Ibid XVIII. cDNA and amino acid sequences
    of members of the storage protein families of
    cottonseed. Plant Molecular Biology.

7. Kyle J. Doolittle RF (1982). A simple method for
    displaying the hydropathic character of a protein. J Mol
    Biol 57:185-132.

Molecular Biology of Plant Growth Control, pages 63–71

PRODUCTION OF A HYBRID PROTEIN BETWEEN E. coli
β-GALACTOSIDASE AND THE PEA PROTEIN SPECIFIED BY
THE EARLY AUXIN-REGULATED mRNA IAA4/5[1]

Paul W. Oeller and Athanasios Theologis

Department of Biological Chemistry, Washington University
School of Medicine, St. Louis, Missouri 63110

ABSTRACT   The complementary DNA sequence IAA4/5 of
an early indoleacetic acid-regulated mRNA in elongating pea stem tissue has been inserted in-frame
into the E. coli lacZ gene of the λgt11 expression
vector. Thermal induction of the λgt11 recombinant
lysogens in the presence of isopropyl-β-D-thio-
galactospyranoside results in the accumulation of a
hybrid of β-galactosidase and a protein specified
by the inserted DNA. The hybrid protein constitutes
5-10% of the total E. coli protein, is heat stable
and in contrast to β-galactosidase is highly insoluble. The molecular weight of the hybrid polypeptide by SDS-PAGE[2] is about 21 Kd larger than that
of β-galactosidase alone.

INTRODUCTION

Auxin is known to regulate various aspects of plant
growth and development (1); however, its primary mechanism
of action is poorly understood. Recent advances in recombinant DNA technology have allowed investigation into whether
auxin has the capacity to act at the transcriptional or post-
transcriptional level concomitant with the initiation of cell

[1]This work was supported in part by National Institutes
of Health grant GM 35447-01 and in part by National Science
Foundation grant DCB-8421167 to A.T.
[2]Abbreviations: IAA, indoleacetic acid. SDS-PAGE,
sodium dodecyl sulfate-polyacrylamide gel electrophoresis.
IPTG, isopropyl-β-D-thiogalactopyranoside. Kd, kilodaltons.

elongation (2). Various laboratories have isolated complementary DNA sequences to specific mRNAs by low auxin concentration (20 µM) in elongating pea and soybean tissue (2).
The response is the fastest known for any plant growth regulator (10-15 min) and is highly specific for auxins. Since
the induction occurs in the presence of protein synthesis
inhibitors (2), the prospect arises that it may be a primary
transcriptional activation directed by auxin.

The function of the polypeptides coded by the early
auxin-regulated mRNAs is unknown. The elucidation of the
biochemical role that the auxin-regulated gene products play
in plant cell growth would be greatly facilitated by the
availability of antibodies to the proteins specified by the
cDNA clones. Herein, we report the production of a hybrid
protein in E. coli using the λgt11 expression system (3,4).
The hybrid protein consists of E. coli β-galactosidase and a
protein from pea specified by the early auxin-regulated mRNA
IAA4/5 (5). Monospecific antibodies directed toward the
hybrid encoded by the pea DNA will be used to purify the
protein, to localize it subcellularly, and to determine the
kinetics of its accumulation during auxin-induced cell
growth in pea stem tissue.

MATERIALS AND METHODS

Enzymes.

T4 DNA ligase, E. coli DNA polymerase I (large fragment) and DNA restriction endonucleases were from New
England Biolabs. Phosphatased λgt11 arms were purchased
from Promega Biotec.

Phage and Bacterial Strains.

λgt11 phage and E. coli strains Y1088 and Y1089 have
been described (6). In vitro-packaging extracts were prepared from strains BHB2671 and BHB2673 as described (7,8).
In vitro-packaged phages were amplified by plating on Y1088
(4). Growth of λ phage was on solid medium and λ DNA was
purified according to Davis et al. (9).

Construction of λgt11 Recombinants Containing the IAA4/5
cDNA.

The recombinant plasmid PIAA4/5 (5) was digested with
EcoRI and the 0.81-Kb cDNA insert was purified by agarose

gel electrophoresis and adsorption to Schleicher and Schuell NA-45 paper. The cDNA was partially digested with nuclease BAL-31 to produce random junction points. The ends were made flush with Pol I, and EcoRl linkers were added. The linkered DNA was subcloned into the EcoRl site of the λgt11 expression vector (3) after previous removal of excess EcoRl linkers. Alternatively, the EcoRl ends of the cDNA were made blunt with T4 DNA polymerase. EcoRl linkers of various lengths (octamers, decamers, dodecamers) were added, and the linkered DNA was subcloned into λgt11. The DNA was packaged in vitro (7,8) and plated on Y1088. Phage stocks from plaque-purified λgt11 recombinants were prepared and stored at 4°C. A detailed experimental protocol will be published elsewhere.

Preparation of Recombinant Antigen from IAA4/5-λgt11 Recombinant Lysogen.

The IAA4/5-λgt11 recombinants were used to lysogenize the E. coli strain Y1089 (4). Lysogens which grew at 30°C and not at 42°C were selected. After growth to $A_{600}=0.5$ in L broth, a lysogen of IAA4/5-λgt11 was induced by a temperature shift (44°C, 30 min) and the production of the fused protein was initiated by the addition of IPTG to 10 mM and incubation at 37°C. Cells were collected by centrifugation 3 hrs after IPTG addition and were frozen in liquid N_2. The E. coli proteins were solubilized by boiling in high SDS buffer, analyzed by SDS-PAGE (10), and visualized by staining with Coomassie brilliant blue. A detailed experimental protocol will appear elsewhere.

RESULTS AND DISCUSSION

λgt11 Expression-Host System.

As a first step in obtaining antisera to the proteins coded by the auxin-regulated mRNA IAA4/5, the corresponding cDNA (5) was introduced into E. coli λgt11 expression vector in order to produce a hybrid protein with E. coli β-galactosidase. The choice of the appropriate expression vector is a prerequisite for successful expression of foreign DNA in E. coli. The first problem is that most foreign DNA does not contain the transcriptional and translational control signals required for expression in E. coli. Consequently, the foreign DNA must be placed under the control of an E. coli promoter that is efficiently recognized by E. coli RNA polymerase and an E. coli ribosomal binding site that can be

used by the bacterial translational machinery. The second
problem in expressing foreign DNA is that unusual polypeptides are efficiently degraded in E. coli (3). The severity
of the problem differs with each antigen; some foreign
proteins are quite stable and some appear highly unstable.
The instability of foreign antigens can often be reduced by
fusing the antigen to a stable host protein and by using
host mutants deficient in specific proteases. Fusion of
unstable foreign antigens to the carboxyterminus of the
stable E. coli β-galactosidase has been shown to enhance the
stability of the fusion proteins (3). More importantly, the
stability of the fusion product of β- galactosidase and the
eucaryotic antigen can be markedly increased in lon mutants
of E. coli (11). The third major problem with foreign
antigen synthesis in E. coli is that the presence of unusual
proteins is occasionally harmful to the cell. A solution of
this problem has been to ensure a transient production of
the foreign antigens: thus the expression of the DNA encoding the foreign protein is repressed during early log-
phase growth of the host cell culture. Near the end of this
period, when the transcriptional and translational apparatus
are still fully active, the expression of the foreign protein is induced and satisfactory levels of the antigen are
produced before cells become inviable.

The above concepts have been incorporated into the
λgt11 expression-host system (3) which was designed by Dr.
Rick Young in Dr. R. Davis' laboratory to improve the levels
to which foreign antigens can accumulate in E. coli. The
λgt11 expression vector (Fig. 1) was constructed to permit
insertion of foreign DNA into the β-galactosidase structural
gene lacZ under the control of the lac operator. λgt11 has
a single EcoRl cleavage site located within the lacZ gene,

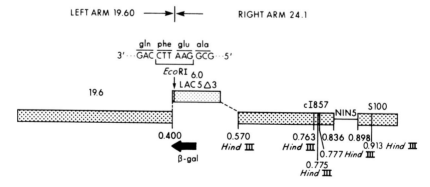

FIGURE 1.   λgt11 expression cloning vector (3).

53 base pairs upstream from the β-galactosidase translation termination codon. It can accommodate up to 7.2 Kb of insert DNA and produce a temperature-sensitive repressor (cI857) which is inactive at 42°C and contains an amber mutation (S100) which renders it lysis-defective in hosts which lack the amber suppressor supF. Because the site of insertion for foreign DNA in λgt11 is within the structural gene for β-galactosidase, foreign DNA sequences in this vector are expressed as fusion proteins with β-galactosidase.

Construction and Preparation of Recombinant Antigen from IAA4/5-λgt11 Recombinant Lysogens.

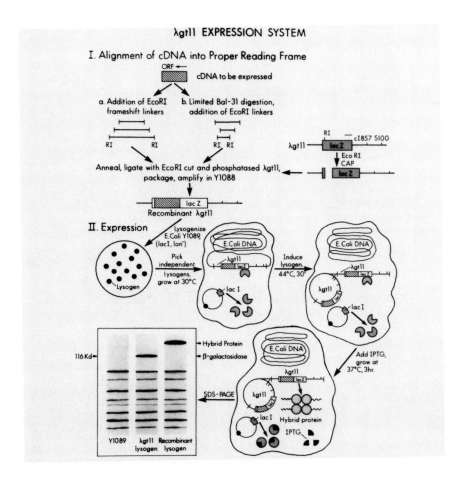

FIGURE 2. λgt11 expression system. See text for details.

The general scheme for preparing a hybrid protein between E. coli β-galactosidase and a protein specified by any cDNA using the λgt11 expression system is shown in Figure 2. It generally involves two steps: first, construction of the recombinant λgt11 phage and second, expression of the hybrid protein specified by the inserted DNA. Proper expression of foreign DNA in λgt11 recombinant lysogens depends on the orientation and reading frame of the inserted DNA with respect to those of lacZ. Thus, one-sixth of the λgt11 recombinants containing a specific cDNA will produce β-galactosidase fused to the protein of interest. To insure proper alignment of the cDNA into proper reading frame, the cDNA insert either is subjected to limited BAL-31 digestion followed by the addition of EcoRl linkers, on its ends or made blund, and EcoRl linkers of various lengths (octamers, decamers, dedecamers) are added. The modified cDNA insert is now ready to be ligated into EcoRl, cut and phosphatased λgt11 vector (Figure 2). The DNA is packaged in vitro and amplified using the E. coli strain Y1088. The λgt11 recombinants are subsequently used to lysogenize the E. coli strain Y1089. The lysogens are thermally induced at 44°C for 30 min and further grown for 3 hrs at 37°C in the presence of 10 mM IPTG. The production of hybrid polypeptides is analyzed by SDS-PAGE (10). E. coli β-galactosidase is a tetramer of identical subunits, each with a MW of 116 Kd (12). The hybrid β-galactosidase synthesized also is a tetramer and its subunits each have a MW of 116 Kd plus the MW of the protein specified by the cloned DNA. The high MW of the hybrid protein permits easy separation of it from the majority of E. coli proteins which have MWs less than 100 Kd.

Figure 3 shows the successful expression of the hybrid protein between β-galactosidase and the protein specified by the pea cDNA IAA4/5 (5) in E. coli following the scheme shown in Figure 2.

The size of the hybrid protein subunit is approximately 21 Kd larger than that the of β-galactosidase subunit (compare lane 3 with lane 2 in Figure 3) which is in close agreement with the observation that the MW of the primary translation product is ∿25 Kd. The pea hybrid portion is smaller than the primary translation product by 4 Kd because the cDNA is not full length (5). For reasons that have been discussed elsewhere, we believe the hybrid portion corresponds to the translational product number 4 (5). The λgt11 recombinant containing the IAA4/5 insert in opposite orientation does not express any hybrid protein (compare lane 4 with lane 3, in Figure 3). Furthermore, the addition of IPTG is

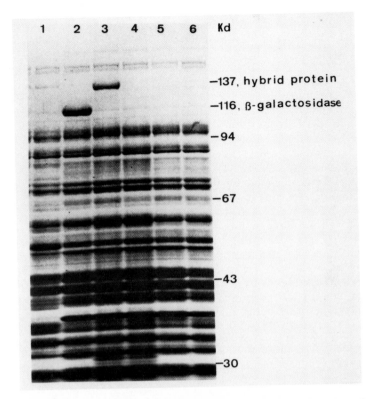

FIGURE 3. SDS-polyacrylamide gel analysis of polypeptides obtained from various λgtll recombinant lysogens. Protein samples (50 μl, corresponding to 1 ml of bacterial culture with A$_{600}$=0.75, were loaded on 6 to 16% polyacrylamide gradient with 4% stacking gel, subjected to electrophoresis and stained with Coomassie brilliant blue (10). Lane 1, E. coli Y1089. Lane 2, λgtll lysogen. Lane 3, IAA4/5-λgtll recombinant lysogen. Lane 4, as lane 3 but with the cDNA insert in opposite orientation. Lanes 5 and 6, as lanes 2 and 3 but without IPTG. Molecular weight of reference proteins are shown on the right x 10^{-3}; carbonic anhydrase, 30; ovalbumin, 43; bovine serum albumin, 67; phosphorylase b, 94.

an absolute requirement for the expression of β-galactosidase and its hybrid (compare lanes 5 and 6 with lanes 2 and 3 in Figure 3). The hybrid protein, constituting 5-10% of the total E. coli protein, is highly insoluble at pH 7.0 compared to β-galactosidase (12). It is markedly resistant to proteolytic degradation, accumulating in nearly equimolar

concentrations as with β-galactosidase. Direct comparison of
the solubility of the various hybrid β-galactosidase proteins
produced with the λgt11 expression system by others shows a
great variation. Our hybrid protein is very insoluble compared to those produced with drosophila (13,14) or mammalian
cDNAs (15,16). Presently we do not know whether this is due
to the hydrophobicity of the protein coded by the IAA4/5
cDNA or due to aggregation of the hybrid protein. Furthermore, the additional 4 Kd in the primary translation product
may lend increased solubility to the hybrid protein.

Since the fused proteins prepared from drosophila
(13,14) and mammalian cDNA (15,16) are as soluble as β-
galactosidase, conventional purification techniques (17) for
the latter have been used to purify these hybrid proteins.
We are currently developing methodology to purify large
quantities of the insoluble hybrid protein between galactosidase and the pea protein specified by the early auxin-
regulated mRNA IAA4/5 in order to raise antibodies to the
hybrid protein moiety.

This report is the first successful demonstration of
the production of a hybrid protein between β-galactosidase
and a plant protein (pea) using the λgt11 expression system.

REFERENCES

1. Thimann KV (1969). The Auxins. In Wilkins MB (ed): "The
    Physiology of Plant Growth and Development," New York:
    McGraw-Hill, p 2.
2. Theologis A (1986). Rapid gene regulation by auxin. Ann
    Rev Plant Physiology 37:407.
3. Young RA, Davis RW (1983). Efficient isolation of genes
    by using antibody probes. Proc Natl Acad Sci USA 80:1194.
4. Huynh TV, Young RA, Davis RW (1985). Construction and
    screening cDNA libraries in λgt10 and λgt11. In Gover,
    D (ed): "DNA Cloning Techniques: A Practical Approach,"
    Oxford: IRL Press, p 49.
5. Theologis A, Huynh TV, Davis RW (1985). Rapid induction
    of specific mRNAs by auxin in pea epicotyl tissue. J Mol
    Biol 183:53.
6. Young RA, Davis RW (1984). Yeast RNA polymerase II
    genes: Isolation with antibody probes. Science 222:778.
7. Scherer G, Telford J, Baldari C, Pirrotta V (1981).
    Isolation of cloned genes differentially expressed at
    early and late stages of drosophila embryonic develop-
    ment. Dev Biol 86:438.

8. Maniatis T, Fritsch EF, Sambrook, J (1982). Molecular Cloning, Cold Spring Harbor Laboratory, Cold Spring Harbor, New York.

9. Davis RW, Botstein D, Roth JR (1980) Advanced Bacterial Genetics, Cold Spring Harbor Laboratory, Cold Spring Harbor, New York.

10. Laemmli UK (1970). Cleavage of structural proteins during the assembly of the head of bacteriophage T4. Nature 227:680.

11. Young RA, Davis RW (1984). Immunoscreening λgt11 recombinant DNA expression libraries. In Setlow J, Hollaender A (eds): "Genetic Engineering," Vol. 7. New York: Plenum Press.

12. Zabin I (1980). β-galactosidase and the lactose operon. In Sigman DS, Brazier MA (eds): "The Evolution of Protein Structure and Function,"New York: Academic Press, p 49.

13. White RA, Wilcox M (1984). Protein products of the bithorax complex in drosophila. Cell 39:163.

14. Beachy PA, Helfand SL, Hogness DS (1985). Segmental distribution of bithorax complex proteins during drosophila development. Nature 313:545.

15. Gallin WJ, Prediger EA, Edelman GM, Cunningham BA (1985). Isolation of a cDNA clone for the liver cell adhesion molecule (L-CAM). Proc Natl Acad Sci USA 82:2807.

16. Schwarzbauer JE, Paul JI, Hynes RO (1985). On the origin of species of fibronectin. Proc Natl Acad Sci USA 82:1424.

17. Fowler AV, Zabin I (1983). Purification, structure, and properties of hybrid β-galactosidase proteins. J Biol Chem 258:14354.

Molecular Biology of Plant Growth Control, pages 73–84
© 1987 Alan R. Liss, Inc.

ROLE OF ABSCISIC ACID AND RESTRICTED WATER

UPTAKE DURING EMBRYOGENY IN BRASSICA[1]

Ruth R. Finkelstein[2], Alice J. DeLisle,
Anne E. Simon[3] and Martha L. Crouch

Department of Biology, Indiana University
Bloomington, Indiana 47405

ABSTRACT   We have identified changes in
stage specific gene expression and developmental
potential during embryogeny and attempted to
correlate these change with potential regulatory
signals both in situ and in vitro. We have
concentrated on the effects of ABA and restricted
water uptake because both of these cues are a
normal part of the seed environment. Sensitivity
to these cues was assayed in terms of their
ability to mimic normal development i.e. to
maintain storage protein synthesis. During
development in situ endogenous ABA is highest
during the reserve accumulation phase, when

[1]This work was supported by NSF grant PCM 80-03803 and
USDA grant 80-CRCR-1-1022 to M. L. Crouch, NSF Postdoctoral
Fellowship in Plant Biology PCM-8412387 to A. J. Delisle,
and NIH Training Grant to the Molecular, Cellular, and
Developmental Biology Program at Indiana University, which
supported R. R. Finkelstein.

[2]Present address: MSU-DOE Plant Research Laboratory,
Michigan State University, East Lansing, MI 48824

[3]Present address: Department of Biology, University
of California at San Diego, La Jolla, CA 92093

storage protein mRNA transcription and accumulation are highest. After the embryos begin to lose water, both endogenous ABA and ABA sensitivity decrease and the embryos gradually acquire the capacity for normal seedling growth. Throughout the later stages of embryogeny (early cotyledon to maturity) embryos cultured on high osmoticum accumulate levels of storage protein and its mRNA similar, to those in embryos developing in situ. Comparison of the kinetics of induction of storage protein mRNA accumulation by ABA or osmotic treatment showed that the osmotic effect was more rapid and therefore possibly more direct. Endogenous ABA in osmotically treated embryos remained low, indicating that high ABA levels are not required for the observed effects on gene expression. These results are consistent with the hypothesis that ABA is important in regulating embryogeny prior to the desiccation phase, possibly by restricting water uptake, but that desiccation itself is required to complete embryo maturation.

INTRODUCTION

Immature embryos acquire the capacity to germinate very early during seed development in many flowering plants, often during the first third of development (1). It is therefore assumed that some factors(s) in the seed must normally inhibit this precocious germination, while allowing the rest of embryo development to occur. Two plant growth regulators have been implicated in the non-toxic inhibition of precocious germination: high levels of abscisic acid (ABA) and low levels of water. The evidence for inhibition by ABA is that a) mutations which result in low levels or insensitivity to ABA often cause vivipary (2, 3), b) inhibitors of ABA biosynthesis may result in vivipary (4), c) exogenous ABA at physiological concentrations will reversibly inhibit the germination of isolated immature embryos (5), and d) ABA levels are very high during seed development (6). However, restricting the rate of water uptake directly by including high concentrations of osmotica in the culture medium can substitute for the effects of exogenous ABA on isolated embryos (1). The osmotic potential of developing embryos is quite low (7), yet water uptake in the seed is much slower during embryo development than it is during germination and seedling growth. In fact, late in embryogeny there is a net loss of water. Therefore, both ABA and restricted water uptake may play a role in inhibiting precocious germination in situ.

We have been investigating the relative roles of ABA and restricted water uptake during seed development in Brassica napus (rapeseed) (8-11). Our approach has been to first correlate ABA and water content changes during development with particular processes characteristic of embryonic and germinative or postgerminative development, and then to determine how exogenous ABA and high osmotica effect those processes in vitro in embryos of different ages.

RESULTS AND DISCUSSION

Pattern of Embryo Development in the Seed

From zygote to mature quiescent embryo takes approximately 60 days in Brassica napus cv. Tower under controlled conditions (8). The morphology of the embryo is relatively simple. After fertilization, the globular embryo grows by cell division to form a root-shoot axis and cotyledons (stage 1). At about 25 days post anthesis (dpa), cell division ceases and subsequent growth is by cell expansion as storage proteins and lipids accumulate in most cells of both the axis and cotyledons (stage 2). The shoot apex produces no visible leaf primordia. By 42 dpa, chlorophyll is lost, the mature embryo enters developmental arrest, and the endosperm is reduced to a thin layer of crushed cells (stage 3). Germination (G) ensues when the seed is imbibed.

We have measured several parameters of embryo development and seedling development, and they are plotted in Figure 1 as a function of days post anthesis for embryogeny or days after imbibition for seedling development. The three broad stages of embryogeny described in the preceeding paragraph are marked at the top of the figure. Embryo dry weight and water content (Fig. 1A, adapted from 8) both increase steadily after the end of cell division until about 40 dpa, when there is a net

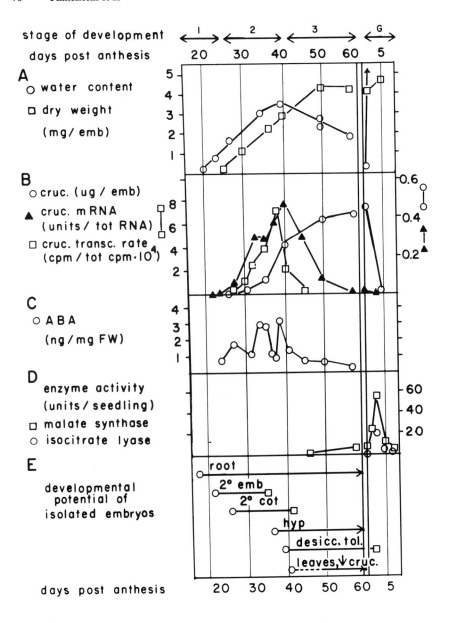

FIGURE 1. Development of B. napus embryos in seeds. Parameters plotted and methods used for quantitation are described in the text.

decrease in water per embryo. The relative amount of water decreases through embryo development from 90% at 21 dpa to 30% at 60 dpa. Dry weight continues to accumulate until 50 dpa. During germination, imbibition is completed by 12 h (12), and water uptake continues rapidly during seedling growth.

Embryo dry weight accumulation is due primarily to deposition of storage lipids and proteins. We followed the accumulation of the major storage protein family, cruciferin, and the accumulation and transcription of cruciferin mRNA (Fig. 1B). Cruciferin protein is first detected by rocket immunoelectrophoresis at 25 dpa, increases rapidly during stage 2, and levels off at seed maturity. It is stored in the seed until germination, when it is completely degraded within 5 d (8). Cruciferin mRNA is first detected on Northern blots at 21 dpa, peaks at 40 dpa, and is barely detectable in mature embryos (10). It is undetectable following germination. Transcription rates were measured in a run-on transcription assay. Nuclei were isolated from embryos and incubated in an $\underline{in\ vitro}$ transcription mixture containing ^{32}P-GTP (13). $\overline{RNA}$ was isolated (14) and the amount of incorporation of ^{32}P-GTP into cruciferin RNA relative to total RNA was determined by DNA excess filter hybridization to the cruciferin cDNA clone, PC1 (15). Transcription of cruciferin genes relative to total transcription has a pattern similar to the relative cruciferin mRNA levels, except that transcription rate begins to decline a few days before the mRNA levels decrease. In contrast, actin transcription rates remain constant relative to total transcription over this period (data not shown).

Levels of ABA are high throughout embryo development (Fig. 1C). Endogenous ABA was measured by radioimmunoassay according to the procedure of Weiler (16), using rabbit anti-ABA-human serum albumin serum (Miles-Yeda Ltd.). The assay was corroborated by GC-MS at two developmental stages. Endogenous ABA levels rise 3- to 4-fold during stage 2, the storage protein accumulation phase (25-39 dpa), reaching peaks at 35 and 38 dpa (Fig. 1C, adapted from 10). The highest level is 3.3 ng/mg FW, which is

approximately 20 μ M. Embryo ABA levels are quite high
throughout the period measured; even the lowest levels are
3-4-fold higher than those in subtending leaves (0.3 ng/mg
FW).

After germination of mature dry seeds, the seedlings
produce a battery of enzymes which break down the storage
lipids and proteins. We measured the activity of malate
synthase, MS, (17) and isocitrate lyase, ICL, (18), two
glyoxylate cycle enzymes which are necessary to convert
lipids to sugars (Fig. 1D). During embryo development, MS
activity first appears around 45 dpa and increases
gradually to 9 units/seed at maturity. Upon germination,
MS activity increases rapidly to about 70 units/seedling
within 3 days, and then declines. Although ICL activity
shows a similar pattern of induction following imbibition
of dry seeds, reaching a peak of 25 units/seedling after 3
days, ICL activity does not reach detectable levels in
developing seeds prior to germination.

Finally, we examined the ability of embryos at
different stages of development to form normal seedlings
when allowed to germinate precociously on a basal culture
medium (19). As the embryos mature they gradually acquire
the ability to form normal seedlings (Fig. 1E). Embryos
from heart stage (15 dpa) on will show root outgrowth in
culture (root). Upon precocious germination, the youngest
embryos do not initiate the formation of primordia at the
shoot apex. Instead, secondary embryos (2° emb) form all
over the surface of the original embryo (Simon and Crouch,
unpublished observations). When 30-40 dpa embryos are
excised and allowed to germinate the shoot apex becomes
active, but forms cotyledons instead of leaves (2° cot).
Storage proteins continue to accumulate in these
"seedlings" even in newly formed root and cotyledon
tissues. Seedlings developing from embryos younger than 36
dpa do not show hypocotyl elongation, but following
germination of 36-40 dpa embryos the hypocotyl elongates
(hyp) even though the apex is still forming cotyledons. At
39-40 d, the embryos become desiccation tolerant. They
lose this property within hours of imbibition; first the
root, then the hypocotyl, and finally the cotyledons become
sensitive to desiccation (12). Between 41-46 dpa, the apex
acquires the capacity to form true leaves. Often, one or
two cotyledons will form, followed by organs that resemble
chimaeras between cotyledons and leaves, and then leaves

with normal morphology. By 46 dpa, only true leaves will form at the apex. Following precocious germination of 41-46 dpa embryos there is a gradual net decrease in storage protein. The normal rapid degradation pattern is seen when embryos older than 50 dpa are germinated. Thus there is a sequential acquisition during embryogeny of the characteristics associated with germination and seedling growth of a mature seed.

Effects of ABA and high osmoticum on storage protein and message accumulation in cultured embryos.

Either ABA or high osmoticum can stimulate synthesis of the embryo specific storage protein cruciferin in embryos cultured at 27 dpa. To determine whether either or both of these cues could substitute for the seed environment in regulating development throughout embryogeny we compared levels of cruciferin and its message in embryos of various ages cultured on either 10 μM ABA or 12.5% sorbitol with those seen during normal development in seeds. Prior to 40 dpa both ABA and sorbitol stimulate cruciferin accumulation to higher levels than in embryos developing in seeds, while embryos cultured on basal medium germinate and show a decrease in storage protein concentration (Fig. 2A). After 40 dpa both ABA and osmoticum still inhibit storage protein turnover relative to the basal treatment, but fully mature dry seeds will not accumulate more storage protein in response to these cues. Although pre-40 dpa embryos cultured on ABA have higher cruciferin message levels than those on basal medium, only the sorbitol treatment maintained cruciferin message accumulation at levels at or above those in seeds (Fig. 2B). This suggests that ABA may be enhancing storage protein synthesis at the level of translation or protein stability. After 40 dpa ABA is no longer effective at maintaining cruciferin message levels above those in embryos on basal medium, but the cruciferin message levels in sorbitol treated embryos are similar to those in embryos developing in seeds. These results indicate that ABA is effective in maintaining embryonic gene expression only in predesiccation stage embryos (younger than 40 dpa) while high sorbitol approximates the seed environment at all stages tested.

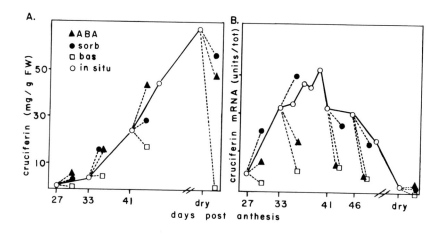

FIGURE 2. Effects of ABA and osmoticum on accumulation of cruciferin and its mRNA. Cruciferin (A) was measured by rocket immunoelectrophoresis and its mRNA (B) was quantitated by dot blots of total RNA hybridized to a cloned ^{32}P-labeled cDNA probe. ($\bigcirc$) connect levels found during development in seeds. (-----) indicate changes occurring after 3 d in culture on 10 μM ABA ($\blacktriangle$), 12.5% sorbitol ($\bullet$) or basal medium ($\square$).

Relationships between ABA and osmotic treatment.

Osmotic effects on a variety of plant tissues appear to be mediated by increased endogenous ABA. However embryos cultured 3 d on high sorbitol do not germinate and continue to accumulate high levels of cruciferin and its message without a concommitant increase in endogenous ABA (11). Working from the assumption that a more direct effect on gene expression would be more rapid, we compared the timing of reinduction of cruciferin mRNA accumulation on 10 μM ABA or 10.9% sorbitol. Embryos (27 dpa) were cultured on basal medium for 10 h, allowing cruciferin mRNA levels to decrease, then transferred to ABA or sorbitol and harvested at various times after transfer. The sorbitol treatment consistently produced a more rapid increase in cruciferin message accumulation, although both ABA- and sorbitol-treated embryos had similar levels of cruciferin

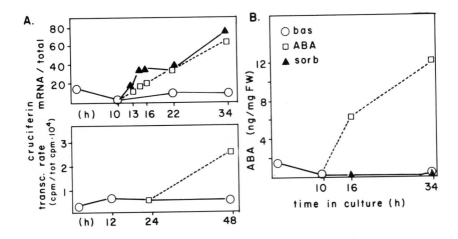

FIGURE 3. Reinduction of cruciferin mRNA accumulation and transcription (A) and ABA levels (B) in response to ABA or osmoticum. Embryos were cultured as described in the text. Cruciferin mRNA levels were measured by dot blots and are expressed in arbitrary units relative to total RNA (10). Transcription rates for cruciferin were determined by run-on transcription assays with isolated nuclei and are expressed relative to total transcription (14). ABA levels were measured by radioimmunoassay (16).

message at the end of the period (Fig. 3A). Measurement of transcription rates in 25 dpa embryos subjected to similar treatments (12 h on basal, then 1 d on 1 μM ABA) showed that at least part of the increase in cruciferin message during the ABA treatment was due to increased transcription (Fig. 3A). Once again the low levels of endogenous ABA in the sorbitol-treated embryos (Fig. 3B) show that high ABA is not required even transiently for the osmotic effects on cruciferin message accumulation. These experiments do not exclude the possibility that ABA is concentrated at its site of action in osmotically treated embryos, or that an ABA metabolite is actually required for its effects on expression. However, they are consistent with the hypothesis that ABA acts via restricted water uptake which affects gene expression in some unknown manner. Support for this hypothesis comes from work with germination of mature rapeseed which has demonstrated that restriction of water uptake by ABA is slower than that due to osmotic treatment, and that ABA and osmotic suppression of germination is synergistic, suggesting that they may be acting through a common effector (20).

Summary

We have characterized normal embryogeny in terms of both expression of various stage-specific proteins and changes in developmental potential, reflected in the pattern of development following precocious germination and seedling growth. To analyze the roles of ABA and restricted water uptake in regulating embryogeny we have looked for correlations between the presence of these cues and maintenance of embryonic development, both in seeds and in culture. ABA is highest from 30-40 dpa when transcription rate and accumulation of cruciferin message is highest. These embryos also respond to exogenous ABA by continuing embryonic development, consistent with a regulatory role for ABA at this stage. In contrast embryos which have entered the desiccation phase have decreased endogenous ABA and reduced sensitivity to applied ABA. High osmoticum can approximate the seed environment for embryos ranging from 27 dpa to maturity, and appears to regulate the observed changes in gene expression and developmental potential more directly than ABA.

REFERENCES

1. Norstog K (1979). Embryo culture as a tool in the study of comparative and developmental morphology. In Sharp WR, Larsen PO, Paddock EF, Raghavan V (eds): "Plant Cell and Tissue Culture, Principles and Applications," Columbus, OH: The Ohio State University Press, p 179.

2. Brenner ML, Burr B, Burr F (1977). Correlation of genetic vivipary in corn with abscisic acid concentration. Plant Physiol Suppl 63:36.
3. Robichaud CS, Wong J, Sussex IM (1980). Control of in vitro growth of viviparous embryo mutants of maize by abscisic acid. Dev Genet 1:325.
4. Fong F, Smith JD, Koehler DE (1983). Early events in maize seed development. Plant Physiol 73:899.
5. Walbot V (1978). Control mechanisms for plant embryogeny. In Clutter M (ed): "Dormancy and Developmental Arrest: Experimental Analysis in Plants and Animals," New York: Academic Press, p 114.
6. Black M (1983). Abscisic acid in seed germination and dormancy. In Addicott FT (ed): "Abscisic Acid," New York: Praeger Publishers, p 113.
7. Yeung EC, Brown DCW (1982). The osmotic environment of developing embryos of Phaseolus vulgaris. Z Pflanzenphysiol 106:149.
8. Crouch ML, Sussex IM (1981). Development and storage protein synthesis in Brassica napus L. embryos in vivo and in vitro. Planta 153:64.
9. Crouch ML, Tenbarge K, Simon A, Finkelstein R, Scofield S, Solberg L (1985). Storage protein mRNA levels can be regulated by abscisic acid in Brassica embryos. In van Vloten Doting L, Groot GSP, Hall TC (eds): "Molecular Form and Function of the Plant Genome," New York: Plenum Publishing Corp, p 555.
10. Finkelstein RR, Tenbarge KM, Shumway JE, Crouch ML (1985). Role of ABA in maturation of rapeseed embryos. Plant Physiol 78:630.
11. Finkelstein RR, Crouch ML (1986). Rapeseed embryo development in culture on high osmoticum is similar to that in seeds. Plant Physiol, in press.
12. Schopfer P, Plachy C (1984). Control of seed germination by abscisic acid II. Effect on embryo water uptake in Brassica napus L. Plant Physiol 76:155.
13. Goldberg R, Walling L, personal communication.
14. Marzluff WF, Huang RCC (1984). Transcription in isolated nuclei. In Hames BD, Higgins SJ (eds): "Transcription and Translation, A Practical Approach," Oxford: IRL Press, p 89.

15. Simon AE, Tenbarge KM, Scofield SR, Finkelstein RR, Crouch ML (1985). Nucleotide sequence of a cDNA clone of Brassica napus 12S storage protein shows homology with legumin from Pisum Sativum. Plant Molecular Biology 5:191.
16. Weiler E (1979). Radioimmunoassay for the determination of free and conjugated abscisic acid. Planta 144:255.
17. Miernyk JA, Trelease RN, Coinski JS Jr. (1979). Malate synthase activity in cotton and other ungerminated oilseeds. Plant Physiol 63:1068.
18. Cooper TG, Beevers H (1969). Mitochondria and glyoxysomes from castor bean endosperm. Enzyme constituents and catalytic capacity. J Biol Chem 244:3507.
19. Finkelstein RR, Crouch ML (1984). Precociously germinating rapeseed embryos retain characteristics of embryogeny. Planta 162:573.
20. Schopfer P, Plachy C (1985). Control of seed germination by abscisic acid II. Effect on embryo growth potential (minimum turgor pressure) and growth coefficient (cell wall extensibility) in Brassica napus L. Plant Physiol 77:676.

Molecular Biology of Plant Growth Control, pages 85–95
© **1987 Alan R. Liss, Inc.**

TRANSCRIPTIONAL REGULATION OF AUXIN
RESPONSIVE GENES

Tom Guilfoyle and Gretchen Hagen

Department of Botany, University of Minnesota
St. Paul MN 55108

ABSTRACT   Four independent cDNA clones
corresponding to mRNAs that rapidly increase
in amount following auxin application have
been isolated from a soybean cDNA library.
The mRNAs to these cDNA clones have been
characterized in detail and their increases
in concentration after auxin treatment have
been shown to be regulated at the level of
transcription. Increased transcription
rates on genes encoding two of the
auxin-responsive mRNAs are not only
increased by auxins but also by other
agents including $CdCl_2$. Direct
application of 2,4-D or $CdCl_2$ to
isolated nuclei in the absence or presence
of an S100 extract does not induce
transcription on the auxin-responsive genes.

INTRODUCTION

In recent years, a number of laboratories have
reported that applied auxins rapidly alter gene
expression in intact and excised plant organs
(reviewed in ref. 1). The function of the gene
products rapidly synthesized in response to auxin
application remain to be identified; however, the
mechanism(s) involved in the synthesis of the
auxin-induced gene products can be elucidated
without understanding the function of the gene
products. cDNA cloning has facilitated the
detection of a number of auxin-responsive mRNAs in

soybean (2-4) and pea (5) which code for
polypeptides that increase in concentration after
auxin treatment (1,5). The increase in polypeptide
concentrations (e.g., in vitro translation products)
is correlated with a several- to many-fold increase
in corresponding mRNA concentrations (2,3,5). The
increase in mRNA concentrations following auxin
treatment could result from increased transcription
rates on the genes encoding the auxin-responsive
mRNAs, increased processing rates for mRNA
precursors, increased mRNA transport rates from
nucleus to cytoplasm, or decreased mRNA turnover
rates (i.e., increased mRNA stability). By
employing in vitro nuclear "run-off" transcription
assays (3,6), we demonstrate that the auxin-induced
accumulation of at least four auxin-responsive mRNAs
results, at least in part, from increased
transcription rates on the corresponding
auxin-regulated genes.

RESULTS

Auxin-responsive mRNAs.

        Four cDNA clones (i.e, pGH1, pGH2, pGH3, and
pGH4) were selected from a cDNA library prepared
from poly(A$^+$) RNA which was purified from
2,4-D-treated soybean seedlings (3). The mRNAs
which correspond to these cDNA clones have been
characterized (3,6), and some of this
characterization is summarized below. The size in
kb for mRNAs to pGH1, pGH2, pGH3, and pGH4 is 1.7,
1.0, 2.4, and 1.0, respectively. mRNAs
corresponding to pGH1-4 increase in concentration
within 10-30 minutes following 2,4-D application to
excised or intact hypocotyls (i.e., elongating or
basal regions) and plumules. Maximum steady-state
mRNA levels are achieved within 2-4 hours following
auxin treatment. Other hormones, including ethylene
do not cause an increase in the auxin-responsive
mRNAs, but a number of other auxins including IAA
and 2,4,5-T induce increases in mRNA levels similar
to 2,4-D. These results indicate that applied
auxins cause a rapid increase in specific mRNAs,

that organs of intact seedlings and different excised organs or regions of organs respond in a similar manner to auxin, and that a variety of auxins, but not ethylene, induce the auxin response.

Transcriptional Regulation of Auxin-responsive mRNAs.

To investigate whether the increase in auxin-responsive mRNAs following auxin treatment results from increased transcription rates on their genes, we conducted in vitro "run-off" transcription experiments with isolated soybean nuclei (3,6). Since the increases in auxin-responsive mRNAs could be demonstrated with a variety of excised or intact organs and organ sections, we chose to work with plumules since this organ provides nuclei of high purity (Fig. 1A) in high yield (i.e., 50-100-fold more nuclei can be purified per equivalent fresh weight of plumules compared to hypocotyls), of uniform size and shape (i.e., compared to hypocotyl nuclei; Fig. 1B) with no detectable breakage, and with highly reproducible total transcription rates for RNA polymerases I, II, and III. Nuclei isolated from excised plumules which were treated for 30 minutes with 100 µM 2,4-D transcribe genes corresponding to the auxin-responsive mRNAs at 10-100-fold greater rates than nuclei isolated from untreated excised plumules (6). Nuclei isolated from 2,4-D-treated (i.e., sprayed) plumules of intact seedlings also display increased transcription rates on the auxin-responsive genes compared to untreated plumules; however, the relative increases in transcription rates are less than that observed with excised plumules. Transcription rates on the auxin-responsive genes increase very rapidly after addition of auxin to excised plumules (6). At least half maximal transcription rates for pGH1-4 genes are achieved within 15 minutes after auxin application, and for pGH1 and pGH4, near maximal transcription rates are observed within this interval. A several-fold increase in transcription rate is observed for pGH3

A

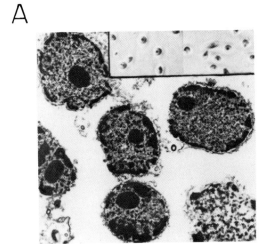

B

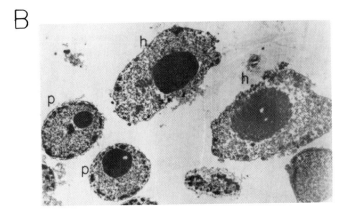

FIGURE 1. Isolated soybean nuclei used for in vitro transcription. Nuclei were purified from plumules or hypocotyls as described by Hagen and Guilfoyle (6). (A) Electron micrograph and light micrograph (insert) of plumule nuclei. (B) Electron micrograph showing a comparison of nuclei isolated from plumules and hypocotyls. Note the uniformity of size and shape of plumule nuclei compared to the hypocotyl nuclei.

within 5 minutes after applying auxin. Once the maximum transcription rate is reached, this rate is maintained for at least the next 1-2 hours, provided that auxin is continually present in the incubation medium. These results indicate that the transcriptional responses to applied auxin can be extremely rapid, similar to those observed with animal hormones (7), and that such rapid responses are within the time frame required to regulate most growth and developmental transitions in plants.

To determine what concentration of auxin is required to alter transcription rates on auxin-regulated genes, we monitored transcription rates in nuclei isolated from excised plumules which were treated with 2,4-D concentrations ranging from nM to mM (6). With all four genes corresponding to pGH1-4, a linear or nearly linear increase in transcription rate is obtained over several logs of increasing 2,4-D concentration. With pGH2, pGH3, and pGH4, the transcriptional response was not saturated at the 1 mM 2,4-D, the highest concentration tested. Only pGH1 showed a saturable dose response, and in this one case, the optimal 2,4-D concentration was 100 µM. As little as 10 nM 2,4-D caused increased transcription rates with pGH3, while 100 nM 2,4-D was required to detect a transcriptional response with pGH1 and pGH2, and µM 2,4-D was required with pGH4. Although these dose responses are striking in some respects (e.g., linear increase in response over several orders of log increase in concentration with no saturation), such dose responses are not uncommon for growth and developmental responses induced by plant growth substances (8,9).

To determine the specificity of the auxin transcriptional response, we tested a variety of auxins and nonauxin analogs for their ability to increase transcription rates on the auxin-regulated genes (see ref. 6 for details). While pGH1 and pGH3 transcription was highly specific for auxins, transcription on pGH2 and pGH4 genes were not entirely specific for auxins.

We have also tested a variety of other compounds and environmental parameters for possible transcriptional activation of auxin-regulated genes.

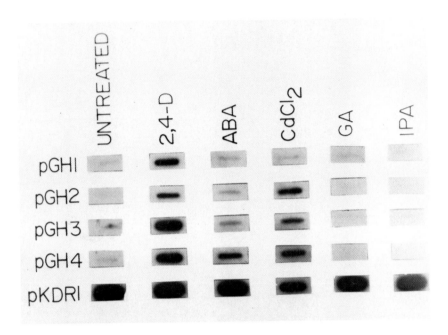

FIGURE 2.   Induction of transcription on genes
corresponding to pGH1-4 following treatments of
plumules with 2,4-D and other agents.   Plumules were
preincubated (6) for 12 hours in the absence of
auxin or other agents and then incubated (6) for 2
hours with the compound indicated.   Nuclei were
isolated and incubated in a transcription cocktail,
and ^{32}P-labeled RNA (10^7 cpm) was hybridized
to M13 single stranded probes which were
complementary to mRNAs of pGH1-4 (6).   Each compound
was present in the plumule incubation mixture at 100
μM.   ABA: abscisic acid; GA: gibberellic acid; IPA:
isopentenyladenosine.   pKDR1 is a soybean ribosomal
DNA clone provided by Dr. R. T. Nagao and J. L. Key.

Neither heat shock nor cold shock affects transcription rates on auxin-responsive genes when administered in the absence of auxin; however, prior heat shock largely prevents the auxin transcriptional response when auxin is subsequently applied. The protein synthesis inhibitors chloramphenicol and cycloheximide do not induce transcription of these genes when auxin is withheld from the incubation medium. Gibberellic acid has no effect on transcription rates of pGH1-4 genes whether applied alone or in combination with 2,4-D (Fig. 2). The cytokinin, isopentenyladenosine, which blocks auxin-induced hypocotyl elongation in soybean (10), does not induce transcription when supplied by itself (Fig. 2), nor does it block auxin-regulated transcription when supplied along with auxin. Ethylene, whose synthesis may be stimulated by applied auxins (11), does not induce transcription of the auxin-regulated genes. In contrast to studies with the above hormones, abscisic acid appears to produce a slight increase in the transcription rate for the gene corresponding to pGH4 and possibly pGH2 (Fig. 2); however, the increase in transcription rate achieved with abscisic acid is several-fold less than that observed with auxin. In addition to the abscisic acid effects on pGH2 and pGH4 genes, we have found that the heavy metal $CdCl_2$ induces the transcription of pGH2 and pGH4 and possibly pGH3 (Fig. 2). With this heavy metal, the induction of pGH2 transcription is as great or greater than the induction by auxins, and the induction of pGH4 transcription is nearly 50% the level of induction achieved with auxins. Interestingly, other heavy metals (e.g., Zn, Co, Ag, and Cu) do not increase transcription rates on any of the auxin-regulated genes. It is interesting to note that a soybean heat shock mRNA has been described which is also increased in concentration in response to 2,4-D and cadmium (12). Furthermore, it is worth noting that stress-induced ethylene biosynthesis can be induced by 2,4-D and cadmium (13).

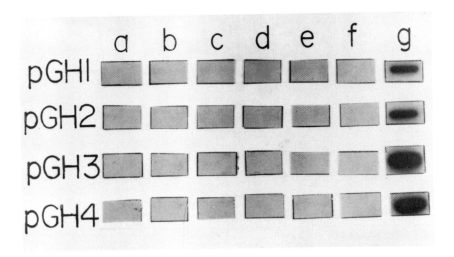

FIGURE 3.   Incubation of untreated soybean
plumule nuclei with 2,4-D or CdCl$_2$.   Nuclei were
isolated from freshly excised plumules as described
by Guilfoyle et al. (15) and incubated for 2 hours
in a transcription cocktail containing (a) no
additions, (b) 100 µM 2,4-D, (c) 100 µM CdCl$_2$,
(d) S100 alone, (e) S100 plus 100 µM 2,4-D, or (f)
S100 plus 100 µM CdCl$_2$.   Lane (g) shows
transcription in nuclei isolated from excised
plumules which were treated in vivo for 2 hours with
100 µM 2,4-D.   In each case, 10 cpm of nuclear
transcripts were hybridized to strand specific
probes of pGH1-4 (6).   The S100 was prepared as
described (14) from excised plumules which were
incubated for 2 hours in the presence of 100 µM each
of 2,4-D and CdCl$_2$.

Direct Application of 2,4-D or $CdCl_2$ to Isolated Nuclei.

In contrast to the above studies which demonstrate that in vivo application of auxins and $CdCl_2$ results in rapid increases in transcription rates on some or all of the auxin-responsive genes (i.e., pGH1-4), in vitro application of these agents to isolated nuclei fails to alter the transcription rates of these genes (Fig. 3). Furthermore, the addition of S100 extracts, which might contain auxin or cadmium binding proteins, along with auxin or $CdCl_2$ fails to increase the transcription rates on genes corresponding to pGH1-4. Similar experiments conducted with animal nuclei have suggested that hormones can activate hormone-responive genes in purified nuclei incubated with S100 extracts (14). The lack of transcriptional activation with plant nuclei could be due to a variety of reasons including inefficient initiation of RNA chains, inactivativation of gene and/or hormone binding proteins, degradation of RNA polymerase II or transcription factors, etc.

DISCUSSION

Now that it has been demonstrated that auxins or cadmium can increase transcription rates on genes corresponding to pGH1-4, it will be important to identify those regions of these auxin-responsive genes which confer the ability to respond to auxin or cadmium. It will be interesting to determine whether a variety of auxin-responsive genes have common regulatory domains. At the protein level, it will be important to gain insight into the nature of the polypeptides that interact with the auxin regulatory DNA sequences as well as the function of the polypeptides that are encoded by the auxin-responsive mRNAs.

ACKNOWLEDGMENTS

This research was supported by grant

PCM-8208496 from the National Science Foundation. We are indebted to M. Lindberg for providing the micrographs of nuclei and to Drs. R. T. Nagao and J. L. Key for providing the pKDRl clone.

REFERENCES

1. Guilfoyle TJ (1986). Auxin regulated gene expression in higher plants. CRC Crit Rev Plant Sci: In Press.
2. Walker JC, Key JL (1982). Isolation of cloned cDNAs to auxin-responsive poly(A$^+$) RNAs of elongating soybean hypocotyl. Proc Natl Acad Sci USA 79:7185.
3. Hagen G, Kleinschmidt A, Guilfoyle TJ (1984). Auxin-regulated gene expression in intact soybean hypocotyl and excised hypocotyl sections. Planta 162:147.
4. Walker JC, Legocka J, Edelman L, Key JL (1985). An analysis of growth regulator interactions and gene expression during auxin-induced cell elongation using cloned complementary DNAs to auxin-reponsive messenger RNAs. Plant Physiol 77:847.
5. Theologis A, Huynh TV, Davis RW (1985). Rapid induction of specific mRNAs by auxin in pea epicotyl tissue. J Mol Biol 183:53.
6. Hagen G, Guilfoyle TJ (1985). Rapid induction of selective transcription by auxins. Mol Cell Biol 5:1197.
7. Ucker DS, Yamamoto KR (1984). Early events in the stimulation of MTV RNA synthesis by glucocorticoids: novel assays of transcription rates. J Biol Chem 259:7416.
8. Kende H, Gardner G (1976). Hormone binding in plants. Ann Rev Plant Physiol 27:267.
9. Trewavas A (1981). How do plant growth substances work? Plant Cell Environ 4:203.
10. Vanderhoef LN, Key JL (1968). Inhibition by kinetin of cell elongation and RNA synthesis in excised soybean hypocotyl. Plant Cell Physiol 9:343.
11. Abeles FB (1966). Auxin stimulation of ethylene evolution. Plant Physiol 41:585.

12. Czarnecka E, Edelman L, Schoffl F, Key JL (1984). Comparative analysis of physical stress responses in soybean seedlings using cloned heat shock cDNAs. Plant Mol Biol 3:45.

13. Fuhrer J (1982). Ethylene biosynthesis and cadmium toxicity in leaf tissue of beans (Phaseolus vulgaris L.). Plant Physiol 70:162.

14. Tata JR, Baker BS (1985). Specific switching on of silent egg protein genes in vitro by an S-100 fraction in isolated nuclei from male Xenopus. EMBO J 4:3253.

15. Guilfoyle TJ, Suzich J, Lindberg M (1986) Synthesis of precursor tRNAs and 5S rRNA in nuclei isolated from wheat embryos. Submitted for publication.

Molecular Biology of Plant Growth Control, pages 97–106
© 1987 Alan R. Liss, Inc.

MODULATION BY ABSCISIC ACID OF GENES ENCODING
β-CONGLYCININ IN DEVELOPING SOYBEAN COTYLEDONS[1]

Elizabeth A. Bray[2] and Roger N. Beachy

Department of Biology, Washington University
St. Louis, Missouri 63130

ABSTRACT The role of the plant growth substance,
abscisic acid (ABA), in the regulation of soybean
storage protein accumulation was investigated using
soybean cotyledon culture. The effect of ABA on seed
maturation was studied by removing the embryonic axes
from the soybean cotyledons. Cotyledons cultured in
10^{-5} M ABA accumulated more of the β-subunit of β-
conglycinin, one of the major storage proteins in soybean seeds, than the controls. The accumulation of the
other two subunits of β-conglycinin was unaffected.
After 3 days in culture, ABA-treated cotyledons accumulated a greater amount of the β-subunit than the
control. In vivo labeling experiments showed that ABA
treatment resulted in increased synthesis of the β-
subunit on each day of culture. The amount of the β-
subunit transcript was correlated with the synthesis
of the subunit. Cotyledons were responsive to ABA
during the stages of development prior to significant
accumulation of the β-subunit. During seed maturation, ABA modulation of the accumulation of the β-subunit may be regulated by the availability of the β-
subunit mRNA.

[1]E.A.B was supported by a research fellowship from
Monsanto Agricultural Products Company.
[2]Present address: Department of Botany and Plant
Sciences, University of California, Riverside, CA 92521.

INTRODUCTION

The role of abscisic acid (ABA) during cotyledon maturation is being investigated to determine ABA's involvement in seed development. ABA is known to prevent precocious germination in developing cultured embryos (1, 2, 3). ABA is also thought to promote maturation of the embryo through the expression of genes that are associated with embryo maturation (1, 2, 4). Changes in ABA concentration that occur during seed development are correlated with increased fresh weight of the seed and dormancy of the seed (5, 6). The effects of ABA during maturation of the cotyledon are addressed in the experiments reported in this paper. Cotyledons were cultured after the axes had been removed. Therefore, the effects of ABA on cotyledon maturation were investigated without also studying the effect of ABA on prevention of precocious germination.

The application of ABA to cultured soybean cotyledons results in increased accumulation of the β-subunit of β-conglycinin (7). The accumulation of the other two subunits of β-conglycinin, the α'- and α-subunits, are not altered by the ABA treatment. The addition of fluridone, an inhibitor of carotenoid biosynthesis that has been shown to decrease the amount of ABA in plant tissues, decreased the accumulation of the β-subunit in cultured soybean cotyledons (7). Alternate methods of increasing the concentration of ABA in the cotyledons, such as increasing the osmotic concentration of the medium, also resulted in increased accumulation of the β-subunit in the cotyledons (7). In vitro translation and hybridization reactions showed that β-subunit mRNA increased two-fold in ABA-treated cotyledons compared to the control. The amount of the β-subunit transcript was decreased by the fluridone treatment.

Although it is known that ABA modulates β-subunit accumulation, the mechanism of action is unknown. In this paper, we describe the results of recent experiments that address the relationship between ABA and seed development. The timecourse of accumulation of the β-subunit, the stage of seed that is optimally sensitive to ABA and the concentration of ABA required for modulation of gene expression are described.

RESULTS

Timecourse of the ABA Response.

The effect of ABA on cultured cotyledons has been analyzed thus far after a 5-day culture period. In order to determine how many days were required to elicit the ABA effect, cotyledons were removed from culture after 1, 3, 4, and 5 days and the proteins were analyzed (Fig. 1). After 3 days, an increased amount of β-subunit accumulated in the cotyledons cultured in ABA compared to the control. The cotyledons cultured in fluridone did not accumulate the β-subunit at any time during the culture period.

The treatment effects on protein accumulation could be due to differences in protein synthesis or stability of the proteins in response to ABA. To determine if ABA affected protein synthesis, the cotyledons were pulse labeled at different times during culture (Fig. 2). The labeled proteins were similar to those shown to accumulate. A greater amount of β-subunit was synthesized in the ABA treatment than the control on days 3, 4, and 5. Little β-subunit was synthesized in the fluridone treatment.

The accumulation of the β-subunit transcript was measured by RNA hybridization reactions to determine if

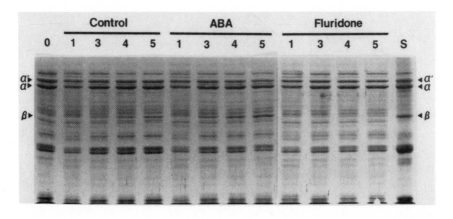

FIGURE 1. Cotyledons from seeds of 9 mm in length were cultured in control, ABA (10^{-5}) or fluridone $(10^{-6}$ M) medium. Proteins were extracted after 1, 3, 4, or 5 days in culture, separated by SDS-PAGE, and stained with Coomassie brilliant blue.

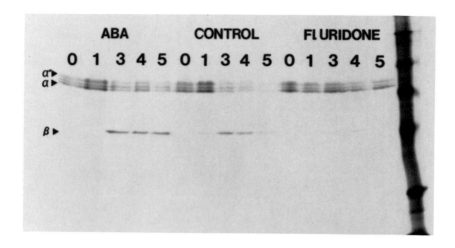

FIGURE 2. Cotyledons were cultured for 1, 3, 4, or 5 days in 10^{-5} M ABA or in control medium and then were labeled with ^{3}H leucine (250 µCi) for 6 hours. The proteins were extracted, immunoprecipitated with β-conglycinin antiserum, separated by SDS-PAGE, and the labeled proteins were detected by fluorography.

the amount of the transcript correlated with synthesis and accumulation of the β-subunit protein. The amount of the β-subunit transcript increased throughout the developmental timecourse in cotyledons that were treated with ABA (Fig. 3). The amount of the β-subunit transcript decreased after 3 days in culture in the control. Little β-subunit transcript was detected in fluridone-treated cotyledons. The amount of α'- and α-subunit mRNA decreased when the cotyledons were cultured. The accumulation of the β-subunit transcript is correlated with the synthesis of the protein. These results suggest that modulation of β-subunit accumulation by ABA is regulated by the availability of the β-subunit transcript.

The concentration of ABA in the cotyledons decreased throughout the culture period (Table 1). The fluridone treatment consistently resulted in a decreased concentration of ABA and the ABA treatment resulted in a small increase in the ABA concentration of the cotyledons compared to the controls.

TABLE 1

ABA CONCENTRATION OF COTYLEDONS CULTURED
IN CONTROL, ABA (10^{-5} M), OR FLURIDONE
(10^{-6} M) MEDIUM FOR 1, 3, 4, or 5 DAYS[a]

Days	Control	ABA	Fluridone
		ng ABA/mg fresh wt	
1	0.62 ± 0.06	0.77 ± 0.12	0.30 ± 0.04
3	0.43 ± 0.05	0.39 ± 0.07	0.18 ± 0.06
4	0.31 ± 0.01	0.45 ± 0.16	0.22 ± 0.05
5	0.29 ± 0.02	0.36 ± 0.05	0.24 ± 0.06

[a]ABA was analyzed by radioimmunoassay (7).

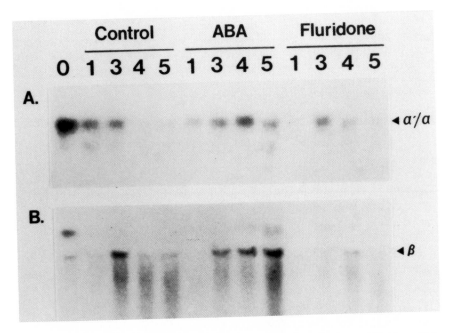

FIGURE 3. Cotyledons were cultured in control, ABA (10^{-5} M) or fluridone (10^{-6} M) medium for 1, 3, 4, or 5 days. RNA was isolated from the cotyledons, separated on an agarose gel containing formaldehyde, and transferred to nitrocellulose. The RNA was hybridized with ^{32}P-labeled pGmg91, which contains a gene encoding the β-subunit (B). After regeneration, the RNA was rehybridized with pGmg17.1, which contains an α'-subunit (A).

Effect of ABA on Cotyledons of Different Developmental Stages.

Cotyledons were cultured from seeds that were from 6 mm to 12 mm in length (Fig. 4). The amount of β-subunit in cultured cotyledons increased with the age of the cotyledon. The earliest stage of development (6 mm in length, stage H as described by Meinke <u>et</u> <u>al</u>. (8)) from which cotyledons could be routinely cultured accumulated a greater amount of the β-subunit in response to ABA in comparison to the control. Cotyledons from the oldest stage of development (11 to 12 mm in length, stage M and N) did not accumulate increased amounts of the β-subunit in response to ABA. At these stages of development, the β-subunit had already accumulated in the cotyledons (8).

Cotyledons from seeds that were 6 to 10 mm in length contained approximately equal concentrations of ABA after 5 days in culture without added ABA (Table 2). Seeds larger than 10 mm had a greater concentration of ABA. The addition of ABA to the culture medium caused an increase in the concentration of ABA in all of the cultured cotyledons when compared to those cultured in the absence of ABA.

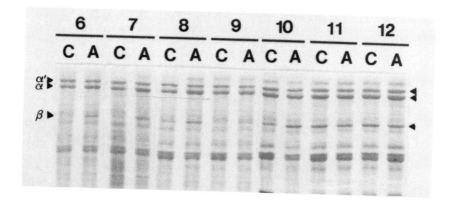

FIGURE 4. Cotyledons were cultured from seeds that were 6 mm to 12 mm in length for 5 days in control medium (C) or in medium containing 10^{-5} M ABA (A). Proteins extracted from these cotyledons were separated by SDS-PAGE and stained with Coomassie brilliant blue.

TABLE 2

THE CONCENTRATION OF ABA IN COTYLEDONS CULTURED
FROM SEEDS OF DIFFERENT LENGTHS WITH OR WITHOUT ABA

| | ABA Concentration[a] | |
Seed Length	Control	ABA
mm	ng ABA/mg fresh wt	
6	0.24 ± 0.02	0.35 ± 0.07
7	0.23 ± 0.04	0.33 ± 0.05
8	0.22 ± 0.02	0.36 ± 0.05
9	0.27 ± 0.02	0.32 ± 0.04
10	0.28 ± 0.02	0.43 ± 0.12
11	0.34 ± 0.12	0.37 ± 0.01
12	0.38 ± 0.01	0.57 ± 0.01

[a] ABA was measured by radioimmunoassay (7).

Concentration of ABA Required for the Response.

To determine the concentration of ABA required to observe changes in β-subunit expression, cotyledons were cultured in different concentrations of ABA. The cotyledons were treated with fluridone to eliminate the contribution of endogenous ABA at the beginning of the culture period. At the same time, different concentrations of ABA were added to the culture medium (Fig. 5). The ABA-treated cotyledons accumulated a greater amount of the β-subunit than the fluridone-treated cotyledons. The amount of the β-subunit that accumulated in the cotyledons increased from 10^{-6} to 10^{-4} M ABA. Cotyledons treated with 10^{-4} M ABA had a greater accumulation of β-subunit and a higher concentration of ABA than the control (Table 3). 10^{-3} M ABA did not result in an increased accumulation of the β-subunit. Although, the amount of ABA in the cotyledons treated with 10^{-3} M ABA was significantly greater than the other treatments. These results suggest that accumulation of the β-subunit in response to ABA is limited.

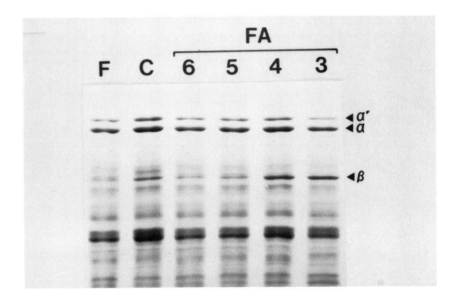

FIGURE 5.   Cotyledons were treated with 10^{-6}, 10^{-5}, 10^{-4}, and 10^{-3} M ABA in the presence of 10^{-6} M fluridone. Proteins were separated by SDS-PAGE and stained with Coomassie brilliant blue.

TABLE 3
THE CONCENTRATION OF ABA IN COTYLEDONS THAT WERE CULTURED
IN INCREASED CONCENTRATIONS OF ABA

Treatment	ABA Concentration[a]
	ng ABA/mg fresh wt
Control	0.17 ± 0.01
ABA 10^{-6} M[b]	0.12 ± 0.01
10^{-5} M	0.21 ± 0.07
10^{-4} M	1.43 ± 0.06
10^{-3} M	98.7 ± 2.8
Fluridone	0.13 ± 0.04

[a] ABA was measured by radioimmunoassay (7).
[b] ABA treatments were given in the presence of 10^{-6} M fluridone.

DISCUSSION

During maturation of cultured cotyledons, the plant growth substance abscisic acid modulates the accumulation of the β-subunit of β-conglycinin. This modulation occurs through increased synthesis of the β-subunit and is correlated with the accumulation of the β-subunit mRNA transcript. Modulation of the accumulation of the β-subunit by ABA can be detected after 3 days of treatment with ABA.

Seeds that have not begun to accumulate, or have a limited accumulation of the β-subunit, respond to ABA in the culture medium by producing an increased amount of the β-subunit in comparison to the control. However, seeds that are in developmental stages in which the β-subunit has begun to accumulate do not accumulate an increased amount of the β-subunit in response to added ABA (Fig. 4). This suggests that there is a limit to the amount of β-subunit that accumulates in response to ABA. A threshold of β-subunit accumulation is also reached when cotyledons are cultured in different sucrose concentrations (7) or in different ABA concentrations (Fig. 5). It is unknown if the limiting step of β-subunit accumulation is at the level of transcription, translation or in the ability of the cell to respond to ABA.

These experiments, where embryo maturation and embryo dormancy can be separated, suggest that maturation of soybean cotyledons and expression of genes during cotyledon maturation are not directly regulated by ABA. If ABA were directly involved in modulation of storage protein expression during seed development, it would be expected that during cotyledon maturation, ABA would modulate the accumulation of all of the proteins that are affected by ABA during prevention of precocious germination (3,6). However, during cultured cotyledon maturation, ABA only modulates the expresson of the β-subunit. The role of ABA during seed development may be an indirect involvement in the regulation of gene expression. Regulation of the β-subunit by ABA may be involved in the plant's response to the environment. The β-subunit is also regulated by other environmental changes such as the addition of methionine to the culture medium (9) or sulphur stress (10).

Although ABA may not be directly involved in seed development, ABA does modulate the expression of the β-subunit. The soybean cotyledon system is being used to study the mechanism of action of ABA. Comparisons between the regulation of the β-subunit and the α'- and α-subunits, which are not affected by ABA, can be used to study the regulatory mechanism.

REFERENCES

1. Crouch ML, Sussex IM (1981). Development and storage-protein synthesis in Brassica napus L. embryos in vivo and in vitro. Planta 153:64-74.

2. Triplett BA, Quatrano RS (1982). Timing, localization and control of wheat germ agglutinin synthesis in developing wheat embryos. Dev Biol 91:491-496.

3. Eisenberg AJ, Mascarenhas JP (1985). Abscisic acid and the regulation of synthesis of specific seed proteins and their messenger RNAs during culture of soybean embryos. Planta 166:505-514.

4. Dure L, Greenway SC, Galau GA (1981). Developmental biochemistry of cottonseed embryogenesis and germination: changing messenger ribonucleic acid populations as shown by in vitro and in vivo protein synthesis. Biochemistry 20:4162-4168.

5. Quebedeaux B, Sweetser PB, Rowell JC (1976). Abscisic acid levels in soybean reproductive structure during development. Plant Physiol 58:363-366.

6. Ackerson RC (1984). Abscisic acid and precocious germination in soybeans. J Exp Bot 35:414-421.

7. Bray EA, Beachy RN (1985). Regulation by ABA of β-conglycinin expression in cultured developing soybean cotyledons. Plant Physiol 79:746-750.

8. Meinke DW, Chen J, Beachy RN (1981). Expression of storage-protein genes during soybean seed development. Planta 153:130-139.

9. Holowach LP, Thompson JF, Madison JT (1984). Effects of exogenous methionine on storage protein composition of soybean cotyledons cultured in vitro. Plant Physiol 74:576-583.

10. Gayler KR, Sykes GE (1985). Effects of nutritional stress on the storage proteins of soybeans. Plant Physiol 78:581-585.

Molecular Biology of Plant Growth Control, pages 107–121
© 1987 Alan R. Liss, Inc.

ETHYLENE INDUCED GENE EXPRESSION IN FRUITS AND STORAGE
ORGANS AND SLICES THEREOF[1]

George G. Laties

Department of Biology, University of California Los Angeles
Los Angeles, California   90024

ABSTRACT   The evocation of ripening and the
respiration climacteric in avocado fruit, as well
as a respiration climax in carrot root, is accompanied by a spate of polysome synthesis and a group
of new mRNAs.   cDNA libraries from ripe avocado and
ethylene treated carrot have yielded a number of
ethylene specific clones subsequently used to assess
the appearance and build-up of ethylene engendered
mRNAs.   Two avocado cellulase clones of near full
length each select a sub-set of three cellulase
mRNAs, together comprising an apparent family of
six or more cellulase genes.   Cellulase processing
in avocado (19) is discussed.   In carrot ethylene
induced transcription ceases in less than a day—
a decline in levels of a cohort of induced mRNAs
occurring soon thereafter.   Reduction of oxygen or
ethylene concentration diminishes respiration rates
at once, while leaving ethylene specific mRNA levels
intact. Avocado slice ripening studies are described
wherein aminoethoxyvinylglycine totally inhibits the
respiration climacteric and ripening.   Carrot slices
show a strong ethylene response that is separable in
time from the wound induced respiration.   A cycloheximide respiration response in carrot slices exceeds
both the wound induced and ethylene induced
respiration.

[1]Current work citing the author was generously
supported by grants from the NSF and USDA Competitive Grants
Program.

INTRODUCTION AND OVERVIEW

Of the myriad ethylene responses of plant tissues
and organs the ripening associated respiration climacteric of many fruits is among the most conspicuous and
best known. In fact, ethylene has been widely viewed as
a ripening hormone, and it is only relatively recently
that the distinction has been pressed between ethylene as
an inducer and ethylene as a mediator of the ripening
process (1,2). The distinction arises in part from the
fact that under given conditions ethylene can evoke a
respiratory climax without ripening concomitants such as
softening, for example, and without autocatalytic ethylene
production (3,4), while under other conditions ripening
related phenomena can be elicited by ethylene without
overt respiratory effects (see 5,6). In any event AVG
(aminoethoxyvinylglycine), an inhibitor of ACC synthase,
the rate limiting enzyme in ethylene production that
converts SAM (S-adenosylmethionine) to ACC (1-amino-
cyclopropane-1-carboxylic acid), inhibits or delays
ripening in a number of climacteric fruits (2), thus
leaving little question of the central role of ethylene
in the ripening process therein. However, tissue
receptiveness, or sensitivity, has increasingly come to
be recognized as a key element in growth hormone response,
due largely to the proselytizing fervor of A. J. Trewavas
(7). In this vein a dual role has been assigned ethylene
in fruit ripening wherein low endogenous ethylene levels
in time bring the fruit to a responsive threshold, a
condition where perhaps the concentration and/or accessibility of ripening-related ethylene binding sites is such
that ethylene initiates the respiration climacteric
together with associated ripening phenomena (1).

A striking example of dual ethylene responsiveness is
found in immature avocado fruit where a so-called pulse
(e.g. 2 days) of propylene, an ethylene analogue, causes
a sharp respiration climax—unaccompanied by autocatalytic
ethylene synthesis or any sign of ripening—that subsides,
to be followed days later by a true climacteric entailing
autocatalytic ethylene production and full ripening. The
more mature the fruit on harvest, the shorter the interval
between what may be dubbed the pseudoclimacteric and the
true climacteric (3). In pulsing early fruit with
ethylene, the time to the onset of the climacteric is
inversely related to the duration of the pulse (Laties and
Buse, unpublished), an affirmation that the events of the

pseudoclimacteric influence the true climacteric—presumably by hastening the response threshold. When exogenous ethylene is given continuously on the other hand—even to early season fruit—the respiration climacteric and ripening seemingly proceed concomitantly (Fig. 1). In mature fruit the only vestige of duality is the frequently observed "shoulder" of the respiration time-course (Fig. 2 (see 15)). Presumably, exogenous ethylene, at concentrations in excess of preclimacteric endogenous levels, so accelerates the advent of the receptive state that attainment thereof and the initiation of the actual climacteric appear to be one event.

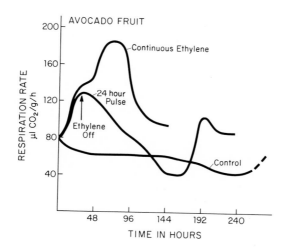

FIGURE 1. Comparison of effect of ethylene pulse and continuous ethylene on respiration behavior of intact early season avocado fruit. Ethylene: 10 μl/l in air. 22°C

Where ripening and the respiration climacteric occur together it has never been clear whether the respiration burst is the cause or the consequence of ripening events. Although ripening is often perceived as the near side of senescence, and senescence connotes degradation and disorganization, there is clear evidence of nucleic acid and protein synthesis alike in the course of the climacteric (5). Is the respiration climacteric, therefore, an "on-call" response to ATP utilization attending a spate of biosynthesis—a homeostatic last hurrah as proposed by Romani (8)—or does the respiration climacteric reflect the biosynthesis of one or more enzymes that directly or indirectly relieve rate controlling bottlenecks in glycolysis and/or electron transport? In this connection it is noteworthy that the climacteric in bananas (9) and avocado (10) is characterized by glycolysis cross-over points where $F-1,6-P_2$ levels rise, and PEP levels drop— suggesting that glycolysis is stimulated. Since there is no significant build-up of lactate or alcohol during the climacteric, however (10), aerobic respiration must keep up with glycolysis.

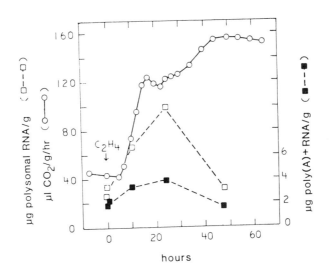

FIGURE 2. Effect of continuous ethylene on respiration climacteric and polysome concentration in mature fruit. Ethylene: 10 $\mu l/l$ in air. 25°C (15).

Quite recently fructose-2,6-P_2 has been discovered as an extraordinarily effective positive modulator of $F-1,6-P_2$ formation in higher plants by the enzyme PP_i-PFK, and a negative effector of $F-1,6-P_2$ ase (see 11). $F-2,6-P_2$ rises sharply in Jerusalem artichoke tubers (and in potato tuber tissue) in response to slicing (12), but more to the point,

has been found to rise in carrot roots in response to
ethylene (13). Withal, does it make sense simply to turn
on an upstream substrate mobilization tap (viz. stimulate
glycolysis), the result of which is to raise respiration to
no specific purpose? I think not, but rather favor the
view that ethylene induces genes that code for enzymes that
implement ripening, as well as for one or more enzymes that
serve to release constraints on respiration in glycolysis
and/or the electron transport chain. In this way ripening
specific metabolic requirements are fulfilled at the same
time that provision is made to sustain the "on-call" energy
demand that attends the spate of ripening related mRNA and
protein synthesis. Other genes that must be turned on or
activated in climacteric fruit are those mediating ethylene
synthesis—in particular the rate-limiting ACC synthase,
recently isolated and purified from tomato pericarp tissue
by Bleecker et al. (14).

Our ideal goal would be to identify and characterize
the many ethylene responsive genes controlling ripening
and the respiration climacteric and to learn how and when
they are influenced by ethylene. Our more realistic and
profoundly circumscribed early task has been to establish
first, a generic effect of ethylene on translation activity
as indicated by polysome proliferation, and secondly,
specific effects on gene expression as manifested by the
appearance and buildup of ethylene specific mRNAs.
Following the characterization of a cDNA clone of one such
specific mRNA in avocado as that for cellulase, a seeming
family has subsequently been identified.

RESULTS

Avocado Fruit Ripening.

Ethylene induced ripening of avocado is attended by
at least a threefold increase in polysome prevalence in
conjunction with the respiration climacteric (Fig. 2; see
15). An array of ethylene induced mRNAs appear and build
up with time through 48 h, while another group diminish
or disappear—as judged by in vitro translation of poly (A)+
mRNA and radioautograms of the translation products
following 1D or 2D PAGE. Since new messages comprise but
a very small fraction of all the mRNA, and total RNA does
not change during the climacteric (16), the proliferation
of polysomes must depend on a spate of overall translation

initiation. The time course of ethylene induced messenger buildup is much the same in all cases, with the onset of a sharp rise at about 10 h and the attainment of a maximal level at about 26 h (15). A cDNA library from poly(A)$^+$ of ripe fruit yielded ethylene specific clones one of which, pAV5, was found to code for cellulase as judged by hybrid selection in conjunction with immunoprecipitation of the in vitro translation product of the hybrid released message with antibody to pure cellulase (17). The pAV5 clone, containing but 600 base pairs, was subsequently used by Rolf Christoffersen (UC Santa Barbara) to select yet another cellulase clone, pAV-6, from the original library, the latter clone comprising some 1800 base pairs, while at the same time Mark Tucker (Molecular Plant Biology, UC Berkeley) starting afresh, and with pAV5 as an initial probe, isolated a cellulase clone, pAV3-63, of some 2200 base pairs. The latter clone has a Pst I restriction site 1200 bp from one end, and a HINC II site 1500 bp from the same end. When the HINC II site of pAV5 is aligned with that of pAV3-63, it is evident that pAV5 lies wholly within pAV3-63. It would seem both pAV-6 and pAV3-63 represent full or almost full messages.

Upon in vitro translation of poly(A)+ from ripe avocado, followed by 2-D PAGE of the products, some 5 or 6 53 kDa polypeptides are seen, while Western immunoblots of ripe extracts show 3 or more spots (18). Do the multiple 53 kDa autographic images of in vitro translation products represent a cellulase gene family or the products of a group of unrelated messages? Do the immunovisualized 53 kDa polypeptides of avocado extracts represent a gene family or post-translational processing intermediates of the product of a single gene? The evidence favors a family. When poly(A)+ from ripe fruit is hybridized with pAV-6 or pAV3-63 at low stringency some 6 53 kDa products are obtained on translation in each case. On the other hand, at high stringency each of the foregoing clones select but 3 mRNAs, and in combination, 6. Further, immunoprecipitation of in vitro translation products of mRNA hybrid-selected at low stringency with either clone shows some 6 products (Fig. 3; DeFrancesco, Tucker and Laties, unpublished). Thus at least 2 subsets of 3 cellulase genes each are suggested. Time course studies with low stringency hybrid selected cellulase mRNA indicate its presence as early as 8 h. In 2-D PAGE gels, furthermore, but one polypeptide translation product is evident at 8 h, whereas several are seen at 31 h—the implication being that at least one

member of one gene subset is manifested earlier than any
members of the other (DeFrancesco and Laties, unpublished).
<u>Processing of avocado cellulase</u>. Bennett and
Christoffersen (19) recently followed the course of
cellulase post-translational processing by establishing
that the enzyme first appeared as a membrane-associated
entity (56.5 kDa) and later as the free enzyme (54.2 kDa).

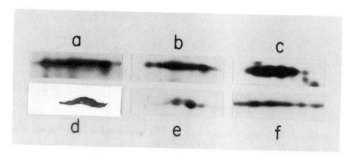

FIGURE 3. Hybrid select cellulase mRNA translation
products and immunoprecipitated translation products.
Autoradiographic images all 53 kDa polypeptides from 2-D
PAGE gels. <u>Hybrid select</u>: a) pAV6, 45°; b) pAV3-63, 45°,
d) pAV6, 55°, e) pAV3-63, 55°, b) pAV-6 + pAV3-63, 55°.
Immunoprecipitation: c) pAV-6, 45°.

Further, they compared the intact and deglycosylated forms
of the membrane-bound and mature free enzyme with the
in vitro translation product of cellulase mRNA. They
adduced evidence for a leader sequence in establishing
that the in vitro translation product (54 kDa) was larger
than the deglycosylated mature enzyme (52.8 kDa). That
the intact mature enzyme and in vitro translation product
were essentially the same size was due to the coincidence
that the leader sequence removed in processing was the
same mass as the glycan moiety added in maturation.
Subsequent sequencing of the amino terminus of the
cellulase protein taken together with the DNA sequence
of the 5' end of the pAV-6 cellulase clone, substantiated
the reality, and established the composition of the leader
sequence (Christoffersen and Bennett, unpublished).

Ripening of Avocado Slices

A study of avocado slice ripening was undertaken in the
expectation that slices might prove a prototype for fruit
ripening more amenable to experimental intervention. Whereas
peeled whole fruit have been shown to ripen much as intact
fruit—except sooner (20)—the respiration climacteric of
slices was shown to bear an inverse relation to thickness
(21). The latter relationship is unexpected if it is assumed
that the contribution of the wound induced respiration is
either nil (21) or diminishes fractionally with slice
thickness. We have learned that whereas the ethylene induced
respiration climacteric in intact avocado fruit occurs as
readily in air as in oxygen (22), the slightest water barrier
impairs ripening in slices, and necessitates the use of
oxygen in the gas phase (Tucker and Laties, unpublished).
Thus, so much as a wetted lens paper on the surface of a
2 mm thick slice prevents ripening even in oxygen, while
submerged slices shaken in oxygenated solution do ripen—
the difference being the unstirred liquid layer in the
former case. The sensitive dependence of ripening—and
presumably of ethylene responsiveness—on oxygenation may
well explain the cited inverse relationship of ripening and
slice thickness (21), since in the experiments in question
slices were wrapped in lens paper designed to serve as a
wick to present bathing medium to the tissue, and it is
surprising they ripened at all. Slices aged on screens in a
stream of moist air ripen in 4 to 5 days following a peak
of endogenous ethylene synthesis at about 60 h. Cellulase
mRNA appears during the course of the climacteric in slices
much as in fruit (Buse and Laties, unpublished). When AVG
is delivered each day in a 25 μl droplet spread on the
surface of a 15 mm diameter avocado disk, ethylene production, the respiration climacteric and ripening are totally
suppressed. Slices submerged in oxygenated solution become
swollen and gray as they ripen, and fail to yield
hybridizable cellulase mRNA. Tissue deterioration is in
fact a consequence of ripening, and the failure to demonstrate cellulase mRNA would seem to be an artifact of the
degraded condition of the tissue. Respiration profiles
with slices are less revealing than with intact fruit.
That is, an ostensible wound-induced rise grades into the
respiratory climacteric with ambiguous demarcation.
Endogenous ethylene production is a better criterion for
the course of the climacteric, and may explain why ripening
proceeds more rapidly in slices and in peeled fruit than

does endogenous ripening in intact fruit (20). The wounding entailed in slicing and peeling presumably hastens endogenous ethylene synthesis (cf. Discussion). In any event, with the appropriate experimental precautions slices promise to provide the hoped-for prototype for ripening studies.

Ethylene Induced Gene Expression in Carrot Roots.

    Intact carrot roots are considerably more impervious to oxygen than avocado skins (22), and accordingly both the ethylene induced respiration climax (the term climacteric being reserved herein for the ripening-associated respiration surge in fruit), and polysome prevalence show a synergistic dependence of ethylene action on oxygen concentration (23). cDNA clones of 3 ethylene responsive genes were selected as probes of the course of appearance and proliferation of ethylene induced mRNAs in carrot roots. The 3 mRNAs in question behaved as a cohort inasmuch as they rose with time in response to ethylene from essentially nil to maximal levels at 21 h, and dropped to half maximal levels at 24 h (24). Nuclear runoff experiments with the same probes demonstrated a sharp transcription maximum at 18 h, with transcription dropping to zero at 27 h. When slices of ethylene treated carrot were prepared at succeeding intervals and pulsed with ^{35}S-methionine, a 42 kDa polypeptide, a product of the gene represented by one of the ethylene specific clones (p10H1), was found to increase well beyond the time when transcription ceased and the mRNA level dropped.
    Since from the outset of our studies we have sought to establish whether the ethylene induced respiration climax depends upon the synthesis of particular mRNAs and their protein products, we examined the effect on ethylene specific mRNA levels of ethylene withdrawal or oxygen replacement by air at 9 or 15 h following the onset of ethylene treatment, times when specific mRNA levels were still rising. Whereas ethylene withdrawal or oxygen diminution caused an immediate sharp drop in respiration, the levels of ethylene specific mRNAs remained steady at the values obtaining at the time the gas environment was changed (25). Since in an unchanged ethylene/O_2 environment augmentation of ethylene induced mRNA proceeds for at least 21 h, it is evident that specific mRNA synthesis depends on ethylene, whereas mRNA maintenance does not. Further, whereas the attainment of the respiration climax may depend

on mRNAs synthesized in response to ethylene, respiration
rate per se does not reflect the levels of such mRNAs. It
would seem the "on-call" explanation of the climax is
problematic, at least where mRNA synthesis is concerned,
although a relation to protein synthesis is not ruled out.
    Ethylene responses in carrot slices. There is a
plethora of literature on wound induced respiration in
storage organ slices, including carrot (26; see 27), but
essentially no indication that ethylene influences slice
respiration except for a report on the metabolism of mitochondria from ethylene treated slices of sweet potato (28).
Thus it came as a considerable surprise when Günter Kahl,
on extending his studies of wound induced gene expression
in carrot slices to the possible effect of ethylene, perceived a dramatic response to ethylene well after the
wound response had subsided (Fig. 4; Kahl and Laties,
unpublished). In O_2 rather than air, in the presence of

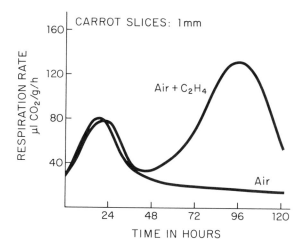

FIGURE 4.   Wound and ethylene induced respiration in
carrot slices.   Slices, 2 mm.   Ethylene 2 μl/l in air. 22°C.

ethylene, an enhanced wound induced respiration merged
directly into an elevated ethylene elicited climax. Of
further interest in these studies was the observation that
cycloheximide alone caused a respiration climax in slices

considerably in excess of the wound induced or ethylene
response. The cycloheximide response is entirely reminiscent
of the simulation by protein synthesis inhibitors of auxin
induced mRNA synthesis in pea epicotyl tissue described by
Theologis et al. (29). We are currently under-using an
extensive λgt11 cDNA expression library simply to screen for
wound induced genes (Kahl, Wood and Laties, unpublished).
With hybridization to wound specific clones, and in vitro
translation polypeptide profiles, studies are under way to
compare wound induced gene expression with ethylene and
cycloheximide induced gene expression. Following upon the
creation of an ethylene specific cDNA expression library and
the ultimate acquisition or preparation of antibodies to
proteins of interest, more meticulous studies will be
possible.

DISCUSSION

A consideration of the role of ethylene in ripening
and the attendant respiratory upsurge in fruit—simulated
by an ethylene induced respiratory climax in bulky storage
organs—inchoately addresses but a small part of the myriad
physiological and biochemical processes influenced by
ethylene (30). Nevertheless, sufficient questions are
raised to promise prolonged study in this area alone. A
major point of interest remains the dual role of ethylene:
viz. its foreshortening of the lag period, or so-called
green life of fruit—the period between harvest and the
onset of the ripening climacteric—and elicitation of the
climacteric when a sensitivity threshold has been attained.
The two events are readily separated in early fruit (e.g.
in avocado and banana) in response to an ethylene pulse
that evokes a pseudoclimacteric followed in due course by
the true climacteric (Fig. 1; (3)); in fully mature avocado
a shoulder in the respiration time-course, the last remnant
of the pseudoclimacteric, is evident when ripening is
initiated by exogenous ethylene (Fig. 2; 15), but less so,
if at all, when ripening proceeds endogenously. Are the
early, preparatory events subsumed in the overall climacteric in the latter case? Studies under way to characterize the pseudoclimacteric should permit an answer to
this question. It would be of great interest to know
whether there is a difference in primary ethylene binding
sites (31) for the two events in question, and whether

the K_M for ethylene in the two cases is the same or different.
Does readiness to respond reflect the dissipation of an
antagonist of ethylene action—a proposition suggested by
the behavior of avocado, where ripening is not initiated
until the fruit is picked (see 5)—or is readiness programmed
by endogenous biology where, for example, the attainment of
a sensitivity threshold entails synthesis of, or access to,
ethylene binding sites. Perhaps a clue may be sought in a
study of ethylene recalcitrance. Thus, in potato tubers a
considerable respiration response to ethylene is followed
by a period of several days of ethylene insensitivity.
Subsequently the tissue responds to ethylene again (32).
Is ethylene binding capacity lost during the refractile
period?

The need for O_2 concentrations in excess of that required
to sustain respiration in order to effect maximal ethylene
responsiveness has long been recognized (33), and discussed
again herein (see 25). It is perhaps unexpected that avocado
(and carrot) slices, in responding to ethylene, are more
sensitive to O_2 concentration than is the intact fruit. The
explanation is to be had in the liquid injection of surface
intercellular gas spaces on cutting or submersion. Nevertheless when avocado slices are adequately oxygenated slices
ripen endogenously much sooner than intact fruit, as do
peeled or bruised fruit (20). Thus, oxygen diffusion is not
at issue here, and the answer may lie in endogenous ethylene
production. However, both in slices and peeled fruit the
lag time to autocatalytic ethylene production is shortened
as well as ripening time, leaving the question of whether
wounding may shorten the lag time independently of ethylene
formation.

When potato tubers are treated with ethylene, fresh
slices therefrom are CN resistant, in contrast to slices
from untreated tubers (34). The role of ethylene in this
instance may relate to membrane integrity, much as suggested
in early classical ripening studies by F. F. Blackman (35)
and/or to effects on components of the mitochondrial transport chain. In this connection Laura DeFrancesco has found
changes in avocado mitochondrial protein in response to
ethylene (unpublished). As noted in the introduction, a
burgeoning field of future investigation into the molecular
biology of ethylene-influenced respiration control will
involve the genesis and regulation of enzymes that not only
regulate the rate of glycolysis, but determine the interrelationship of glycolysis and gluconeogenesis (11).

REFERENCES

1. McGlasson WB (1985). Ethylene and fruit ripening. HortSci 20(1):51.
2. Yang SF (1985). Biosynthesis and action of ethylene. HortSci 20(1):41.
3. Eaks I (1980). Respiratory rate, ethylene production, and ripening response of avocado fruit to ethylene or propylene following harvest at different maturities. J Am Soc Hort Sci 105(5):744
4. Brady CJ, O'Connell PBH (1976). On the significance of protein synthesis in ripening banana fruits. Aust J Plant Physiol 3:301.
5. Biale JB, Young RE (1981). Respiration and ripening in fruits—retrospect and prospect. In Friend J, Rhodes MJC (eds): "Recent Advances in the Biochemistry of Fruits and Vegetables," Academic Press, p 1.
6. Rhodes MJC (1980). The maturation and ripening of fruits. In Thimann KV (ed): "Senescence in Plants," CRC Press, p 158.
7. Trewavas A (1982). Growth substance sensitivity: the limiting factor in plant development. Physiol Plant 55:60.
8. Romani R (1984). Respiration, ethylene, senescence, and homeostasis in an integrated view of post harvest life. Can J Bot 62:2950.
9. Young RE, Salminen S, Sornorwichae P (1975). Enzyme regulation associated with ripening in banana fruit. In "Facteurs et Régulation de la Maturation des Fruits," Colloques Internationaux du CNRS:238:161.
10. Solomos Th, Laties GG (1974). Similarities between the actions of ethylene and cyanide in initiating the climacteric and ripening of avocados. Plant Physiol 54:506.
11. Black CC, Smyth DA, Wu M-X (1985). Pyrophosphate-dependent glycolysis and regulation by fructose 2,6-bisphosphate in plants. In Ludden PW, Burris JE (eds): "Nitrogen Fixation and CO_2 Metabolism," Elsevier Publ. Co., p. 361.
12. Van Schaftingen E, Hers H-G (1983). Fructose 2,6-bisphosphate in relation with the resumption of metabolic activity in slices of Jerusalem artichoke tubers. FEBS Lett 164:195.
13. Stitt M, Cséke C, Buchanan B (1986) Ethylene-induced increase in fructose-2,6-bisphosphate in plant storage tissues. Plant Physiol 80:246.

14. Bleecker AB, Kenyon WH, Kende H (1986). 1-aminocyclo-propane-1-carboxylate (ACC) synthase: purification and production of monoclonal antibodies. Abstract Am Soc Plant Physiol Louisiana State Univ, June 1986.
15. Tucker ML, Laties GG (1984). Interrelationship of gene expression, polysome prevalence, and respiration during ripening of ethylene and/or cyanide-treated avocado fruit. Plant Physiol 74:307.
16. Christoffersen RE, Warm E, Laties GG (1982) Gene expression during fruit ripening in avocado. Planta 155:52.
17. Christoffersen RE, Tucker ML, Laties GG (1984) Cellulase gene expression in ripening avocado fruit: The accumulation of cellulase mRNA and protein as demonstrated by cDNA hybridization and immunodetection. Plant Mol Biol 3:385.
18. Tucker ML, Christoffersen RE, Woll L, Laties GG (1985) Induction of cellulase by ethylene in avocado fruit. In Roberts J, Tucker G (eds): "Ethylene and Plant Development," Butterworths, p 163.
19. Bennett AB, Christoffersen RE (1986). Synthesis and processing of cellulase from ripening avocado fruit. Plant Physiol, in press.
20. Ben-Yehoshua S, Robertson RN, Biale JB (1963) Respiration and internal atmosphere of avocado fruit. Plant Physiol 38:194.
21. Tingwa PO, Young RE (1974) The effect of tonicity and metabolic inhibitors on respiration and ripening of avocado. Plant Physiol 54:907.
22. Theologis A, Laties GG (1982). Potentiating effect of pure oxygen on the enhancement of respiration by ethylene in plant storage organs: a comparative study. Plant Physiol 69:1031.
23. Christoffersen RE, Laties GG (1982) Ethylene regulation of gene expression in carrots. Proc Natl Acad Sci USA 79:4060.
24. Nichols S, Laties GG (1984). Ethylene-regulated gene transcription in carrot roots. Plant Mol Biol 3:393.
25. Nichols S, Laties GG (1985). Differential control of ethylene-induced gene expression and respiration in carrot roots. Plant Physiol 77:753.
26. Laties GG (1978). The development and control of respiratory pathways in slices of plant storage organs. In Kahl G (ed): "Biochemistry of Wounded Plant Tissues," Berlin: de Gruyter, p 421.

27. Kahl G (ed) (1978). "Biochemistry of Wounded Plant Tissues," Berlin: de Gruyter, 680 pp.
28. Makimoto N, Asahi T (1981). Stimulation by ethylene of mitochondrial development in wounded sweet potato root tissue. Plant Cell Physiol 22:1051.
29. Theologis A, Huynh TV, Davis RW (1985). Rapid induction of specific mRNAs by auxin in pea epicotyl tissue. J Mol Biol 183:53.
30. Lieberman M (1979) Biosynthesis and action of ethylene. Ann Rev Plant Physiol 30:533.
31. Sisler EC (1982). Ethylene binding in normal, rin, and nor mutant tomatoes. J Plant Growth Regul 1:219.
32. Reid MS, Pratt HK (1972). Effects of ethylene on potato tuber respiration. Plant Physiol 49:252.
33. Burg SP, Burg EA (1965). Ethylene action and the ripening of fruits. Science 48:1190.
34. Day DA, Arron GP, Christoffersen RE, Laties GG (1978) Effect of ethylene and carbon dioxide on potato metabolism. Plant Physiol 62:820.
35. Blackman FF (1954). "Analytic Studies in Plant Respiration." London: Cambridge University Press.

Molecular Biology of Plant Growth Control, pages 123–132
© 1987 Alan R. Liss, Inc.

BENZYLADENINE REGULATION OF THE EXPRESSION OF TWO
NUCLEAR GENES FOR CHLOROPLAST PROTEINS [1]

Susan Flores and Elaine M. Tobin

Department of Biology, UCLA
Los Angeles, California 90024

ABSTRACT. When white-light grown Lemna gibba plants
are placed in the dark for 6 days, levels of mRNA
encoding the major chlorophyll a/b-binding protein
(LHCP) and the small subunit (SSU) of ribulose-1,5-
bisphosphate carboxylase (RuBisCO) decline to less than
10% of the initial amounts. We found that treatment of
plants in darkness with a cytokinin, benzyladenine
(BA), causes an increase in the abundance of both these
mRNAs relative to total RNA.   BA pretreatment also
amplifies the increase in these mRNAs resulting from a
1-min red light pulse. The effect of BA treatment was
somewhat specific, since not all mRNAs examined show
increased levels after treatment. Pretreatments with BA
for less than 8 h do not lead to an increase in the
abundance of LHCP RNA relative to total RNA, either in
darkness or after a red light treatment, suggesting
that the observed increase after longer treatment is
not a primary response to cytokinin.   Analysis of both
nuclear RNA levels and in vitro transcription products
from nuclei isolated from BA treated and non-treated
plants indicates that BA regulates the expression of
LHCP and SSU at a post-transcriptional level.

INTRODUCTION

Chloroplast development is stimulated in many higher
plants by both light perceived by phytochrome, and by
cytokinins (1,2).   One aspect of phytochrome regulation of

[1] This work was supported by NIH grants GM-23167 (E.M.T)
and NRSA GM-09656 (S.F.).

chloroplast development is the control of the expression of nuclear genes for chloroplast proteins (3). It has now been shown in several species that the levels of mRNA encoding the small subunit (SSU) of ribulose-bisphosphate carboxylase (RuBisCO) and a polypeptide of the major light-harvesting chlorophyll a/b-binding protein (LHCP) are under phytochrome control (4). In this laboratory it was demonstrated that, in Lemna gibba, phytochrome regulates the abundance of these mRNAs at the level of transcription (5).

In contrast, less is known concerning the regulation of chloroplast development by cytokinins, and possible interactions with the phytochrome system. It has, however, been shown that cytokinins regulate the expression of genes for LHCP (6) and both subunits of RuBisCO (7), in tobacco cells and detached pumpkin cotyledons, respectively and that this regulation occurs at the mRNA level.

L. gibba, a small aquatic monocot, can be maintained in darkness if sucrose is provided; however, both red light and an exogenous cytokinin are required for growth (8). We have used the apparent cytokinin deficiency of dark-treated Lemna as a model system to study one aspect of the role of cytokinins in chloroplast development. Specifically, we have characterized the effect of benzyladenine, a synthetic cytokinin, on the expression of two nuclear-encoded chloroplast proteins, SSU and LHCP.

METHODS

Lemna gibba L. G-3 plants were grown aseptically under continuous illumination as previously described (9). For dark, red light (R), and benzyladenine (BA) treatments, plants were transferred to a dark growth chamber for 7 d. One minute R treatments were as described (9) and were given 2 h before harvest. For BA treatments, 0.5 µM BA was added to the medium 24 h before harvest, unless otherwise stated. Procedures used for the isolation of total and nuclear RNA, and for Northern and slot blot analysis, are described in detail elsewhere (9).

For analysis of run-off transcription, nuclei were isolated using a modification of the method of Watson and Thompson (10). Fresh tissue was homogenized in a hexylene glycol buffer using a Waring blendor fitted with razor blades, and nuclei then separated on Percoll gradients. Isolated nuclei were incubated at 30° for 20 min in a medium containing 50 mM $(NH_4)_2SO_4$, 2 mM $MnCl_2$, 0.2 mM each of ATP,

GTP, CTP, 10μM UTP, and 200-300 μCi of [³²P]UTP (3000Ci/mmol, ICN). Labelled RNA was isolated (11), and hybridized to cloned DNA sequences blotted to nitrocellulose (5). Transcript abundance was estimated from densitomiter scans of the autoradiogram.

RESULTS AND DISCUSSION

When white light grown <u>Lemna</u> plants are placed in total darkness for one week, the level of RNA encoding LHCP declines relative to total RNA. Treatment of plants with BA 24 h before harvest, or R 2 h before harvest, results in an increase in LHCP mRNA above the dark level. Figure 1 depicts the relative levels of LHCP mRNA in plants kept in total darkness for 1 week and then given either R, BA, or R plus BA treatments before harvest. The levels of LHCP mRNA were estimated from densitometer scans of autoradiograms of

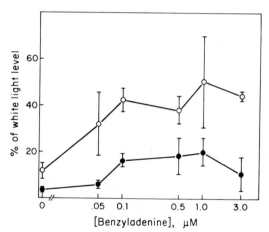

FIGURE 1. Levels of LHCP mRNA after dark, R, and BA treatments. Total RNA was isolated from <u>L. gibba</u> plants kept in total darkness for 7 d and given the indicated amount of BA 24 h before harvest (closed circles); or from plants given the same dark and BA treatments plus 1-min R 2 h before harvest. The RNA was then blotted to nitrocellulose and probed with pAB30, a <u>L. gibba</u> genomic clone encoding a polypeptide of LHCP (12). The average values from two experiments, with the range indicated, are shown. Reproduced with permission from (9).

slot blots of total RNA from treated plants, and were
normalized relative to LHCP mRNA levels in total RNA from
plants grown in continuous white light. In the absence of
BA or R treatment, the amount of LHCP mRNA is only about 5%
of the level in light-grown plants (closed circles, 0 BA).
R alone results in about a 3-fold increase (open circles, 0
BA). BA alone, in the range from 0.1 to 1.0 μM, causes
about a five-fold increase in LHCP mRNA (closed circles).
Similarly, in the presence of an optimal concentration of
BA, there is about four times more LHCP after a R pulse than
if the light treatment is given in the absence of the
hormone. The increase in the amount of LHCP mRNA observed
when both treatments are given is more then the sum of the
increase due to either factor alone.

In order to estimate how rapidly BA might be acting on
dark-treated plants to increase LHCP mRNA levels, the time

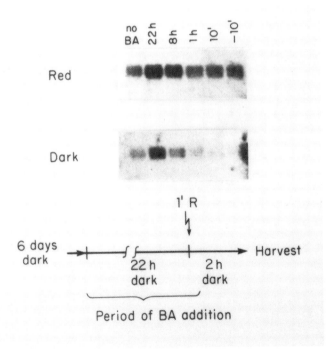

FIGURE 2. Northern blots of total RNA extracted from
plants placed in the dark for 1 week and given 0 or 0.5 μM
BA at varying times before harvest, as indicated. Top: 1-
min R was given 2 h before harvest. Bottom: Dark controls.
Filters were probed with pAB30 (12)

of BA addition was varied.   Figure 2 (top) shows a Northern
blot   analysis   of total RNA from plants given 1-min R   2   h
before   harvest,   as in Fig.   1.   However,   BA was added at
various   times ranging from 22 h before the light pulse   (as
in Fig.   1) to 10 min after the pulse.   It is apparent that
addition   of BA for 1 h or less before the light pulse   does
not   cause a stimulation over the no   BA   controls;   however
when the BA is present for 8 h, the stimulation is nearly as
great   as   at   24 h.   Similarly,   in plants kept   in   total
darkness (Fig.   2, bottom), BA must be present for at least 8
h   before   harvest for an increase in LHCP   mRNA   levels   to
occur.   Interestingly, the kinetics of the BA effect seem to
be   different   for plants kept in total darkness and   plants
receiving a R pulse;   for the latter the effect seems to   be
nearly   maximal   at   8 h while for the plants   receiving   no
additional illumination there is a substantial increase seen
between 8 and 24 h.

    We   were   particularly   interested   in   determining   if
pretreatment with a cytokinin also influenced the expression
of   genes   encoding SSU since it was previously   shown   that
phytochrome regulates the levels of both SSU and LHCP   mRNAs
in Lemna at the level of transcription (5).   Fig.   3, panel
A, shows a Northern blot analysis of total RNA isolated from
dark-treated   plants provided with 0 (lanes 1 and 2) or   0.5
μM   (lanes 3 and 4) BA 24 h before harvest and kept in total
darkness for that period (lanes 1 and 3), or given 1-min R 2
h   before   harvest (lanes 2 and 4).   The SSU   message   also
increases in response to BA in the dark, and the increase in
response   to   R   is larger in the presence   of   BA   than   in
untreated   plants.   In   a   similar   experiment,   when
incorporation   of [^{35}S]-methionine into protein was examined
by   SDS-PAGE and fluorography,   synthesis of SSU was   barely
detectable   unless   plants   received both   R   and   cytokinin
treatment [data not shown, see (3)].

    The   observation that BA treatment increases the levels
of   LHCP   and SSU RNAs relative to total   RNA,   as   seen   in
previous   figures,   could   be   explained   by   two   different
mechanisms.   This   result might be observed if BA treatment
led to qualitative changes in the mRNA population,   but also
would be seen if BA treatment caused a general shift in   the
proportion   of   mRNA to ribosomal RNA.   To test the   latter
possibility,   we examined the effect of BA and R   treatments
on levels of additional nuclear-encoded mRNAs.   From Fig. 3,
(B,C)   it   can   be seen that BA treatment has no   effect   on
levels   of   sequences   hybridizing   to   pLg106,   an
uncharacterized L. gibba cDNA (Fig.   3B) or to RNA encoding

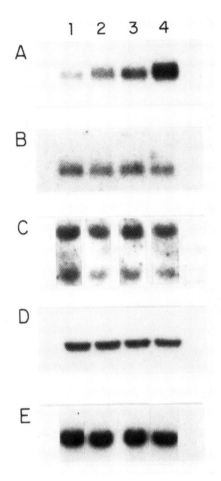

FIGURE 3A-E.   Northern blot analysis of total RNA extracted from plants placed in the dark for 1 week and given the following treatments: lane 1, no additions or light treatments; lane 2, 1-min R was given 2 h before harvest; lane 3, 0.5 µM BA was given 24 h before harvest, no light treatment; lane 4, 0.5 µM BA was added 24 h before harvest and a 1-min R pulse was given 2 h before harvest. Blots were probed with the L. gibba cDNAs pLgSSU1 (A) or pLg106 (B), a chicken cDNA encoding β-actin (C), or cloned chloroplast DNA fragments encoding the 32 kDa herbicide-binding protein (D) or the large subunit of RuBisCo (E). Reproduced with permission from (9).

β-actin (Fig. 3C). The former message increases in abundance when white-light grown plants are placed in the dark (data not shown), and thus clearly represents an mRNA synthesized during the course of the experiment.

Because numerous reports have linked cytokinins to chloroplast development, we were interested to see if the enhanced levels of two nuclear-encoded mRNAs which were observed after BA treatment were correlated to changes in chloroplast mRNA levels. We used probes for mRNAs encoding two chloroplast proteins: the 32 kDa herbicide-binding protein (Fig. 3D) and the large subunit of ribulose-bisphosphate carboxylase (Fig. 3E). No effect of BA pretreatment was seen on levels of either mRNA. Unlike SSU and LHCP messages, substantial amounts of the chloroplast-encoded transcripts remain even after plants have been in darkness for one week, and small changes would not be detectable above the high background. However, it is clear that any increase relative to the total amount of message is insignificant.

The changes in steady-state SSU and LHCP mRNA levels which were observed as a result of BA treatment could be due to either alterations in rates of transcription or changes in RNA life-time. One approach to distinguishing these possibilities is to examine events occurring in the nucleus, to reduce the contribution of cytoplasmic processes. Nuclei were isolated from plants placed in the dark for one week and given 1-min R 2 h before harvest, and RNA isolated from these nuclei. It was found that the addition of 0.5 uM BA 24 h before harvest had no effect on relative levels of LHCP sequences in nuclear RNA, as estimated from slot blot analysis [data not shown, see (8)]. Furthermore, isolated nuclei were used to synthesize P-labelled RNA in a run-off transcription system. In this assay, nuclei are allowed to extend transcripts in vitro in the presence of nucleotides, including labelled UTP. Since it is not believed that re-initiation of transcription by RNA polymerase occurs in such a system, the amount of label incorporated into a particular transcript in vitro is believed to reflect the extent of RNA polymerase engaged in transcription in vivo, at the time the nuclei were isolated. When run-off transcripts were synthesized by nuclei isolated from plants placed in the dark and given R and BA treatments, as in previous experiments, there were only small differences in levels of SSU and LHCP sequences in products of nuclei isolated from plants receiving BA (Table 1, D+/D and R+/R). However, a R treatment resulted in a several fold-increase in LHCP

TABLE 1

THE EFFECT OF BA AND R ON RELATIVE RATES
OF TRANSCRIPTION IN ISOLATED NUCLEI

	D+/D–	R+/R–	R–/D–	R+/D+
LHCP	1.5	1.4	3.6	3.3
SSU	0.9	1.2	1.4	1.4

TABLE 1.   Relative abundance of LHCP and SSU sequences
in run-off transcripts from nuclei isolated from L. gibba
plants given BA or R treatments.   Plants were placed in
darkness for 1 week before isolation of nuclei.   (D) plants
were given no additional light treatment;   (R) plants
received 1-min R 2 h before harvest.   (+) indicates the
addition of 0.5 μM BA 24 h before harvest.   [^{32}P]-labelled
transcripts were isolated as described (11) and equal cpm
hybridized to LHCP and SSU sequences blotted to
nitrocellulose.   Relative amounts were estimated from
densitometer scans of the autoradiogram of the blot.   The
results of 2 experiments were averaged.

transcript levels (Table 1, R+/D+, R/D), in the presence or
absence of BA.   It is interesting to note that the effect of
R on the relative amount of SSU transcription in the green
plants used in these experiments was less than the R
stimulation of LHCP gene transcription, and less than might
be expected from the R stimulation of SSU gene transcription
in etiolated plants (5).   In other experiments we observe a
greater effect of R on SSU transcript levels in green
tissue, however the lack of BA effect has been consistent.
We have also noted variability in the response of steady-
state SSU mRNA levels to R, and in those experiments where
little response to R is seen, the background level of SSU
mRNA in dark-treated plants is high.

CONCLUSION

In green <u>Lemna</u> fronds transferred to darkness, it appears that benzyladenine and phytochrome act independently to regulate levels of mRNA encoding SSU and LHCP. Our data is consistent with the hypothesis that phytochrome regulation occurs primarily at the level of transcription, and BA acts primarily post-transcriptionally. Since a low level of transcription of these two messages remains in dark-treated plants (data not shown), a decreased rate of turnover in the presence of BA would lead to increases in steady state message level even in the absence of R. Similarly, the nuclear run-off experiments show that R stimulates transcription in the absence of BA. However, the greater stimulation of steady-state mRNA levels by R after BA treatment is expected if mRNA degradation is reduced. This model is consistent with the observation that BA and R effects on mRNA levels are more nearly multiplicative than additive. It also provides an explanation for the observation that BA must be present longer to affect LHCP mRNA levels in darkness than to enhance mRNA levels after a R pulse: since transcription rates are several-fold higher after a R pulse, but only about slightly higher after BA treatment (Table 1), LHCP mRNA will accumulate more rapidly after R.

The physiological significance of the cytokinin requirement in dark-grown <u>Lemna</u> remains unclear. Although white-light grown fronds placed in darkness remain green and appear healthy, the cessation of growth indicates at least partial dormancy. It is possible that the effect of cytokinin in green tissue is more relevant to the role of this growth regulator in processes of senescence and dormancy rather than in the phenomenon of greening. The answer to this question awaits more thorough characterization, at a molecular level, of the effects of cytokinins on the various phases of chloroplast development. The fact that BA must be present for over an hour to see an effect on LHCP and SSU mRNA levels suggests that we are not looking at a primary response to cytokinin. However, it is hoped that identification of the molecular mechanisms involved in the apparent post-transcriptional regulation of specific mRNAs may eventually lead to the identification of early events.

REFERENCES

1. Mohr H (1984). Phytochrome and chloroplast development. In Baker NR, Barber J (eds): "Chloroplast biogenesis," Amsterdam: Elsevier, p 305.
2. Parthier B (1979). The role of phytohormones (cytokinins) in chloroplast development. Biochem Physiol Pflanzen 174:173.
3. Tobin EM, Silverthorne J, Flores S, Leutwiler LS, Karlin-Neumann GA (1986). Regulation of the synthesis of two chloroplast proteins encoded by nuclear genes. In Fox JE, Jacobs M (eds): "Molecular Biology of Plant Growth Control," New York: Alan R. Liss, in press.
4. Tobin EM, Silverthorne J (1985). Light regulation of gene expression in higher plants. Ann Rev Plant Physiol 36:569.
5. Silverthorne J, Tobin EM (1984). Demonstration of transcriptional regulation of specific genes by phytochrome. Proc Natl Acad Sci USA 81:1112.
6. Teyssendier de la Serve B, Axelos M, Péaud-Lenöel C (1985). Cytokinins modulate the expression of genes encoding the protein of the light-harvesting chlorophyll a/b complex. Plant Mol Biol 5:155.
7. Lerbs S, Lerbs W, Klyachko NL, Romanko EG, Kulaeva ON, Wollgiehn R, Parthier B (1984). Gene expression in cytokinin- and light-mediated plastogenesis of Cucurbita cotyledons: Ribulose-1,5-bisphosphate carboxylase/ oxygenase. Planta 162:289.
8. Hillman WS (1957). Nonphotosynthetic light requirement in Lemna minor and its partial satisfaction by kinetin. Science 126:165.
9. Flores S, Tobin EM (1986) Benzyladenine modulation of the expression of two genes for nuclear-encoded chloroplast proteins in Lemna gibba. Planta, in press.
10. Watson JC, Thompson WF (1986) Purification and restriction endonuclease analysis of plant nuclear DNA. Methods Enzymol 118:57.
11. Groudine M, Peretz M, Weintraub H (1981) Transcriptional regulation of hemoglobin switching in chicken embryos. Mol Cell Biol 1:281.
12. Kohorn BD, Harel E, Chitnis PR, Thornber JP, Tobin EM (1986) Functional and mutational analysis of the light-harvesting chlorophyll a/b protein of thylakoid membranes. J Cell Biol 102:972.

Molecular Biology of Plant Growth Control, pages 133–143
© 1987 Alan R. Liss, Inc.

ETHYLENE REGULATION OF PLANT DEFENSE GENES [1]

Joseph R. Ecker and Ronald W. Davis

Department of Biochemistry
Stanford University School of Medicine
Stanford, California   94305

ABSTRACT   One of the earliest detectable events to   occur
during the plant-pathogen interaction is a rapid increase
in the key ethylene biosynthetic   enzyme, ACC   synthase.
The subsequent increase in ethylene production, which
follows such a biological stress, may be a signal for
plants   to   erect   defense   mechanisms against invading
pathogens. These observations have led us to examine the
effect of ethylene on two plant defense response genes:
1) phenylalanine ammonia-lyase (PAL), the first enzyme in
phenylpropanoid biosynthetic   pathway   and 2) hydroxyproline rich glycoprotein (HRGP), the major   protein   compon-
nent of the plant cell wall. Northern blot analysis of
RNA from ethylene-treated carrot roots reveals a marked
increase in the level of both PAL mRNA and certain HRGP
mRNA species compared to control air-treated roots.   In
order to further examine the regulation of these defense
response genes, chimeric genes have been constructed and
introduced into plant cells using electroporation.

INTRODUCTION

As in animal systems, resistance of plants to disease not
only involves static processes, but also involves inducible defense mechanisms. The enzyme phenylalanine ammonia-lyase (PAL)
plays a central role in this response. PAL, the first enzyme
in the pathway of phenylpropanoid compounds in plants, catalyzes the elimination of ammonia and hydrogen from phenylala-

[1]This   work   was   supported   by   the   National Science
Foundation Research Grant DBC 8402390 and by the United States
Department of Energy Research Grant DE-FG013-84ER13265.

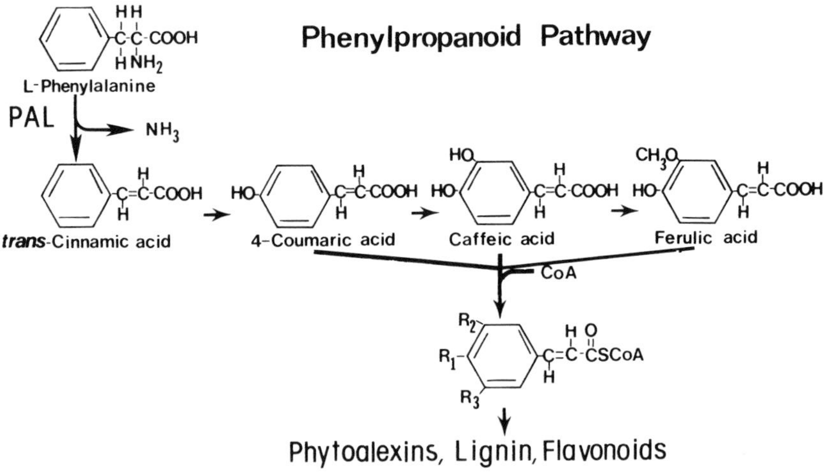

Phenylpropanoid Pathway

L-Phenylalanine

PAL ⟶ NH₃

trans-Cinnamic acid   →   4-Coumaric acid   →   Caffeic acid   →   Ferulic acid

CoA

Phytoalexins, Lignin, Flavonoids

FIGURE 1.   Phenylpropanoid biosynthetic pathway.

nine to form cinnamic acid, the precursor of several compound
involved in the defense of plants againsts pathogens (Figur
1).   Increases in phenolic compounds occur during stress o
tissue invasion and may be responsible for   active plant de
fense.   Induction of PAL enzyme activity occurs in respons
to pathogenic infection (1), treatment with ethylene (2), U
light (3), wounding (2) and to a host of other biotic an
abiotic stimuli (1).

The regulation of PAL enzyme activity in plants has bee
most intensively studied in relationship to defense response
(1). Treatment of plants or plant cell cultures with activ
fungal cultures or elicitors induces PAL mRNA (4,5) whic
results in the synthesis of phytoalexins (5) and lignin (6)
These compounds are products of the phenylpropanoid pathwa
and their accumulation results from the coordinate increase i
phenylpropanoid biosynthetic enzymes. Induction of PAL enzym
enzyme activity correlates with increased resistance to patho
genic infection (7).   Inhibition of PAL activity with the spe
cific inhibitor L-α-aminooxy-β-phenylproprionic acid results
in greatly reduced levels of phenolic compounds and correlate
with an increase in susceptibility of plants to infection (8)
The induction of PAL and several phenylpropanoid biosyntheti
enzymes by fungal infection or elicitor treatment occurs via a
increase in the rate of gene transcription (3).

Several other defense response proteins including hydroxy
proline-rich glycoproteins (HRGPs)(9,10,11), peroxidase (12)
chitinase, β-1,3-glucanase (13), proteinase inhibitors

(14) and pathogenesis-related (PR) proteins (15) are induced as a result of pathogen attack, elicitor or ethylene treatment. HRGPs are the major structural proteins of plant cell walls (16). The accumulation of HRGPs in response to stress or tissue invasion during infection occurs in many plant systems and is correlated with the expression of disease resistance. Consistent with these results, rapid accumulation of HRGP mRNAs occurs in response to fungal infection (10). The exact role of HRGPs in the defense response is not clear, but HRGPs may act as structural barriers, provide a matrices for the deposition of lignin (6) and/or may act as specific agglutinins of microbial pathogens (17).

The earliest detectable event to occur during the plant-pathogen interaction is a rapid increase in ethylene biosynthesis (15). Ethylene, the gaseous plant hormone, is involved in the regulation of numerous plant processes ranging from growth and development to fruit ripening (18). Furthermore,ethylene can regulate the rate of transcription of specific plant genes (19). The molecular mechanisms underlying operation of ethylene action are unknown but the pathway of its biosynthesis in plants has been determined to be from methionine through S-adenosylmethionine and 1-aminocyclopropane-1-carboxylic acid (ACC) (18). A variety of biological stresses can induce ethylene production in plants including wounding (18), viral (15), or fungal infection (9) and by treatment with elicitors (20). The increase in stress ethylene production is caused primarily by an increase in activity of the rate limiting ethylene biosyn- synthetic enzyme, ACC synthase. The induction of ACC synthase is the most rapid of any enzyme documented thus far in response to elicitor treatment (20). We propose that ethylene produced in response to biological stress is a signal for plants to erect defense mechanisms against invading pathogens and report that ethylene causes the rapid accumulation of PAL and certain HRGP mRNAs in carrot root.

RESULTS

In order to study the regulation of the PAL carrot gene in response to ethylene, a cDNA (3) encoding a portion of the parsley PAL mRNA was used as probe in southern blot experiments to determine the extent of homology to carrot genomic DNA. Carrot genomic DNA was digested with several restriction en- zymes and hybridized using stringent hybridization conditions with a radiolabled parsley PAL cDNA. A very simple hybridiza- tion profile indicitative of only 1 or 2 carrot PAL genes is revealed (Figure 2b). In order to obtain the homologous carrot

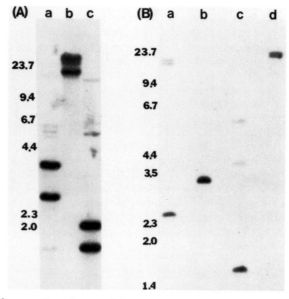

FIGURE 2. Southern blots of carrot DNA digested wit
restriction enzymes (blot A) a)EcoRI b)BamHI c)HindIII or (blc
B) a)BamHI b)EcoRI c)HindIII d)PstI and probed with the a
carrot HRGP gene or b) parsley PAL cDNA.

PAL gene, Sau3A partially-digested carrot DNA was used to construct a genomic library in the λ vector EMBL-3. The parsley
cDNA was used as probe to identify two λ clones from the carrot genomic library which contain the carrot PAL gene (Figur
3). The promoter region of the PAL gene was mapped and se
quenced. The carrot PAL gene was used as probe in norther
blot experiments to examine the effect of ethylene on PAL mRNA
Whole carrot roots were placed in 4 l jars and allowed t
equilibrate in a stream of moist air. After 24 hrs, the root
were flushed at a rate of 100 ml/min with either 10 pp
ethylene in hydrocarbon-free air, 10 ppm ethylene in oxygen o
with hydrocarbon-free air or oxygen alone. At various time
during the treatment period groups of at least four carrots pe
treatment condition were removed and polyadenylated RNA pre
pared. Polyadenlyated RNA was denatured by glyoxal, electro
phoresed on agarose gels and transfered to diazotized paper
or to Genatran 45 nylon blotting membrane (Plasco). As show
in figure 4, ethylene causes the accumulation of PAL mRNA i
carrot roots. Densitometic analysis reveals a ten-fold in
crease in the level of PAL mRNA after 50 hours of ethylen

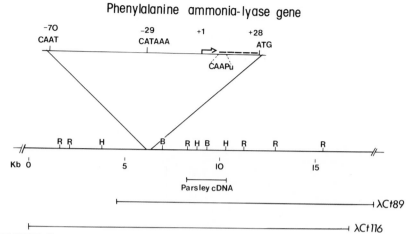

FIGURE 3.   Restriction map of the carrot PAL gene.

treatment. A more complex pattern of regulation of HRGP mRNA
is evident from similar RNA analysis using a carrot HRPG gene
(16) as probe.   Three mRNA species of sizes 1.5, 1.8 and 4.0 kb
are present in air-treated roots (Figure 5).   The 4.0 and 1.8
kb RNAs are much less abundant than the 1.5 kb RNA. After exposure to 10 ppm ethylene, there is a dramatic 50 to 100-fold
increase in the level of the 4.0 and 1.8 kb mRNAs, while the
1.5 kb mRNA decreases in abundance during the same time period.
Several additional ethylene-induced mRNAs homologous to the
HRGP probe are also detected (Figure 5).   Such transcripts
either have low homology to the HRGP gene probe and/or are very
low abundance transcripts. The former possibility seems likely
since southern blot analysis reveals several weakly hybridizing
bands (Figure 2a).   Furthermore, several additional lambda
clones weakly homologous to the HRGP gene probe have been
isolated (unpublished data). Additional studies have revealed
that the ethylene inducibility of HRGP transcripts may be
developmentally controlled (data not shown).
     Treatment of most storage roots and fruits with ethylene
in air causes an increase in the rate of respiration (21).
Furthermore, treatment with ethylene in oxygen, instead of
air, causes a potentiation of the respiration rate (21).   In-
order to determine if the induction of PAL and HRGP mRNAs is
linked to the ethylene-induced rise in respiration, the effect
of ethylene in air or oxygen on both PAL and HRGP mRNAs was
examined. Analysis of RNA prepared from carrot roots so
treated revealed that the ethylene induced rise in PAL and HRGP

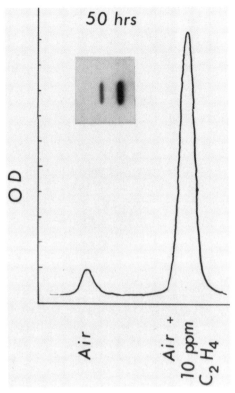

FIGURE 4. Slot blot and densitometer tracing of 50 hr
ethylene-treated carrot RNA probed with the carrot PAL gene.

mRNAs is not potentiated by oxygen (data not shown).

Upon wounding of the carrot root, there is a dramati
increase in accumulation of the 1.5 kb mRNA species relative t
the other HRGP transcripts. Within one hr, the level of th
1.5 kb mRNA increased over 10-fold while the level of the 1.
kb mRNA decreased (data not shown). Similarly, protoplast- in
of carrot cell cultures induces the accumulation of the 1.5 k
HRGP mRNA species (data not shown). Conversely, protoplast
ing of cell cultures results in a decrease of PAL mRNA and tub
ulin mRNA (data not shown). The kinetics of accumulation c
the 1.5 kb HRGP mRNA parallels resynthesis of the cell wall.
An additional approach to understanding the regulation of PA
and HRGP genes is to construct chimeric genes containing th
PAL promotor fused a reporter gene. The carrot PAL ger
promoter was fused to the coding region of the bacterial chlor

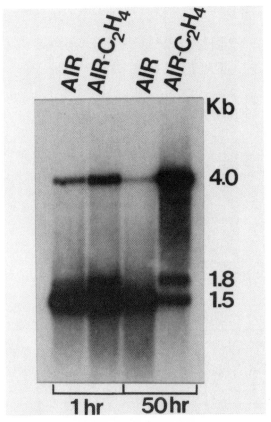

FIGURE 5. Northern blot analysis of 1 or 50 hr ethylene-treated carrot RNA probed with the carrot HRGP gene.

amphenicol acetyltransferase (CAT) gene and the polyadenyla-tion signal of the Agrobacterium nopaline synthase gene (NOS) gene. This and a similar construction containing the NOS gene promoter fused to the CAT gene were introduced into either carrot or tobacco protoplasts by electroporation (22). Relative to the NOS/CAT/NOS plasmid, the PAL/CAT/NOS plasmid is ex-pressed at a higher level in homologous cells (carrot) than in heterologous cells (tobacco) (Figure 6). Studies are in-progress to examine the inducibility of the PAL and HRGP gene promotor fusions in both transiently transfected cells and stably transformed plants.

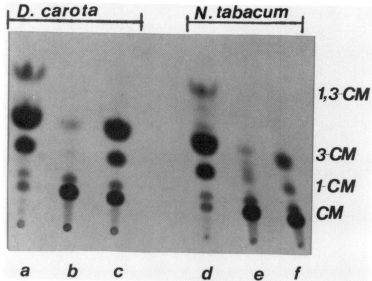

FIGURE 6. CAT enzyme assays from carrot or tobacco cel
electroporated with chimeric genes: a and d) NOS/CAT/N
plasmid, b and e) SON/CAT/NOS (inverted NOS promoter, contr
plasmid), c and f) PAL/CAT/NOS plasmid.

DISCUSSION

The effect of the gaseous plant hormone ethylene on two pla
defense response genes has been examined.   PAL performs
enzymatic role in the response and HRGPs may perform a
structural role in the response.   Ethylene causes a mark
accumulation of 4.0 and 1.8 kb HRGP mRNAs and PAL mRNA in u
wounded carrot roots. Wounding of plants results in inducti
of a different size (1.5 kb) HRGP transcript. The induction
this transcript is very large in stored roots but only sever
fold in rapidly growing roots. These results suggests that t
1.5 kb HRGP mRNA may be the normal size transcript produc
during plant growth. Similarly, the induction of PAL mRNA
ethylene in rapidly growing roots is not as great as in matu
or stored roots. Presumably, the high basal levels of the
mRNAs in developing plant tissues reflects the requirement
PAL and HRGP proteins for plant growth.
      Treatment of unwounded carrot roots with ethylene induc
the   accumulation of the 1.8 and 4.0 kb HRGP transcripts aft
only 1 hour of exposure to the hormone .   Furthermore, whi

treatment with ethylene causes a dramatic increase in the levels of 1.8 and 4.0 kb HRGP transcripts, it causes a reduction in the level of the 1.5 kb HRGP transcript. These results indicate that the 1.5 and 1.8 kb HRGP transcripts are differentially regulated. The 1.5 kb HRGP mRNA is inducible by wounding but not by 10 ppm ethylene, while the 1.8 and 4.0 kb HRGP mRNAs are induced by ethylene. Both the 1.5 and 1.8 kb transcripts have been mapped to the same HRGP gene (16). It has been proposed that each transcript has its own promoter; the 1.8 kb transcript promoter being upstream to the 1.5 kb transcript promoter. If this is the case, then the reduction in abundance of the 1.5 kb transcript during ethylene treatment may be explained by promoter occlusion. Alternatively, ethylene may regulate the differential splicing of HRGP mRNAs. Studies are in progress to more precisely map these two transcripts. Sequencing of primer-extended cDNA products and comparison to the genomic sequence should allow an unambiguous answer. Furthermore, it will be of interest to examine and compare the primary structure of the HRGP proteins encoded by the 4.0 and 1.5 kb mRNAs. Structural differences between these two proteins may give us a clue as to their roles in the development or defense of the plant.

Plants can produce defense compounds in response to ethylene. The compound 3-methyl-6-methoxy-8-hydroxy-3,4-dihydroisocoumarin (6-methoxymellein), an isocoumarin phytoalexin, is produced in large (mg/g tissue) quantaties in carrot roots in response to either fungal infection (23) or ethylene (2). Carbon dioxide, an antagonist of ethylene action, inhibits the induction (2). Accumulation of 6-methoxymellein in carrot is clearly an example of a defense response induced by ethylene. Furthermore, treatment of melon (*Cucumis melon*) with with elicitor from either fungal or plant origin or infection with *Colletotrichium lagenarium* results in ethylene production and a dramatic increase in HRGPs (9). In the presence of an inhibitor of ethylene biosynthesis, inhibi- tion of both the elicitor-induced ethylene and elicitorinduced HRGP occurs. Analogous to elicitor, addition of the immediate precursor of ethylene, ACC, induces ethylene production and the accumulation of HRGP (9). These results suggest that the induction of phytoalexins and HRGPs by fungal, viral or bacterial infections, may result from triggering of ethylene production. Moreover, as a gas, the locally produced ethylene is eminently suited to induce defense response genes in both neighboring and somewhat dis- stant plant tissues in advanse of pathogen movement. To this end, Showalter etal(10) have recently reported that infection of bean (*Phaseolus vulgaris*) hypocotyls with a fungal pathogen (*Colletotrichum lindemuthianum*) not only induces defense response mRNAs in cells at the infection site but also at uninfected sites distantly located. Thusfar, increases in the levels of the plant defense-related enzymes, ACC synthase, PAL, lignin biosynthetic enzymes, peroxidase, HPGPs, chitinase, β-1,3-glucanase, and PR-proteins have been shown to occur during pathogen attack or in response to ethylene. Taken together, these results strongly support the proposal that ethylene produced in response to biological stress is a signal for plants to erect defense mechanisms against potential pathogens.

REFERENCES

1. Hahlbrock K, Grisebach H (1979) Enzymic controls in the biosynthesis of lignin and flavonids. Ann Rev Plant Physio: 30:105.
2. Chalutz E (1973) Ethylene-induced phenylalanine ammonia-lyase activity in carrot roots. Plant. Physiol. 51 1033.
3. Chappell J, Hahlbrock K (1984) Transcription of plant defense genes in response to UV light or fungal elicitor. Nature 311:76.
4. Edwards K, Cramer CL, Bolwell GP, Dixon RA, Wolfgang S, Lamb CJ (1985) Rapid transient induction of phenylalanine ammonia-lyase mRNA in elicitor-treated bean cells. Proc Nat Acad Sci USA 82:6731.
5. Darvill AG, Albersheim P (1984). Phytoalexins and their elicitors: A defense against microbial infection in plants. Ann Rev Plant Physiol 35: 243.
6. Vance CP, Kirk Tk, Sherwood RT (1980). Lignification as a mechanism of disease resistance. Ann Rev Phytopath 18: 259.
7. Cramer CL, Bell JN, Ryder TB, Bailey JA, Schuch W, Bolwell C Robbins MP, Dixon RA, Lamb CJ (1985) Co-ordinated synthesis of phytoalexin biosynthetic enzymes in biologically-stressed cells of bean (Phaseolus vulgaris L.) EMBO 4:285.
8. Moesta P, Grisebach H (1982) L-2-Aminooxy-3-phenylproprioni acid inhibits phytoalexin accumulation in soybean with concomitant loss of resistance against Phytopthora megasperma. Physiol Plant Path 21:65.
9. Roby D, Toppan A, Esquerre-Tugaye MT (1985). Cell surfaces i plant microorganism intetactions. V. Elicitors of fungal and of plant origin trigger the synthesis of ethylene and of cel wall hydroxyproline-rich glycoprotein in plants. Plant Physiol 77:700.

10. Showalter AM, Bell JN, Cramer CL, Bailey JA, Varner J, Lamb CJ (1985). Accumulation of hydroxyproline-rich glycoproteins mRNAs in response to fungal elicitor and infection. Proc Natl Acad Sci USA 82: 6551.
11. Bolwell GP, Robbins MP, Dixon RA (1985). Metabolic changes in elicitor-treated bean cells: Enzymatic responses associated with rapid changes in cell wall components. Eur. J. Biochem. 148: 571.
12. Hammerschmidt R, Lamport DTA, Muldoon EP (1984) Cell wall hydroxyproline enhancement and lignin deposition as an early event in the resistance of cucumber to Cladosporium cucumerium. Physiol Plant Path 24:43.
13. Boller T (1985) Induction of hydrolases as a defense reaction against pathogens. In Key JK, Kosuge T (eds): "Cellular and Molecular Biology of Plant Stress," New York: Alan R. Liss, p 247.
14. Ryan C (1978) Proteinase inhibitors in plant leaves: a biochemical model for natural plant prtection. TIB 5:148.
15. Van Loon LC (1983) Mechanisms of resistance in virus infected plants. In JA Bailey , BJ Deverall (eds) "The Dynamics of Host Defense", New York: Academic Press.
16. Chen J, Varner JE (1985). An extracellular matrix protein in plants:characterization of a genomic clone for carrot extensin. EMBO 4: 2145.
17. Mellon JE, Helgeson JP (1982) Interaction of a hydroxyproline rich glycoprotein fron tobacco callus with potential pathogens. Plant Physiol 70:401
18. Yang SF, Hoffman NE (1984) Ethylene biosynthesis and its regulation in higher plants. Ann Rev Plant Physiol 35:155.
19. Nichols SE, Latis GG (1984). Ethylene-regulated gene transcription in carrot roots. Plant Mol Bio 3: 393.
20. Chappell J, Hahlbrock K, Boller T (1984). Rapid induction of ethylene biosynthesis in cultured parsley cells by fungal elicitor and its relationship to the induction of phenylalanine ammonia-lyase. Planta 161: 475.
21. Theologis A, Latis GG (1982). Potentiating effect of pure oxygen on the enhancement of respiration by ethylene in plant storage organs: A comparative study. Plant Physiol 69: 1031.
22. Fromm M, Taylor LP, Walbot V (1985) Expression of genes transfected into monocot and dicot plants cells by electroporation. Proc Natl Acad Sci USA 82:5824.
23. Harding VK, Heale JB (1981). The accumulation of inhibitory compounds in the induced resistance response of carrot root slices to Botrytis cinerea. Physiol Plant Path 18:7.

Molecular Biology of Plant Growth Control, pages 145–155
© 1987 Alan R. Liss, Inc.

CONSTRUCTION AND TESTING OF VECTORS FOR THE CLONING OF
PLANT GENES BY PHENOTYPIC COMPLEMENTATION[1]

Neil Olszewski, Daniel Voytas, Jim Hu, Joanne Chory, and
Frederick Ausubel

Department of Genetics, Harvard Medical School, and
Department of Molecular Biology, Massachusetts General
Hospital, Boston, Massachusetts 02114

ABSTRACT We have constructed cosmid-cloning vectors for
the cloning of plant genes by phenotypic complementation
of mutations. These vectors contain either the right
border region or both the right and left border regions
from the T-DNA of the Ti plasmid of Agrobacterium
tumefaciens and dominant selectable markers for use in
plant cells. These constructions were made in two
different mobilizable broad-host-range plasmids derived
from RK2 (pRK290; Ditta et al. 1980 Proc. Nat. Acad.
Sci. USA 77: 7347–7351 and pTJS75; Schmidhauser and
Helinski 1985 J. Bact. 164: 446–455). These vectors
transform plants. Genomic libraries of Arabidopsis
thaliana have been constructed, and those in pRK290-
based vectors are stable in A. tumefaciens. The
libraries constructed in the pTJS75-based vectors are
not. The pRK290-based vectors containing genomic DNA
inserts have been shown to transform plants indicating
the probable transfer of the genomic DNA. Finally, we
discuss strategies by which these libraries can be used
to clone plant genes which code for enzymes involved in
gibberellic acid biosynthesis.

INTRODUCTION

Agrobacterium tumefaciens is a gram-negative soil
bacterium which is a pathogen of dicotyledonous plants.
During the infection process, A. tumefaciens transfers a
portion of its Ti-plasmid, the T-DNA, into the plant nuclear

[1]This work was supported by a grant from Hoechst AG.

genome (1). The T-DNA is flanked by 23 bp imperfect-direct repeats (2-5). These repeats are cis-acting signals which approximately delimit the DNA sequences to be integrated into the plant genome. If the right imperfect repeat is deleted, transformation does not occur. The left repeat is not required for transformation. The right repeat directs the vectorial transfer of DNA into the plant genome and sequences adjacent to the right repeat affect the efficiency of transformation (6-10). The Ti-plasmid also contains a vir region which must be present in either cis or trans with the right border for T-DNA transfer (11,12). These features allow plant transformation strategies in which foreign DNA is placed adjacent to the right border on a broad-host-range replicon. When plants are infected by A. tumefaciens containing this construct, the vir region which is present in trans directs the transfer of the foreign DNA into the host plant's genome.

Cloning genes by phenotypic complementation of characterized genetic mutations is commonly employed in both yeast and bacteria; however, this strategy has not yet been employed in the cloning of plant genes. To utilize this strategy for plants, one must be able to perform the following: (1) generate genomic libraries from wild-type plants; (2) transfer to, and express these libraries in, mutant plant cells; (3) select plants or plant cells in which complementation has occurred (plants or plant cells which show the wild-type phenotype); and (4) recover or identify the clone which is responsible for the complementation.

CLONING VECTOR FEATURES

We have constructed a series of cosmid-cloning vectors which contain the features necessary for use in strategies to clone plant genes by phenotypic complementation. Maps of these vectors are shown in Figure 1. These vectors contain either the right border region or both the right and left border regions of the A. tumefaciens T-DNA in two different broad-host-range mobilizable plasmids (13,14). Both of these plasmids are derivatives of the mobilizable broad-host-range RK2 plasmid, and carry a gene which confers tetracycline resistance in bacteria. In addition, these vectors contain: (1) a cos site; (2) unique restriction sites in which genomic libraries can be constructed; (3) a ColE1 replicon; (4) a supF tRNA gene; and (5) a dominant selectable marker for use in plants. The A. tumefaciens T-DNA border sequences provide cis-acting signals which are

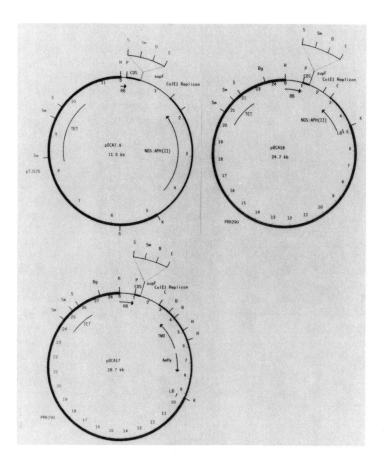

FIGURE 1.   Map of the pOCA7.9, pOCA17 and pOCA18
vectors. The genetic loci indicated in this figure are: RB,
the right border repeat from the pTiC58 Ti-plasmid; COS, the
cos region of phage lambda; supF, a bacterial supF tRNA
gene; ColE1, the ColE1 plasmid origin of replication;
NOS::APH(II), a chimeric gene which is expressed in plants
and confers resistance to kanamycin (19,20); TMO and AmHy,
genes from the T-DNA of the pTiC58 which are expressed in
plants to produce enzymes which synthesize indoleacetic
acid; LB, a synthetic oligonucleotide containing the 25 bp
left border repeat of the pTiB6S3 Ti-plasmid; TET, a bacterial tetracycline resistance gene; pTJS75 and pRK290, mobilizable broad-host-range plasmids (13,14). The restriction
sites indicated are: B, BamHI; Bg, BglII; C, ClaI; E, EcoRI;
H, HindIII; K, KpnI; P, PstI; S, SalI; Sm, SmaI.

required for transformation. In pOCA17 and pOCA18, DNA
located between the borders will be transferred into the
plant genome. pOCA7.9, which contains only the right T-DNA
border, was designed such that DNA is transferred
vectorially from these sequences (Fig. 1). Genomic
libraries for use in the cloning of plant genes by
phenotypic complementation have been constructed in the
unique restriction sites which are located between the right
border and the plant selectable marker. This assures that
transformed cells which express the plant selectable marker
also contain the genomic clone.

Phenotypic complementation strategies require a means by
which particular sequences responsible for reversing a
mutant phenotype can be identified. This is especially true
if genomic libraries are transferred in bulk (such as by co-
cultivation with protoplasts). To accomplish this, the
ColE1 replicon and supF gene have been included in the
vectors to allow the recovery of a portion of the
complementing clone by "reverse-cloning" (15). In this
strategy, genomic DNA is isolated from plant cells which
show the complemented phenotype. This DNA is digested with
a restriction endonuclease which does not cut in the region
containing the ColE1 replicon, the supF gene, and the
restriction site in which the library has been constructed.
The digested DNA is then self-ligated under dilute
conditions, where circularization rather than concatamer-
ization will occur. The ligated DNA is transformed into E.
coli where the supF tRNA is used as a selectable marker.
Clones obtained in this manner should contain at least a
portion of the genomic clone which conferred the
complemented phenotype. This portion of the genomic clone
can then be used as a hybridization probe to recover the
genomic clone which is responsible for the complementation.

TRANSFORMATION OF PLANTS

To test the stability of pOCA7.9, pOCA17 and pOCA18 in
A.tumefaciens, they were mated from E. coli into A.
tumefaciens strain LBA4404 (11,16) by a three-way mating
with E. coli (14). The Ti-plasmid of LBA4404 contains a
deletion of the entire T-DNA region, but the vir region is
present and functional. All of these vectors were stable
in LBA4404 (data not shown).

pOCA17 carries two genes from the T-DNA of the pTiC58
Ti-plasmid (Fig. 1) that function as a selectable marker in
plant cells. These genes are homologous to two octopine Ti-
plasmid genes (17) which are expressed in transformed plant

cells and code for enzymes which produce auxin (18). Expression of these genes in the plant causes the growth of a rooty teratoma (7). When LBA4404 containing pOCA17 is inoculated onto Kalanchoë diagremontiana stems, rooty teratomas are produced (Fig. 2). LBA4404 alone is avirulent (Fig. 2). This result indicates that Agrobacterium containing pOCA17 transformed Kalanchoë stem cells.

A chimeric Nos::APH(3')II gene, which confers resistance to kanamycin in plants, is the dominant selectable marker in pOCA7.9 (19) and pOCA18 (20) (Fig. 1). When LBA4404 containing pOCA7.9 is co-cultivated with tobacco leaf discs (21), shoots resistant to 500 ug/ml of kanamycin are obtained. These shoots can be rooted on 100 ug/ml kanamycin (not shown). Hybridization of ^{32}P-pOCA7.9 to genomic DNA from these plants indicated that pOCA7.9 DNA was present in the genomes of all plants tested; genomic DNA isolated from plants which were not co-cultivated contained no sequences which hybridized to pOCA7.9 (Fig. 2). This result indicates that pOCA7.9 can transform plant cells and confer a kanamycin resistant phenotype. Following co-cultivation of tobacco leaf discs with LBA4404 containing pOCA18, kanamycin resistant tobacco shoots were obtained (data not shown). Each of these plants will be tested similarly to determine if pOCA18 DNA is integrated into its genome.

CHARACTERISTICS OF GENOMIC LIBRARIES
CONSTRUCTED IN pOCA VECTORS

Genomic libraries of EcoRI partial digests of Arabidopsis thaliana DNA have been constructed in the EcoRI site of pOCA7.9. Similarly, TaqI partial digests were ligated into the ClaI site of pOCA17 and pOCA18. The characteristics of these libraries are shown in Table 1. The number of clones obtained per ug of insert DNA was 30-fold greater in the pOCA17 and pOCA18 libraries than in the pOCA7.9 library. This may not reflect actual differences in the cloning efficiencies of these vectors, however, as the libraries were constructed by different methods (Table 1). Also, since the genomic DNA inserts are larger in the pOCA7.9 library, the cloning efficiency would be expected to be lower. Based on the packaging capacity of lambda phage, the cloning capacity of pOCA7.9 is between 27 and 39 kb and the cloning capacity of pOCA17 and pOCA18 is expected to be between 10 and 23 kb and between 14 and 27 kb, respectively. When the size of the inserts contained in these clones was determined by restriction mapping, all of the genomic clones

contained inserts which were within the expected size range
(Table 1). We examined the stability of these libraries in
E. coli and in all cases, the pOCA17 and pOCA18 libraries
were stable. Ninety-nine percent of the pOCA7.9 clones were
stable in E. coli.

TABLE 1
CHARACTERISTICS OF GENOMIC LIBRARIES

Vector	Colonies/ ug Insert	Insert Size kb[b]	Percent stability[a] E. coli[c]	Agrobacterium[d]
pOCA7.9[e]	6×10^4	28–36	99	30
pOCA17[f]	2×10^6	14–22	100	100
pOCA18[f]	2×10^6	15–25	100	N.T.

[a]Stability was determined by restriction mapping.
[b]The size of the inserts was determined by restriction
mapping following isolation of the clones from E. coli.
At least 40 clones were examined in each case.
[c]Value obtained from examination of at least 40 clones.
[d]Value obtained from examination of at least 10 clones.
[e]The genomic library was constructed in the EcoRI site
of pOCA7.9 using EcoRI partial digests of nuclear
A. thaliana DNA, which had been size selected (25–50 kb)
by isolation from a low melting point agarose gel, by
the method of Ish–Horowicz and Burke (22).
[f]Genomic libraries were constructed in dephosphorylated
ClaI sites of both pOCA17 and pOCA18 using
A. thaliana nuclear DNA which had been partially
digested with TaqI and isolated as above.

The stability of the pOCA7.9 and the pOCA17 libraries in
LBA4404 was also determined (Table 1). In seventy percent
of the pOCA7.9 genomic clones analyzed, deletions had
occurred in the cloned DNA. All of the pOCA17 genomic
clones examined were stable. It is possible that the larger
size of the cloned DNA in the pOCA7.9 library contributes to
insert instability. We feel that this is unlikely, however,
based on the results of Schmidhauser et al. (13) who have
recently reported that pTJS75 is not maintained in
Agrobacterium. Clearly, the instability of libraries

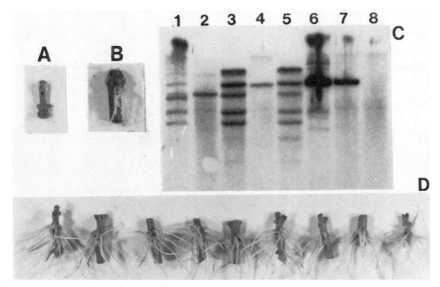

FIGURE 2. Kalanchoë stems infected with LBA4404 containing either pOCA17 or pOCA17 containing genomic clones of A. thaliana DNA, and a blot of DNA from pOCA7.9 transformed tobacco plants. Kalanchoë stems infected with either (A) LBA4404, (B) LBA4404 containing pOCA17, or (D) LBA4404 containing gemonic clones of A. thaliana constructed in pOCA17. C: Tobacco (10 ug/lane) was digested with EcoRI subjected to agarose gel electrophoresis, blotted onto nitrocellulose, and hybridized to ^{32}P-pOCA7.9. Lanes 1-5: DNA isolated from pOCA7.9 transformed tobacco. Lanes 6 and 7: Reconstruction containing one or ten copies of pOCA7.9 per tobacco genome equivalent. Lane 8: Tobacco DNA from uninfected plants.

constructed in pOCA7.9 makes it unsuitable to clone genes by phenotypic complementation. We are in the process of testing the stability of the genomic library constructed in pOCA18. We expect that this library will be stable since it is constructed in the same broad-host-range plasmid (pRK290) as pOCA17.

GENOMIC CLONES IN pOCA17 TRANSFORM PLANTS

Nine different genomic clones of A. thaliana DNA in pOCA17 were mated into LBA4404 and used to infect Kalanchoë

daigremontiana stems. Each of the strains tested produced
rooty teratomas (Fig. 2). This indicates that insertions of
large fragments of genomic DNA (10-23 kb) do not inhibit the
transfer of the downstream selectable marker. Because the
genomic DNA was cloned between the right border and the
plant selectable marker, it also seems likely that the
genomic DNA contained in these clones was transformed into
the Kalanchoë cells. Experiments are in progress to
determine if this transformation occurs without
rearrangement of the DNA.

DISCUSSION

Genomic libraries of A. thaliana DNA constructed in
pOCA17 and pOCA18 are probably suitable for use in cloning
plant genes by phenotypic complementation. A. thaliana was
chosen as the source of DNA for complementation experiments
because of its small genome size ($7x10^7$ bp) compared to that
of most plant species (23,24). It is not certain that the
genes from A. thaliana will be expressed in heterologous
plant species, but several plant genes tested to date have
been shown to be faithfully expressed when they are moved
from one plant species to another (25). Based on the
average insert size obtained in pOCA17 and pOCA18, an A.
thaliana genomic library must contain 16,000 members to have
a 99% probability of having every unique sequence
represented at least once. Since genes transformed into
plants show variability in the level of their expression
(generally lower than normal), it may be necessary to screen
multiple transformants of each genomic clone to obtain ones
in which the expression of the cloned genes is high enough
to observe complementation.

The genomic libraries of A. thaliana DNA constructed in
pOCA17 and pOCA18 will be used to clone genes which code for
enzymes involved in the gibberellic acid (GA) biosynthetic
pathway. The Le gene of peas is believed to be involved in
the conversion of GA_{20} to GA_1 (26). The dwarf pea, Progress
#9, carries this mutation, and reversal of this phenotype
will be used as the criterion for complementation. The
genomic library constructed in pOCA17 will be used to
generate tumors on Progress #9 stems. We are using this
strategy rather than screening pea plants which have been
transformed and regenerated because of the difficulty in
regenerating pea plants. We will also try to complement
two tomato mutations which are believed to be blocked in the
GA biosynthetic pathway (27). In this case, the genomic
library constructed in pOCA18 will be used to transform

dwarf tomato leaf discs by co-cultivation (21), and kanamycin resistant plants will be regenerated. The dwarf phenotype will be reversed in plants where complementation has occurred. In these experiments we will screen the libraries using one clone per leaf disc or stem inoculation, making it possible to immediately identify the complementing clone.

ACKNOWLEDGMENTS

We wish to thank M. Stempowski and J. Smith for providing the left border oligonucleotide and R. Hyde for assisting in the preparation of this paper.

REFERENCES

1. Nester EW, Gordon MP, Amasino RM, Yanofsky MF (1984). Crown Gall: a molecular and physiological analysis. Ann Rev Plant Physiol 35:387.
2. Yadav NS, Vanderleyden J, Bennett DR, Barnes WM, Chilton M-D (1982). Short direct repeats flank the T-DNA on a nopaline Ti-plasmid. Proc Nat Acad Sci USA 79:6322.
3. Simpson RB, O'Hara PJ, Kwok W, Montoya AL, Lichtenstein C, Gordon MP, Nester EW (1982). DNA from A6S/2 tumor contains scrambled Ti-plasmid sequences near its junction with plant DNA. Cell 29:1005.
4. Zambryski P, Depicker A, Kruger K, Goodman H (1982). Tumor induction by Agrobacterium tumefaciens: analysis of the boundaries of T-DNA. J Mol Appl Genet 1:361.
5. Holsters M, Villarroel R, Seurinck J, De Greve H, Van Montagu M, Schell J (1983). An analysis of the boundaries of the octopine TL-DNA in tumors induced by Agrobacterium tumefaciens. Mol Gen Genet 190:35.
6. Holsters M, Silva B, Van Vleit F, Genetello C, De Block M, Dhaese P, Depicker A, Inze D, Engler G, Villarroel R, Van Montagu M, Schell J (1980). The functional organization of the nopaline Agrobacterium tumefaciens plasmid pTiC58. Plasmid 3:212.
7. Joos H, Inze D, Caplan A, Sormann M, Van Montagu M, Schell J (1983). Genetic analysis of T-DNA transcripts in nopaline crown galls. Cell 32:1057.
8. Wang K, Herrera-Estrella L, Van Montagu M, Zambryski P (1984). Right 25 bp terminus sequence of the nopaline T-DNA is essential for and determines direction of DNA transfer from Agrobacterium tumefaciens to the plant genome. Cell 38:455.

9. Shaw CH, Watson MD, Carter GH, Shaw CH (1984). The right hand copy of the nopaline Ti-plasmid 25 bp repeat is required for tumour formation. Nucleic Acids Res 15:6031.
10. Peralta EG, Ream LW (1985). T-DNA border sequences required for crown gall tumorigenesis. Proc Nat Acad Sci USA 82:5112.
11. Hoekema A, Hirsch PR, Hooykaas PJJ, Schilperoort RA (1983). A binary plant vector strategy based on the separation of the vir- and T-region of the Agrobacterium tumefaciens Ti-Plasmid. Nature 303:179.
12. de Framond AJ, Barton KA, Chilton M-D (1983). Mini-Ti: a new vector strategy for plant genetic engineering. Bio/Technology 1:262.
13. Schmidhauser TJ, Helinski DR (1985). Regions of broad-host-range plasmid RK2 involved in replication and stable maintenance in nine species of gram-negative bacteria. J Bacteriol 164:446.
14. Ditta GS, Stanfield S, Corbin D, Helinski DR (1980). Broad host range cloning system for gram-negative bacteria: construction of a gene bank of Rhizobium meliloti. Proc Nat Acad Sci USA 77:7347.
15. Koncz C, Kreuzaler F, Kalman ZS, Schell J (1984). A simple method to transfer, integrate and study expression of foreign genes, such as chicken ovalbumin and α-actin in plant tumors. EMBO J 3:1029.
16. Ooms G, Hooykaas PJJ, Van Veen RJM, Van Beelen P, Regenburg-Tuink TJG, Schilperoort RA (1982). Octopine Ti-plasmid deletion mutants of A. tumefaciens with emphasis on the right side of the T-region. Plasmid 7:15.
17. Willmitzer L, Dhaese P, Schreier PH, Schmalenback W, Van Montagu M, Schell J (1983). Size, location and polarity of T-DNA-encoded transcripts in nopaline crown gall tumors; common transcripts in octopine and nopaline tumors. Cell 32:1045.
18. Thomashow MF, Hugly S, Buchholz WG, Thomashow LS (1986). Molecular basis for auxin-independent phenotype of crown gall tumor tissues. Science 231:616.
19. Herrera-Estrella L, De Block M, Messens E, Hernalsteens J-P, Van Montagu M, Schell J (1983). Chimeric genes as dominant selectable markers in plant cells. EMBO J 2:987.
20. Hain R, Stabel P, Czernilofsky AP, Steinbib HH, Herrera-Estrella L, Schell J (1985). Uptake, integration, expression and genetic transmission of a selectable chimeric gene by plant protoplasts. Mol Gen Genet 199:161.

21. Horsch RB, Fry JE, Hoffmann NL, Eichholtz D, Rogers SG, Fraley RT (1985). A simple and general method for transferring genes into plants. Science 227:1229.
22. Ish-Horowicz D, Burke JF (1981). Rapid and efficient cosmid vector cloning. Nucleic Acids Res 9:2989.
23. Leutwiler LS, Hough–Evans BR, Meyerowitz EM (1984). The DNA of Arabidopsis thaliana. Mol Gen Genet 194:15.
24. Bennett MD, Smith JB (1976). Nuclear DNA amounts in angiosperms. Proc R Soc Lond B 274:227.
25. Nagy F, Morelli G, Fraley RT, Rogers SG, Chua N–H (1985). Photoregulated expression of a pea rbcS gene in leaves of transgenic plants. EMBO J 4:3063.
26. Ingram TJ, Reid JB, Murfet IC, Gaskin P, Willis CL, MacMillan J (1984). Internode length in Pisum: the Le gene controls the 3β-hydroxylation of gibberellin A_{20} to gibberellin A_1. Planta 160:455.
27. Zeevaart JAD (1984) Gibberellins in single gene dwarf mutants of tomato. Plant Physiol Supp 75:186.

Molecular Biology of Plant Growth Control, pages 157–166
© 1987 Alan R. Liss, Inc.

SEARCHING FOR MOLECULAR MECHANISMS INVOLVED
IN FRUIT RIPENING[1]

Peter H. Morgens, Jana B. Pyle,
and Ann M. Callahan

Department of Plant and Soil Sciences
West Virginia University
Appalachian Fruit Research Station
Kearneysville, West Virginia 25430

ABSTRACT In vitro translation studies indicate that
there are both increases and decreases in individual
mature RNA species during tomato fruit ripening. Some
of the changes are localized to either the locular or
pericarp tissue of the fruit. Polygalacturonase,
an abundant cell wall softening enzyme associated
with ripening (1), appears to be synthesized de novo
during ripening as a 55 Kd precursor to the 46 Kd
mature protein. Ripening-associated cDNAs were isolated and their transcripts were characterized with
respect to developmental timing. One ripening-
specific cDNA is associated with a gene family. DNA
sequence rearrangements appear to be associated with
the time of expression of two of the cDNAs, one
ripe-specific and one green-specific.

INTRODUCTION

Our goal is to understand changes in gene expression
that are associated with fruit ripening well enough to
attempt to modify the process by genetic engineering.
Fruit ripening is a complex developmental process that
involves numerous biochemical and physiological changes
that lead to softening as well as changes in flavor and

[1]This work was supported by USDA specific
cooperative agreement No. 58-32U4-3-583 with West Virginia
University.

coloring (2,3). The gaseous hormone ethylene has long been
associated with ripening and exogenous ethylene can
accelerate ripening under certain conditions (2,4,5,6). No
molecular mechanism for either fruit development or the
action of ethylene has been demonstrated.

To probe the ripening process we have looked for
changes in accumulated proteins, in newly synthesized
proteins, and in mRNA populations as indicated by in vitro
translation studies. We have constructed a large cDNA
library, isolated individual ripening specific cDNA clones,
and examined the expression of individual RNA species
through ripening. We have also looked for genome
rearrangements that might be involved in ripening specific
gene expression.

RESULTS

The experiments reported here were done in the tomato
cultivar Lycopersicon esculentum Mill var Pixie. Ripening
stages are define in Table 1. To begin our study
we fractionated the proteins extracted from tomato pericarp
at different stages of ripening by SDS-PAGE (data not
shown). As others have shown previously (4), there are
many changes in the cold protein pattern and the most

TABLE 1
DEFINITIONS OF FRUIT DEVELOPMENT STAGES

Stage No. and name	Characteristics
1 Immature green	Fruit is less than mature size and has no gel material in locular area.
2 Mature green	Fruit is full size and has green gel material in locular area.
3 Very early breaker	Locular material has some yellowish coloration, pericarp is still green.
4 Early breaker	Locular material is orange and pericarp has yellow tinge on blossom end.
5 Late Breaker	Red color is evident on less than half of the outside of the fruit.
6 Ripe	Fruit is fully ripe and colored.

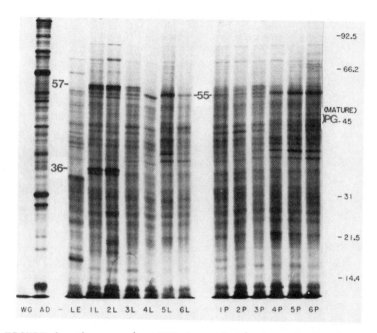

FIGURE 1. Changes in mRNA translation products from
locular (lanes 1L-6L) and pericarp (lanes 1P-6P) tissue in
ripening tomato fruit. Total RNA was isolated from liquid
nitrogen frozen and lyophilized material as described (7)
except that a 2M LiCl precipitation step was added prior to
precipitation with ethanol. RNAs were translated in a
wheat germ extract in the presence of S-35-methionine (8).
The products were separated by SDS-PAGE and detected by
autoradiography (8). Translation products from tomato leaf
RNA (LE), Adenovirus RNA (Ad), and with no exogenous RNA
(WG) added to the wheat germ extract are also shown. The
positions of marker proteins (MW x 10^{-3}) run in a
separate lane on the same gel and the size of the mature PG
are indicated.

prominent of these is the accumulation of polygalacturonase
(PG) that migrates as a polypeptide of 46 Kd. The huge
increase in a 46 Kd protein was not, however, evident in in
vivo labelling experiments. The most prominent band in the
in vivo labelling pattern ran as a polypeptide of
approximately 55 Kd (data not shown). Differences between
pericarp and locular material (in this work, all the fruit
material that is not the pericarp is termed locular) are

also readily evident when examined by SDS-PAGE (Callahan and Morgens, manuscript in preparation).

In Vitro Translations

Figure 1 shows the wheat germ in vitro translated products from each of six stages of pericarp and locular RNA. As both locular and pericarp material ripen there are appearances and disappearances of proteins or changes in their amounts (e.g. the 55 Kd protein in 4L-6L and 4P-6P, and the 57 Kd protein in lanes 1L-3L and 1P-3P). There are also some striking differences between locular and pericarp (e.g. the 36 Kd protein seen in lanes 1L and 2L that is not detectable at all in the corresponding pericarp lanes).

Immunoabsorptions

To test which of the in vitro translation products were antigenically related to PG, we immunoabsorbed the in vitro translation products with a rabbit anti-PG serum (kindly made available by Dr. Frederick Abeles of the USDA) and protein A-sepharose according to the method of Sato et al (9). The major absorbed product is a 54 Kd protein seen primarily in ripe locular (Figure 2, lane 6L) and ripe pericarp (lane 6P). A longer exposed autoradiograph indicates that the 54 Kd protein is also present in stages 4 and 5 of both locular and pericarp material (data not shown). Two minor proteins at 50 Kd and 46 Kd are also absorbed, but these co-migrate with the major bands immunoabsorbed by a rabbit antiserum to total ripe protein (lane anti-total). We feel that these represent highly antigenic proteins that were minor contaminants of the purified PG.

Isolation and Initial Characterization of Ripening Associated cDNA Clones

To study changes in RNA levels directly and to be able to isolate ripening specific controlling regions, we constructed a large cDNA library in the expression vector lambda gt11 using RNA from ripe pericarp (Morgens and Callahan, manuscript in preparation). Ripening associated clones, both ripe-specific and green-specific, were

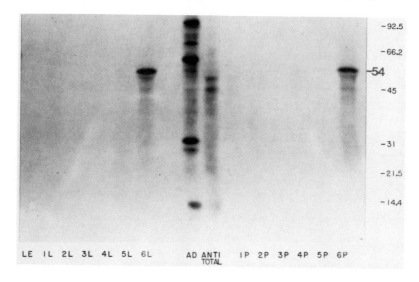

FIGURE 2. Immunoabsorption of mRNA translation
products from ripening fruit with a PG antiserum. RNA
samples were translated as decribed in Figure 1 and the
putative PG precursor was separated by immunoabsorption (9)
and detected as in Figure 1. Translation products from
ripe pericarp RNA that are immunoabsorbed with an antiserum
made to total ripe tomato protein are also shown (lane
anti-total).

isolated by differential hybridization with radioactively
labelled cDNA from ripe versus green fruit. Some of these
clones have been characterized as to their cross-
hybridization with other clones, the length of the RNA
transcript they correspond to, the pattern of RNA expression during ripening, and the number of genes to which they
appear to hybridize. Table 2 summarizes this data.

Developmental Northerns

Figure 3 shows developmental Northerns for the
ripe-specific clone cDNA102 and the green-specific
clone cDNA204. The transcript for cDNA102 is present in
all stages of fruit tissue examined, but it is greatly
increased in the later stages of ripening (lanes 5L-6L and
5P-6P). It is not detectable at all in leaf RNA (lane

TABLE 2

CHARACTERIZATION OF RIPENING ASSOCIATED cDNAs

cDNA #	family size[a]	mRNA length[b]	# of genes comple- mentary to cDNA[c]	Peak expression[b]
			Ripe-specific clones	
102	15	1700	one or two	6L and 6P
103	3	2100	one or two	5L and 4P-5P
104	2	3200	six or more	5L and 5P
106	1	2000	one or two	5L and 4-5P
120	4	1450	one or two	5L and 4-5P
			Green-specific clones	
204	3	800	one or two	1L-2L, 1P
208	1	2000	one or two	1P

[a]Family size was determined by hybridizing the inserts of clones to 80 ripening associated cDNAs.
[b]Determined from Northern analyses (e.g. see Figure 3).
[c]Estimated from number of bands and intensities of Eco RI digested total DNA (e.g. see Figure 4).

LE). The green-specific transcript for cDNA204 is present in large amounts in immature green fruit (lanes 1L and 1P) and mature green fruit (lanes 2L and 2P) and is greatly reduced in later stages of ripening (lanes 4L-6L and 4P-6P). This green fruit specific transcript is present, albeit at low levels, in leaf RNA (lane LE).

Developmental Southerns

To determine whether DNA rearrangements might be associated with changes in levels of gene expression, Eco RI digested total DNA from leaf, mature green fruit, and ripe fruit were hybridized to individual cDNA probes. Figure 4 shows that total DNA from both green and ripe fruit (lanes 2P and 6P, left panel) has a different hybridization pattern with cDNA102 than does total DNA isolated from leaves (lanes LE, left panel). The hybridization pattern for the green-specific cDNA204 probe is different in green fruit total DNA (lane 2P, right panel) than it is in either leaf or ripe fruit total DNA (lanes LE and 6P, right panel).

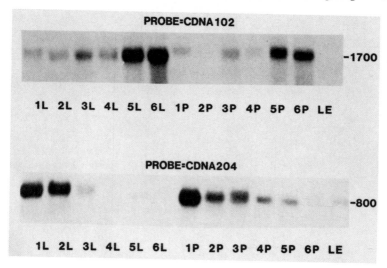

FIGURE 3. Changes in levels of a green-specific and a ripe-specific mRNA in ripening locular and pericarp tissue. Ten micrograms of stage specific RNA (see lane explanation in Figure 1) were electrophoresed through 1.4% agarose containing 5 mM methyl mercury hydroxide as a denaturant (10), electroblotted onto Nytran paper (Schleicher and Schuell, Keane, NH), and hybridized to P32-labeled (11) insert DNA from clones cDNA102 or cDNA204 according to instructions with the Nytran paper. Final washes were 0.5X SSC at 65°C. Size determinations (in nucleotides) indicated on the right were determined from radioactive DNA fragments (HpaI digests of bacteriophage T7) that were visible on the autoradiograph.

DISCUSSION

Our approach to understanding ripening has been to investigate different possible molecular mechanisms that control gene expression. These include post-translational modifications, transcriptional regulation, and regulation of the primary sequence of the genome. Our search suggests the involvement of all these processes. In addition tissue specific gene expression is evident in locular and pericarp material (Figure 1). Fruit ripening is not just a degradative process. New gene expression and/or RNA processing are really involved in this developmental process. This is demonstrated by changing in vitro

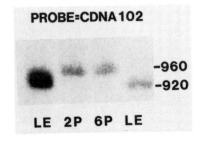
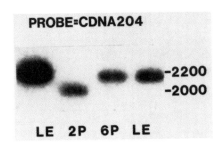

FIGURE 4. Changes in DNA hybridization pattern in
ripening fruit. Approximately 10 micrograms (the left most
lanes clearly had more material than the other three) of
EcoRI restricted total DNA from tomato leaf (lane LE),
mature green tomato pericarp (lane 2P), or ripe tomato
pericarp (lane 6P) were electrophoresed through 1% agarose,
transferred to nitrocellulose, and hybridized with
P32-labeled (11) insert DNA from clones cDNA102 or cDNA204
according to Southern (12). DNA was prepared from
lyophilized material according to Dellaporta (13). Final
wash was in 0.5X SSC at 65°C. Exposure times were
approximately ten days with two intensifying screens at
-70°C. Sizes indicated on the right in base pairs were
derived from radioactive DNA fragments visible on the
autoradiogram. Shifts in fragment size are evident.

translation products (Figure 1 and (14, 15)), the ability
to isolate ripening-specific cDNAs (Table 1 and (16,17)),
and the ability to measure changing levels of specific RNAs
(Table 1, Figure 3, and (16,17)).
     Polygalacturonase (PG) is the best studied of the
enzymes induced during ripening. We think it is
synthesized de novo as a 54-55 Kd protein (Figure 2) and is
processed to the mature 46 Kd protein. The de novo
synthesis result is consistent with previously published
work by Brady et al (5) and Grierson and Tucker (6). The
results agree with the work of Sato et al (9) on the size
of the precursor, but disagree with that of Grierson et al
(14). The discrepency may be due to the addition of 4
micrograms of unlabelled PG to the immunoabsorptions
described by Grierson et al (14). That the unlabelled PG
would compete out the small amount of radioactive PG from
the in vitro translation is indicated by the experiment
presented in Figure 2 of Sato et al (9). We believe that
the 48 Kd putative precursor of PG described by Grierson et

al (14) is a highly antigenic contaminating protein (seen in lane "anti-total" of Figure 2. Slater et al (16) use the same technique of Grierson et al (14) to identify a putative cDNA clone for PG--we think that this identification must be considered suspect until discrepencies are clarified.

We have found changes in DNA sequences in or near the two ripening-associated genes that correlate with time of expression. With the green-specific gene detected by cDNA204, a particular DNA rearrangement correlates well with the time of high expression of the 800-base transcript (Figures 3 and 4). The turn-off of the gene at other times and even in the leaf, however, is not absolute. The ripe-specific clone cDNA102, in contrast, shows a shift in DNA hybridization pattern that does not correlate well with time of high abundance but does correlate with a period of absolute turn off in the leaf (Figures 3 and 4). Thus the DNA rearrangement might make the expression of the gene susceptible to other factors (e.g. ethylene stimulation or mRNA stability). We are currently studying the nature of these rearrangements and isolating the genes.

ACKNOWLEDGMENTS

We thank Dr. Frederick B. Abeles for the PG antiserum and for sharing his laboratory facilities, Dr. William Welker and George Vass for showing us how to grow a lot of tomatoes, Wilbur Hershberger for technical assistance, and the staff of the USDA-Appalachian Fruit Research Station for helping us start a new molecular biology laboratory.

REFERENCES

1. Tucker GA and Grierson D (1981). Synthesis of polygalacturonase during tomato fruit ripening. Euro J Bioch 112:119-124.
2. Sacher JA (1973). Senescence and postharvest physiology. Ann Rev Plant Physiol 24:197-224.
3. Piechulla B, Chonoles Imlay KR, Gruissem W (1985). Plastid gene expression during fruit ripening in tomato. Plant Mol Biol 5:373-384.
4. Grierson D, Slater A, Maunders M, Crookes P, Tucker GA, Schuch W, Edwards K (1984). Regulation of the expression of tomato fruit ripening genes: the

involvement of ethylene. In Roberts JA, Tucker GA (eds): "Ethylene and plant development", London: Buttersworths, p 147.

5. Brady CJ, Macalpine G, McGlasson WB, Ueda Y (1982). Polygalacturonase in tomato fruits and the induction of ripening. Aust J Plant Physiol 9:171-178.

6. Grierson D, Tucker GA (1983). Timing of ethylene and polygalacturonase synthesis in relation to the control of tomato fruit ripening. Planta 157:174-179.

7. Morgens PH, Grabau EA, Gesteland RF (1984). A novel soybean mitochondrial transcript resulting from a DNA rearrangement envolving the 5s rRNA gene. Nuc Acids Res 12:5665-5684.

8. Dure LS and Galau GA (1981). Developmental biochemistry of cottonseed embryogenesis and germination. Plant Physiol 68:187-194.

9. Sato T, Kusaba S, Nakagawa H, Ogura N (1984). Cell-free synthesis of a putative precursor of polygalacturonase in tomato fruits. Plant Cell Physiol 25:1069-71.

10. Bailey JM and Davidson N (1976). Methyl mercury as a reversible denaturing agent for agarose gel electrophoresis. Anal Biochem 70:75-85.

11. Maniatis T, Jeffrey A, Kleid D G (1975). Nucleotide sequence of the rightward operator of phage lambda. Proc Natl Acad Sci 72:1184-1188.

12. Southern EM (1975). Detection of specific sequences among DNA fragments separated by gel electrophoresis. J Mol Biol 98:503-517.

13. Dellaporta SL, Wood J, Hicks JB (1983). A Plant DNA Minipreparation: version II. Plant Mol Biol Report 1:19-21.

14. Grierson D, Slater A, Speirs J, Tucker G.A. (1985). The appearance of polygalacturonase mRNA in tomatoes: one of a series of changes in gene expression during development and ripening. Planta 163:263-271.

15. Speirs J, Brady CJ, Grierson D, Lee E (1984). Changes in ribosome organization and messenger RNA abundance in ripening tomato fruits. Aust J Plant Physiol 11:225-33.

16. Slater A, Maunders M, Edwards K, Schuch W, Grierson D (1985). Isolation and characterisation of cDNA clones for tomato polygalacturonase and other ripening-related proteins, Plant Mol Bio 5:137-147.

17. Mansson PE, Hsu D, and Stalker D (1985). Characterization of fruit specific cDNAs from tomato. Mol Gen Genet 200:356-361.

Molecular Biology of Plant Growth Control, pages 167–176
© 1987 Alan R. Liss, Inc.

PLASTID AND NUCLEAR GENE EXPRESSION DURING TOMATO
FRUIT DEVELOPMENT AND RIPENING

Birgit Piechulla and Wilhelm Gruissem

Department of Botany, University of California,
Berkeley, CA 94720

ABSTRACT Photosynthetic capacity and protein
patterns of green and red tomato fruits have been
compared with transcript levels of respective
genes. Steady-state mRNA levels for photosynthesis-specific proteins and glycolytic enzymes,
encoded by nuclear and plastid genes, have been
determined during tomato fruit development and
ripening. Differential gene expression is correlated with physiological changes and plastid differentiation.

INTRODUCTION

The fruit formation in tomato (Lycopersicon esculentum)
involves three phases: fruit development, ripening, and
senescence. Fruit development is usally triggered by pollination of flowers, followed by the fertilization of the egg
inside the ovule. After pollination, cell division in tomatoes occurs for 7-10 days; further growth is due to cell
enlargement (1). This process is associated with a number of
morphological and physiological changes, some of which are
under the control of hormones (2). Fully developed fruits
(mature) are unripened and green. Fruit ripening is the
second phase and follows soon after the fruit attains a
maximum size. The onset of ripening in tomato is characterized by a significant increase of respiratory activity
(respiratory climacteric, ref.3). During this period the

This work was supported by a research grant from the
Deutsche Forschungsgemeinschaft (Pi 153/1-1) to B.P.
and a grant from NIH (GM 33813) to W.G.

color of the fruit changes as a result of chlorophyll degradation and synthesis of carotenoids (4, 5). The high levels
of starch and organic acids detectable in mature green
fruits decrease and are accompanied by an increase in reducing sugar (6) and aromatic compounds. Enzymes responsible
for the degradation of cell walls (e.g. polygalactorunase)
are newly synthesized and accumulate during the ripening
process, which results in softening of the fruit (7, 8, 9).
At the organelle level chloroplasts differentiate into chromoplasts. The chloroplast/chromoplast transition is one
example of tissue-specific differentiation of plastids found
throughout plant development (10). The ultrastructural changes observed during the chloroplast/chromoplast conversion
in tomato fruits include the disappearance of the thylakoid
membrane system and the degradation of chlorophyll (11, 12).
It is unclear at present if degradation of the photosynthetic apparatus is a consequence of decreased chlorophyll
synthesis, causing increased LHCP turnover, or if inactivation of cab gene expression directly affects chlorophyll
synthesis. During early stages of chromoplast development
lycopene synthesis and accumulation increases (4, 5).

In contrast to the broad knowledge about alterations of
the physiology and morphology in developing and ripening
tomato fruits, very little is known about the regulatory
mechanisms which control these events at the level of transcription, processing of transcripts, translation, and/or
post-translational processes. As a first step to elucidate
possible control mechanisms during the process of fruit
development and ripening, we analyzed the expression of
various plastid and nuclear encoded genes. Genes coding for
subunits of protein complexes which are localized on genetically distinct systems are of particular interest. For
example, the small and large subunit of ribulose-1.5-bis-
phosphate carboxylase as well as thylakoid membrane proteins, such as the light harvesting chlorophyll a/b binding
and proteins of photosystems I and II, are required for the
biogenesis of chloroplasts. The expression of their genes
has to be coordinately regulated. It is therefore possible,
that a communication system exists which includes the
nucleus, plastids and mitochondria.

We are currently involved in studying the regulatory
mechanisms that might occur at the transcriptional level
during tomato fruit development and ripening, which result
in the organ and/or stage-specific expression of several

genes. In this report we present a summary of our recent results on the analysis of steady-state mRNA levels of genes coding for photosynthetic proteins and glycolytic enzymes.

RESULTS and DISCUSSION

Changes during chloroplast/chromoplast differentiation

During ripening in tomato the plastids undergo differentiation, chloroplasts of the green tomato convert into chromoplasts (red fruit). Chloroplasts in developing and mature green tomato fruits resemble leaf chloroplasts, since they contain thylakoid membranes arranged into grana stacks and starch grains (12, 13). Based on their morphological structure fruit chloroplasts were thought to be photosynthetically active. Recently, we demonstrated that a number of photosynthesis-specific proteins are present in mature green tomato fruits (FIG.1, ref. 14). These include the P700 reaction center protein, the light harvesting complex proteins and 14 kD protein of photosystem I, the P680 reaction center protein and the Q_B binding protein of photosystem II, the Fd-NADP-oxidoreductase and plastocyanin of the electron transfer chain, as well as the large subunit of ribulose-1.5-bisphosphate carboxylase. The chloroplast protein complexes are active as was shown by our measurements of the ribulose-1.5-bisphosphate carboxylase activity and the P700 apoprotein and Q (primary electron acceptor) concentrations (Table 1). The PS II/PS I ratio is 1.64 in tomato fruit chloroplasts and therefore similar to ratios determined for chloroplasts of other higher plants (15). However, the Fv/Fo measurements indicate that fruit chloroplasts have a smaller light harvesting antenna size than leaf chloroplasts (14, 16). In contrast to the photosynthetically active chloroplasts of green tomato fruits, no photosynthetic proteins, chlorophyll a and b, and P700 and Q are present in the chromoplasts of red fruits (Table 1).

At the molecular level we were able to demonstrate that the transcripts for plastid encoded photosynthetic proteins (psaA, psbA, B, C, D and rbcL decrease during plastid transition (FIG.2, ref.17). In ripe tomatoes these transcripts are reduced to almost undetectable levels. However, significant mRNA levels for the 32 kD Q_B binding protein (psbA) of photosystem II are still present in red fruits (FIG.2), which is consistent with the identification of this

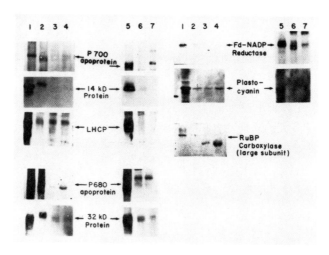

FIGURE 1.   Changes in photosynthetic protein levels
during tomato fruit ripening.
Identification of proteins in leaf and pericarp tissue at
different ripening stages of tomato fruits ('Western
blots'). 20 μg of total protein extract of leaves (lane 1),
mature green fruits (lanes 2 and 5), intermediate fruits
(lane 3 and 6), and ripe fruits (lane 4 and 7) were sepa-
reted on SDS polyacrylamide gels (9-19%). Proteins of soluble (lane 1 to 4) and membrane (lane 5 to 7) fractions were
analyzed with heterologous antibodies. Proteins of photosystem I (P700 apoprotein, 14 kD protein, and light harvesting
complex protein), of photosystem II (P680 apoprotein, Q_B
binding protein), of the electron transport chain (Fd-NADP-
oxidoreductase, plastocyanin), and the stromal ribulose-1.5-
bisphosphate carboxylase (large subunit) were identified.

protein by cross-reactions with specific antibodies (FIG.1).
The 32 kD polypeptide is characterized by its high turnover
rate relative to other photosynthetic proteins. It has been
suggested that the expression of psbA is controlled at the
translational level, and that light regulation of the 32 kD
protein synthesis is independent of the transcript level
(18). It is presently unclear, however, what significance
the presence of the 32 kD polypeptide has for the function
of the chromoplasts.

TABLE 1
PHOTOSYNTHETIC COMPONENTS

	leaf	green fruit	orange fruit	red fruit
RuBPCase activity[*]	100	34.9	2.0	0.8
electron transfer[*]	100	41.4	13.0	n.d.
chl a/b	2.89	2.56	/	/
PSII/PSI	1.36	1.64	/	n.d.
Fv/Fo	1.33	0.8	/	/

n.d. not detectable
*) % activities compared to leaf tissue

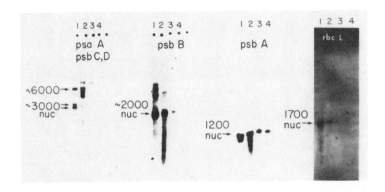

FIGURE 2.   Changes in transcript levels of photosynthetic proteins.
Hybridization of heterologous plastid genes with total RNA isolated from leaves (lane 1), premature green fruits (lane 2), intermediate stage fruits (lane 3), and red fruits (lane 4) ('Northern blots'). Identification of the transcript of the P700 apoprotein of PS I (psaA), the chlorophyll a binding proteins of PS II (psbC and D), the P680 apoprotein of PS II (psbB), and Q_B binding protein of PS II psbA), and the large subunit of ribulose-1.5-bisphosphate carboxylase (rbcL).

Concomitant with the loss of photosynthetic activity during the chloroplast/chromoplast conversion, a number of

unique proteins appear in the chromoplasts (19). No evidence
exists that transcription and/or translation of plastid
encoded genes occurs during ripening, although the presence
of plastid rRNA in chromoplasts indicate that they retain
the capacity of independent protein synthesis (17, 19).
Conversely, a number of ripening-specific, nuclear encoded
proteins have been found to be expressed during the ripening
process.

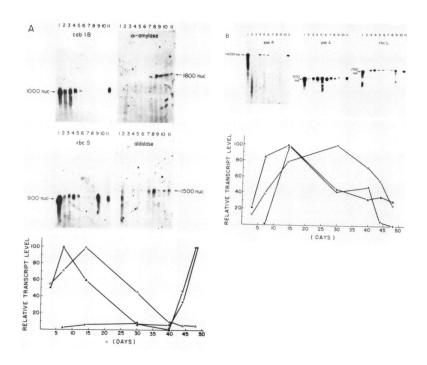

FIGURE 3. Determination of steady-state mRNA levels of
plastid and nuclear encoded genes in total RNA preparations
from leaves (lane 1), fruits of different developmental and
ripening stages (3 days-lane 2, 7 days-lane 3, 14 days-lane
4, 30 days-lane 5, 40 days-lane 6, 44 days-lane 7, 48 days-
lane 8), etiolated seedlings (lane 9), roots (lane 10), and
stems (lane 11).
Panel A: 'Northern blots' and changes in relative transcript
levels (%) of nuclear encoded genes. Hybridizations with
probes coding for the chlorophyll a/b binding protein (cab
from tomato, ▲-▲), the small subunit of ribulose-1.5-bis-
phosphate carboxylase (rbcS from tomato, △-△), the alpha-
(Continued)

Fig. 3 (Continued)
amylase (maize, o-o), and the fructose-1.6-bisphosphate
aldolase (maize, ●-●).
Panel B. 'Northern blots' and changes of the relative transcript levels (%) of plastid encoded genes. Hybridizations
with probes coding for the P700 reaction center protein of
PS I (psaA, spinach, ●-●), the 32 kD Q_B binding protein of
PS II (psbA, tobacco, o-o), and the large subunit of ribu-
lose-1.5-bisphosphate carboxylase (rbcL, tobacco, ▲-▲).

Expression of nuclear genes during fruit formation

There is evidence that de novo synthesis of some proteins is required for the ripening process of tomato (20).
Changes in the mRNA population and ribosome organization hve
been detected (21, 22). The cell wall degrading enzyme
polygalacturonase is a remarkable example which demonstrates
that not only changes in enzyme activity but also increasing
levels of mRNAs occur after the onset of ripening (8, 9). In
addition, increased levels of a number of other enzymes,
such as the beta-galactosidase, pectin esterase, and invertase have been detected (23). In the locule tissue of tomato
fruits a rapid decrease of starch is associated with increased alpha-amylase activities (24). We have shown that
the increase of the alpha-amylase activity in ripening fruit
tissue is correlated with the increase of mRNA (FIG.3A, ref
25). At the same time when the increase of alpha-amylase-
specific mRNA levels can be observed, the rise in the transcripts for another glycolytic enzyme, fructose-1.6-bisphos-
phate aldolase, has been de-tected (FIG.3A). Starch and
sugar degradation in ripening tomato fruits can be correlated with the regulation of the genes for the respective
enzymes. At present, little information is available about
the signals (proteins, sequences, ions, hormones, enviromen-
tal conditions) which are responsible for the regulation of
their gene activity during fruit development and ripening.
It has been suggested that certain developmental programs
and/or fruit stages are necessary for the initiation of such
processes. In contrast, light has been shown to stimulate
the expression of the genes coding for the small subunit of
ribulose-1.5-bisphosphate carboxylase (rbcS) and light
harvesting chlorophyll a/b binding protein (cab). 5' upstream sequences of the small subunit genes are thought to
be involved in the light-inducibility of transcription (26).
The mRNA levels of these two prominent nuclear encoded
chloroplast polypeptides (rbcS and cab) have been determined
in RNA preparations from tomato fruits (FIG.3A, ref.27).

During tomato fruit development and ripening the mRNA transcribed from rbcS and cab genes reach their maximum levels in 7- to 14- day old fruits, but decline to non-detectable levels in the presence of light long before the fruit has reached its maximum size. It appears therefore that gene expression is activated early in fruit development, most likely as a response to light and/or other developmental signals. The mechanisms involved in the inactivation of the genes are not known.

To provide chloroplasts with all components for the formation of photosynthetic complexes to enable photosynthetic light and dark reactions, a coordinated regulation of the expression of nuclear and plastid genes is required. The mRNA levels of several plastid encoded genes coding for the P700 reaction center protein of photosystem I (psaA), the Q_B binding protein (psbA), and the large subunit of ribulose-1.5-bisphosphate carboxylase (rbcL) have been measured at various timepoints during tomato fruit development (FIG.3B) and were compared to the relative transcript levels of the nuclear encoded photosynthetic genes (FIG.3A). Highest levels of the psaA and rbcL transcripts are found in 14 day old fruits while the psbA mRNA level is at its maximum in approximately 25 day old fruits (FIG.3B). The increase and decrease of the steady-state mRNA levels coding for photosynthesis-specific proteins (with the exception of psbA) appears simultaneously. These observations suggest that regulatory mechanisms exist for both nuclear and plastid genes which are highly coordinated during tomato fruit development and ripening. It will be of interest to determine the factors which are required for the control and regulation of such processes in development.

ACKNOWLEDGEMENTS

We thank Dr.M.Sugiura, Dr.K.Shinozaki, Dr.J.H.Bohnert, Dr.W.F.Thompson, Dr.G.Zurawski, Dr.E.Pichersky, and Dr.E.M.Orozco for providing us with plasmid DNAs. We thank Dr.R.Malkin, Dr.A.Barkan, Dr.W.Taylor, and Dr.L.McIntosh for specific antibodies. We also thank K.R.Chonoles Imlay, Dr.B.Moll, Dr.R.E.Glick, and Dr.A.Melis for their help.

REFERENCES

1. Bollard EG (1970). The physiology and nutrition of developing fruits. In Hulme AC (ed): "The biochemistry of fruits and their products," New York: Academic Press, p 387.

2. Nitsch JP (1970). Hormonal factors in growth and development. In Hulme AC (ed): "The biochemistry of fruits and their products," New York: Academic Press, p 427.

3. Biale JB, Young RE (1981). Respiration and ripening in fruits--retrospect and prospect. In Friend J, Rhodes MJC (eds): "Recent advances in the biochemistry of fruits and vegetables," New York: Academic Press, p 1.

4. Raymundo LC, Chichester CO, Simpson KL (1976). Light-dependent carotenoid synthesis in tomato fruit. J Agric Food Chem 24:59.

5. Laval-Martin D, Quennemet J, Monéger R (1975). Pigment evolution in Lycopersicon esculentum fruits during growth and ripening. Phytochemistry 14:2357.

6. Hobson GE, Davies JN (1971). In Hulme AC (ed): "The biochemistry of fruits and their products," New York: Academic Press, p 437.

7. Crookes PR, Grierson D (1983). Ultrastructure of tomato fruit ripening and the role of polygalactorunase isoenzymes in cell wall degradation. Plant Physiol 72:1088.

8. Grierson D, Slater A, Speirs J, Tucker GA (1985). The appearance of polygalactorunase mRNA in tomatoes: one of a series of changes in gene expression during development and ripening. Planta 163:263.

9. Sato T, Kusaba S, Nakagawa H, Ogura N (1985). Polygalactorunase mRNA of tomato: size and content in ripe fruits. Plant Cell Physiol 26:211.

10. Thomson WW, Whatley JM (1980). Development of nongreen plastids. Ann Rev Plant Physiol 31:375.

11. Harris WM, Spurr AR (1969). Chromoplasts of tomato fruits: I. Ultrastructure of low-pigment and high-beta mutants: carotene analyses. Amer J Bot 56:369.

12. Harris WM, Spurr AR (1969). Chromoplasts of tomato fruits: II. The red tomato. Amer J Bot 56:380.

13. Rosso SW (1968). The ultrastructure of chromoplast development in red tomato. J Ultrastructure Res 25:307.

14. Piechulla B, Glick RE, Bahl H, Melis A, Gruissem W (1986). Changes in photosynthetic capacity and photosynthetic protein pattern during tomato fruit ripening. Plant Physiol, submitted.

15. Melis A, Harvey GW (1981). Regulation of photosystem stoichiometry, chlorophyll a and chlorophyll b content and relation to chloroplast ultrastructure. Biochem Biophys Acta 637:138.

16. Laval-Martin D, Tremolières A (1980). Three chloroplast membrane models corresponding to different photosynthetic potentialities in the same plant. Planta 149:34.

17. Piechulla B, Chonoles Imlay KR, Gruissem W (1985). Plastid gene expression during fruit ripening in tomato. Plant Mol Biol 5:373.

18. Fromm H, Devic M, Fluhr R, Edelman M (1985). Control of psbA gene expression: in mature Spirodela chloroplasts light regulation of 32 kD protein synthesis is independent of transcript level. EMBO J 4:291.

19. Bathgate B, Purton ME, Grierson D, Goodenough PW (1985). Plastid changes during the conversion of chloroplasts to chromoplasts in ripening tomato. Planta 165:197.

20. Tucker GA, Grierson D (1982). Synthesis of polygalactorunase during tomato fruit ripening. Planta 155:64.

21. Speirs J, Brady CJ, Grierson D, Lee E (1984). Changes in ribosome organization and messenger RNA abundance in ripening tomato fruits. Aust J Plant Physiol 11:225.

22. Rattanapanone N, Speirs J, Grierson D (1978). Evidence for changes in messenger RNA content related to tomato fruit ripening. Phytochemistry 17:1485.

23. Goodenough PW, Tucker GA, Grierson D, Thomas   T (1982). Changes in colour, polygalactorunase, monosaccharides and organic acids during storage of tomatoes. Biochemistry 21:281.

24. Davies JW, Cocking EC (1967). Protein synthesis in tomato fruit locule tissue: the sites of synthesis and the pathway of carbon into protein. Planta 76:285.

25. Piechulla B, Gruissem W (1986). Coordination of nuclear and organelle gene expression during development and ripening of tomato fruit. Manuscript in preparation.

26. Morelli G, Nagy F, Frayley RT, Rogers SG, Chua N-H (1985). A short conserved sequence is involved in the light-inducibility of a gene encoding ribulose-1.5-bisphosphate carboxylase small subunit of pea. Nature 315:200.

27. Piechulla B, Pichersky E, Gruissem W (1986). Correlation of nuclear and plastid encoded photosynthesis-specific genes during tomato fruit development and ripening. Manuscript in preparation.

Molecular Biology of Plant Growth Control, pages 177–180
© 1987 Alan R. Liss, Inc.

WORKSHOP SUMMARY: METHODS TO STUDY THE MOLECULAR
BIOLOGY OF PLANT HORMONES

Mark Jacobs

Department of Biology
Swarthmore College
Swarthmore, Pennsylvania 19081

In this workshop we focused on two kinds of techniques
used effectively in recent research on the molecular biology
of plant growth control by plant hormones: mutants and
monoclonal antibodies. A brief summary of the applications
of these techniques discussed at the workshop is presented
below.
    R. Firn opened the workshop with a plea to use the
molecular techniques we were about to discuss to study
developmental changes and responses to plant growth substances at the cellular and tissue level, not just the whole
organ level. Dr. Firn pointed out that, traditionally, our
thinking about plant cell growth emerged from the results of
research carried out on the growth of whole isolated organs
in response to plant growth substances. A general increase
in organ elongation rate was interpreted as a general increase in all of the component cells' elongation rates, and
tropic responses were often interpreted, without evidence, as
simply an increase in elongation rate of cells on one side of
an organ with the growth of cells on the other side assumed
to be continuing unchanged. However, recent studies of organ
growth responses at the tissue and cellular levels have indicated a good deal more complexity than was previously thought:
(1) individual cells (e.g. in the apical hook) often follow a
complex pattern of growth rate changes within a developmentally short period of time, (2) different tissues within one
organ (e.g. the epidermal vs. the cortical cells in a hypocotyl) respond to growth substances in different ways, and
(3) directed "growth" of an organ (e.g. phototropism) can
often result from differential reduction or cessation of cell
growth rather than from an increase in cellular growth rates.
    B. Phinney began the discussion of the use of mutants to
study the molecular biology of hormones by describing his
recent experiments using the transposable element, Robertson's
mutator (Mu), to isolate genes controlling dwarfism in maize.

Phinney's group has already characterized five nonallelic
dwarfing genes in maize that control specific and different
steps in the gibberellin biosynthetic pathway leading to the
key bioactive gibberellin of maize, GA_1. They have shown
that shoot elongation in maize is dependent on the level of
GA_1, not on the level of the other gibberellins present in
the maize shoot. In their recent work, they isolated GA-
responding dwarf mutants from Mu lines of maize; these mutants
presumably arose as a result of the insertion of Mu into a
region of the maize genome associated with gibberellin biosynthesis. Since the phenotypes of several of these Mu
dwarfs are reminiscent of the phenotypes of the original
dwarfing genes, the group hopes that ^{32}P-labeled Mu can be
used as a probe to clone the dwarfing genes of maize. The
products from cloning can be used not only to study GA-
controlled regulation of development but also to obtain the
hydroxylases, oxidases and cyclases coded for by the different maize dwarfing genes.

P. King described his group's experiments with auxin-
requiring mutants of Hyoscyamus and Nicotiana. Dr. King
has developed a method for mutagenizing haploid leaf protoplasts in tissue culture, selecting for temperature-sensitive
(ts) auxin auxotrophy, and then growing up shoots with the
same phenotype from the selected clones. In one such clone
of Hyoscyamus muticus, temperatures above 29 C totally inhibit growth unless the cells are supplied with an auxin. The
value of such mutants is that if one can uncover the ts step
one potentially has in hand a gene coding for a product
crucial in the auxin physiology of the cell: biosynthesis,
degradation, receptors, transport or natural antagonist. In
the H. muticus clone, the cells are rescued at restrictive
temperatures by indoleacetic acid (IAA) addition, but analysis
with HPLC/RIA has shown that variant cells can synthesize and
accumulate IAA to wildtype levels. The mutation could thus
be one that affects compartmentation of IAA within cells or
receptors for auxin in certain parts of the cells and these
possibilities are being investigated. Characterizing others
of a series of auxin auxotrophic mutants will allow the
further study of crucial gene products along the auxin biosynthesis and action pathways in plant cells.

M. Estelle discussed hormone-insensitive mutants of
Arabidopsis that he has been investigating in Somerville's
laboratory at Michigan State University. The mutants are
A. thaliana plants mutagenized as seeds with a 3-4 hr treatment in EMS, then allowed to grow and self, and screened for
resistance to 2,4-dichlorophenoxyacetic acid (2,4-D) on petri

dishes (5 μM 2,4-D inhibits wild type root growth but does
not inhibit the root growth of mutants). The resistant plants
(13 out of 300,000 germinating seeds) had stable single gene
recessive mutations and showed dramatic alterations in growth
behavior including a reduction in secondary root initiation,
loss of apical dominance (bushy growth habit) and partial to
complete male sterility. The genetic alteration appears to
be at the same single locus for at least 8 of the mutants;
Estelle calls this gene aar-1 for "altered auxin response."
Since aar-1 mutants exhibit reduced 2,4-D-induced ethylene
biosynthesis from mature leaves and form callus poorly under
standard conditions, the 2,4-D resistance does not appear to
be simply a side consequence of the altered growth habit but
rather both phenomena seem due to a change in a protein involved in some aspect of auxin metabolism or response. Since
the metabolism of radiolabeled 2,4-D is not altered in the
aar-1 mutants, Estelle felt that the aar-1 gene might code
for an auxin receptor. The group's future plans include an
examination of the auxin and ethylene binding components in
both wild type and aar-1 Arabidopsis plants.

R. Morris opened our discussion of the uses of monoclonal antibodies (mAbs) in the study of the molecular biology of hormones by describing the work he has done using
immunoaffinity purification of cytokinins as a tool to
analyze cytokinin levels in heavily contaminated plant or
bacterial extracts. Monoclonal and polyclonal antibodies
have been obtained to a number of cytokinins and linked to
column matrix materials that can be used for single pass
isolation of cytokinins from mixtures. Since different mAbs
have different affinities for different cytokinins, columns
of a desired specificity can be obtained by mixing specific
mAbs. The columns must be operated in the pH range of 6.5
to 8.5, but are resistant to denaturation (e.g. can be run
up to 55 C) and will last for from ca. 2 passes (mAb columns)
up to 10 passes (polyclonal Ab columns). The best agents to
use to elute the cytokinins from the columns are pyridine-
formate (0.1M, pH 2.2), glycine-HCl (0.1M, pH 2.3), methanol
or methanol ether. The usual recovery is greater than 75%.
Such immunoaffinity columns have been used to measure cytokinin levels in developing wheat spikes, wheat grains,
aquatic monocots, and the male pollen buds of Douglas fir.

E. Weiler summarized a great deal of research his group
has done making and using mAbs against several plant growth
regulators: abscisic acid, IAA, gibberellins and cytokinins.
Weiler's focus has been on the use of the mAbs in enzyme-
linked immunosorbent assays (ELISAs) which allow detection of

picogram quantities of the respective plant growth regulators.
Immunoaffinity chromatography using these same mAbs can be
performed before the ELISA assay to clean up the hormone prep
in cases of a badly contaminated starting material. After a
discussion of the uses and advantages of mAbs in plant science,
and of the procedures and possible interferences in the ELISA
assay employing mAbs to detect plant hormones, Dr. Weiler
gave some examples from his research, including the measurement of gibberellin levels in Norway spruce.

M. Jacobs ended the workshop discussion with a brief
description of the indirect immunofluorescence technique
using mAbs that he has used to localize presumptive auxin
transport carriers in dark-grown pea stem tissue. The same
mAbs can be used in immunoprecipitation assays that have
implicated a 77,000 MW protein from pea stem tissues as the
protein, or a subunit of the protein, that may transport
auxin and/or bind the transport inhibitor naphthylphthalamic
acid during polar auxin transport. Jacobs discussed the
problems that arise when using a physiological assay (plant
hormone binding) as a screening assay for mAbs. They include
interference in the assay by some of the large number of
components of the hybridoma feeding medium (the medium into
which the mAb is secreted). He also discussed the interferences possible when performing indirect immunofluorescence
localization experiments in plant stem tissue. Here, a major
problem arises if one uses a control mAb that is produced by
a hybridoma made from the same spleen as the positive mAbs,
but one which was negative in the screening assay. If the
original injections of the mice were of a heterogeneous
protein mixture, a control mAb for the hormone binding site
in the screening assay may still be a positive mAb for some
other, unscreened membrane-bound protein, and thus may give
false positives in the immunofluorescence work. Finally,
in immunoprecipitation experiments, oligoclonal mixtures of
different positive mAbs were much more effective in recog-
nizing and precipitating a protein than were individual mAbs
used alone.

Molecular Biology of Plant Growth Control, pages 181–196
© 1987 Alan R. Liss, Inc.

FUNCTIONAL ANALYSIS OF THE T-DNA ONC GENES*

Dirk Inzé[1], Anders Follin[1], Harry Van Onckelen[2],
Patrick Rüdelsheim[2], Jeff Schell[1,3],
and Marc Van Montagu[1]

[1] Laboratorium voor Genetica, Rijksuniversiteit Gent
   B-9000 Gent (Belgium)
[2] Departement voor Biologie, Universitaire Instelling
   Antwerpen, B-2610 Wilrijk (Belgium)
[3] Max-Planck-Institut für Züchtungsforschung
   D-5000 Köln 30 (F.R.G.)

ABSTRACT Agrobacterium tumefaciens causes crown
gall tumor formation by the introduction of a well
defined DNA segment, called T-DNA, into the plant
nuclear genome. The combined activities of the
T-DNA genes 1 and 2 have an auxin-like effect. We
have isolated Nicotiana tabacum plants transformed
with either gene 1 (called rG1) or gene 2 (called
rG2), as the only T-DNA gene. Both rG1 and rG2
plants grow and develop in a normal fashion.
These observations lead to the following conclusions. First, normal plant cells do not contain a
compound, such as indole-3-acetamide (IAM), which
can be converted to a biologically active auxin by
the gen 2 product. Second, N. tabacum cannot
convert the auxin precursor synthesized by the
gene 1 product into IAA at a physiologically
significant rate. The IAM content of rG1 plants
was about 500–1000 times higher (about 1500 pmol/g
fresh weight) compared to an untransformed control
plant. Cell-free extracts of tobacco tissue
transformed with gene 1 were shown to convert
^{3}H-Trp into ^{3}H-IAM. These data indicate that the

* This work was supported by grants of the "ASLK-Kanker-
fonds", the "Fonds voor Wetenschappelijk Geneeskundig Onder-
zoek" (F.G.W.O. 3.0001.82), and the Services of the Prime
Minister (O.O.A. 12.0561.84) to JS and MVM.

T-DNA gene 1 encodes a tryptophan-2-monooxygenase. A similar enzyme has been described for the plant pathogenic bacterium, Pseudomonas savastanoi. We could show that the P. savastanoi iaaM gene (encoding tryptophan-2-monooxygenase) can mimic the phenotypic effect of gene 1. Furthermore, evidence is provided that the IAM production is one of the factors limiting the host range of P. savastanoi (oleanders and olives) since a wild-type P. savastanoi strain produces large outgrowths on rG1, but not on rG2 plants.

INTRODUCTION

Agrobacterium tumefaciens induces crown gall tumors upon infection of most dicotyledonous plants [1]. Recently, it has been demonstrated that Agrobacterium can also transform or infect some monocotyledonous plants such as Asparagus and Narcissus [2,3]. Crown gall tumors show two new characteristics when compared to normal plant tissues. First, crown galls synthesize unusual compounds, called opines, which are not found in normal plant tissues (for a review, see [4]). Second, crown gall tissue grows in culture on synthetic media lacking plant hormones in the absence of viable bacteria. Normal, untransformed tissue requires the external addition of auxins and cytokinins for in vitro culture [5].

All oncogenic agrobacteria harbor large plasmids, up to 200 kb in size [6]. These so-called Ti plasmids were shown to be essential for tumor induction [7,8]. Crown gall tumor formation is the result of the transfer, stable integration, and expression of a well-defined Ti plasmid segment, the T-DNA, into the plant nuclear genome (for a review, see [9]).

At least three genes of the T-DNA are directly involved in tumor induction (Figure 1). Mutants in gene 4 (Roi⁻, root inhibition) or in gene 1 and/or 2 (Shi⁻, shoot inhibition) respectively produce tumors that sprout roots or many shoots on some host plants [10-13]. In analogy to what is known about plant growth regulators, the effect of gene 4 can be thought of as "cytokinin-like", so that inactivation might result in a low cytokinin to auxin ratio, and hence, to root formation. Similarly, the combined effects of genes 1 and 2 can be thought of as "auxin-like" since mutations in either gene appear to increase the cytokinin to auxin ratio of tumors [14-16], and this might in turn lead to shoot formation [10,17]. Therefore, the undifferentiated appearance of a crown gall tumor and its ability to grow without exogenous hormones, appears to reflect the combined activities of the products of genes 1, 2, and 4.

In this chapter, we report on the genetic analysis of the individual pTiC58 T-DNA genes, especially genes 1 and 2. Part of this work has been published [18-20].

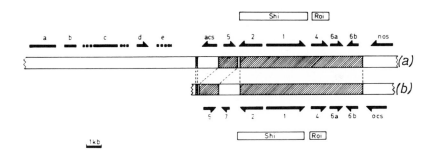

FIGURE 1.   Structure of a nopaline and octopine-type T-DNA. The open boxes with staggered ends represents the T-DNA of the nopaline-type pTiC58 plasmid (a) and the TL-DNA of the octopine-type pTiB6S3 plasmid (b). The hatched parts indicated homologous regions [21]. The thick arrows show the approximate locations, and the orientations of the T-DNA transcripts that have been identified by [22,23]. Shi (shoot inhibition), and Roi (root inhibition) are loci involved in tumor formation [12,13].

RESULTS AND DISCUSSION

Phenotypical Effects of the Individual T-DNA Genes on Plants

One way to analyze the function of the T-DNA genes is to study their individual effect on plant cells. Therefore, we cloned pTiC58 T-DNA genes 1, 2, 4, 5 and 6a/6b separately, and introduced them in the Ti plasmid gene vector, pGV3850 (Fig. 2) [18]. This plasmid is a pTiC58 derivative in which the internal T-DNA HindIII fragments are replaced by pBR322. The pBR sequences can serve as a region of homology for recombination in order to introduce foreign DNAs, in our case the T-DNA genes, between the T-DNA borders [24]. The T-DNA genes 1, 2, 4, and 5 of the octopine-type plasmid pTiB6S3 have been studied in a similar way [Budar et al., in preparation].

We have been able to show that a 2040-bp pTiC58 T-DNA fragment overlapping gene 4 contains all the information required to induce tumorous growth when introduced into plant cells. In addition, the resulting tumors sprout numerous shoots in tissue culture [18]. Recently, it has been demonstrated that gene 4 encodes isopentenyl transferase, an enzyme that catalyzes the first step in cytokinin biosynthesis [26,27].

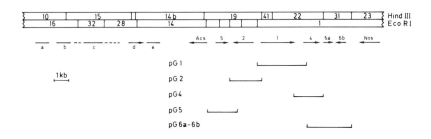

FIGURE 2. Location of the cloned fragments with the pTiC58 T-DNA onc genes used to introduce into Ti plasmid gene vector pGV3850. The HindIII and EcoRI restriction map of the pTiC58 T-DNA is drawn according to [25]. Immediately below the restriction map are the locations of the T-DNA transcripts [23]. The lines under the position of the transcripts indicate the extent of the cloned fragments which were subsequently introduced into pGV3850 according to [24]. The number after "pG" refers to the introduced T-DNA gene, e.g. pG1 contains a fragment with gene 1. The clones with the individual T-DNA genes contain the following approximate number of base pairs upstream of the start and downstream of the stop of translation. pG1, 290 and 750; pG2, 500 and 270; pG4, 670 and 300; pG5, 600 and 330. Clone pG6a-6b contains approximately 2200 bp, and 540 bp respectively, upstream of the start of translation of gene 6a, and gene 6b. The exact position of the T-DNA genes is based on DNA sequence analysis [J. Seurinck, unpublished results].

Neither of the pGV3850 derivatives with gene 1 or gene 2 induced tumors by themselves, whereas a double infection of plants with pGV3850 derivatives containing either gene 1 or gene 2 resulted in the formation of small tumors. These tumors produced roots, particularly in vitro culture. This led to the conclusion that genes 1 and 2 are both necessary to obtain auxin-independent growth, to suppress shoot formation, and to stimulate root formation of plant tissue cultures [18]. Furthermore, we could show that the activity of genes 1 and 2 could be replaced in vivo by addition of α-naphthalene acetic acid, and that the activity of gene 1 could be substituted by α-naphthalene acetamide [13,18]. The results suggest that the product of gene 2 converts α-naphthalene acetamide into an active auxin. Recently, it has been shown biochemically that gene 2 codes for a hydrolase that is able to convert indole-3-acetamide (IAM) into indole-3-acetic acid (IAA) [28,29], and it has been confirmed that the gene 2 product can also use other substrates, such as α-naphthalene acetamide [30]. The genetic data (see also below) indicate that the T-DNA gene 1 is responsible for the synthesis of IAM as a substrate for gene 2. The pGV3850 constructions with genes 5 or 6a/6b did not have any detectable effect on plants.

Tobacco Plants Transformed with only Gene 1 or Gene 2 Are Morphologically Normal

It was important to test whether the products of gene 1 or gene 2 by themselves would have any influence on plant development. Therefore, we isolated N. tabacum plants transformed with either gene 1 or gene 2 (called rG1 and rG2, respectively) [19]. Both these plants were found to grow and develop in a normal way. This lack of phenotypic effect by the presence of the gene 1 or gene 2 products was not due to inactivity of these genes since A. tumefaciens strains only harboring an active gene 2 were able to produce neoplasmic proliferations on rG1 plants, while A. tumefaciens strains carrying only an active gene 1 had a similar effect on rG2 plants [19].

These observations led to the following conclusions. First, the product of gene 2 obviously does not interfere with normal auxin metabolism in plants. In other words, normal plant cells do not seem to contain a compound which can be converted to biologically active auxin by the gene 2 product. In agreement with this, it is generally accepted that IAM, the normal substrate for gene 2, plays no role in auxin biosynthesis in plants [31]. Second, the activity of gene 1 in rG1 plants has no obvious effect by itself. This implies that Nicotiana tabacum cannot convert the auxin precursor (IAM) synthesized by the gene 1 product into IAA at a physiologically significant rate. The above data indicate that the combined activities of the T-DNA genes 1 and 2 are able to override the normal control mechanisms regulating the levels of auxin in plants by introducing a new pathway for the production of this plant growth factor. This new auxin biosynthetic pathway (Fig. 3) is similar to that encoded by the plant pathogenic bacterium, Pseudomonas savastanoi (see further). Unlike tumors induced by Agrobacterium, the Pseudomonas-induced tissue proliferations depend on the continuous production of IAA by the infecting bacteria. Pseudomonas savastanoi produces IAA by means of tryptophan-2-monooxygenase and indole-3-acetamide hydrolyse according to the following pathway : L-tryptophan → IAM → IAA [32-34].

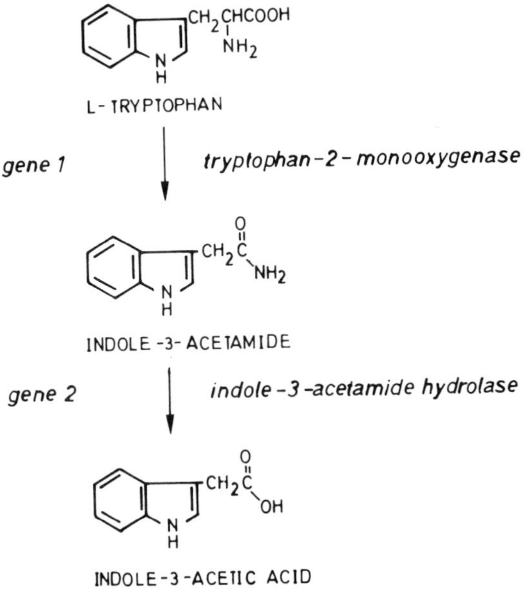

FIGURE 3. Auxin biosynthetic pathway encoded by the T-DNA genes 1 and 2 in crown gall tissues.

Genetic Evidence that the T-DNA Gene 1 Is Functionally Equivalent to the Tryptophan-2-monooxygenase Gene of Pseudomonas
savastanoi

It is now well established that the gene 2 product is
genetically equivalent to the indole-3-acetamide hydrolase
of Pseudomonas savastanoi [28,29]. The following experiments demonstrated that the gene 1 product is functionally
equivalent to the tryptophan-2-monooxygenase of P. savastanoi [19]. The plasmid pLUC2 expresses the P. savastanoi
iaaM gene (encoding tryptophan-2-monooxygenase) in E. coli
[35]. When the E. coli strain K514(pLU2) was inoculated on
decapitated rG2 plants, a burst of root and callus formation
could be observed after 8—10 days (Fig. 4). No such proliferations were noted when the same E. coli strain was inoculated, on either rG1 or normal plants. Similarly, a
pronounced response on rG2 plants could be observed when the
iaaM gene was expressed under its own promoter in Agrobacterium (data not shown). The observed effect can be explained by assuming that E. coli K514 (pLU2) produce and
release IAM which is taken up by the plant cells at the site
of inoculation and converted into active IAA by the activity
of the gene 2-encoded enzyme in rG2 plants.

If this interpretation is correct, one would expect the
root and callus proliferation to be dependent on the active
growth of the IAM-producing bacteria at the infection site.
Consistent with this expectation, it was observed that
tissues isolated from the proliferations did not grow with-

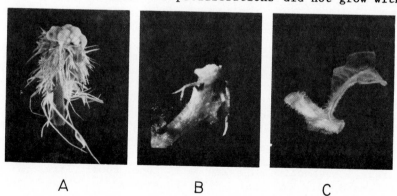

A   B   C

FIGURE 4. Phenotypic response of decapitated and in-
fected rG2 Nicotiana tabacum plants with E. coli
K514(pLUC2), (A); A. tumefaciens C58C1Rif^R(pGV3850::pG1),
(B); and no bacteria, (C).

out plant growth factors in axenic culture conditions. It was also observed that even 4 weeks after inoculation of the rG2 plants, viable E. coli K514(pLUC2) could be reisolated.

Finally, the supernatants of an overnight-grown K514(pLU2) culture contained about 81 pmol/ml IAM whereas less than 8 pmol/ml were detected in the supernatant of a similarly grown K514 culture. The bacteria were grown in LB medium [36] at 37°C and the IAM contents were analyzed as described [20].

From this data it appears that the gene 1 product is functionally interchangeable with the P. savastanoi trypto-phan-2-monooxygenase. Recently, significant amino acid sequence homology has been been found [37] between the pre-dicted sequence of the tryptophan-2-monooxygenase of P. savastanoi and the deduced product of the T-DNA gene 1 of the octopine-type plasmid pTiA6NC from Agrobacterium tume-faciens. Strong homology was found in the 25-amino-acid se-quence of the putative FAD-binding region of tryptophan-2-monooxygenase.

Biochemical Evidence That Gene 1 Encodes a Tryptophan-2-monooxygenase

If the T-DNA gene 1 encodes a tryptophan-2-monooxy-genase activity, one would expect to find elevated IAM levels in tissue transformed with this gene. Therefore, we analyzed the endogenous IAM and IAA concentrations in trans-formed tobacco tissues using different reversed-phase HPLC systems as described [20]. The results are summarized in Table 1.

The results show that SR1 crown gall tissues transform-ed with A. tumefaciens C58 T-DNA mutant pGV3132, and there-fore containing gene 1 but not gene 2, accumulated large amounts of IAM. An endogenous level of 17160 pmol IAM/g fresh weight was found 12 days after subculture. In compa-rison, untransformed SR1 callus and the SR1 3845 line, defective for gene 1, but with the amidohydrolase activity encoded by gene 2, contained 31.0 and 10.9 pmol IAM/g fresh weight, respectively. The wild-type tumor line SR1 C58 contained only 120 pmol IAM/g fresh weight, probably due to the indole-3-acetamide hydrolase encoded by gene 2.

The IAM content of the rG1 plant, which contain an active gene 1, was about 500–1000 times higher compared to rG6a/6b and the untransformed N. tabacum cv. W38 control plant.

TABLE 1
ENDOGENOUS IAM AND IAA CONCENTRATIONS IN DIFFERENT
TRANSFORMED TOBACCO CELL LINES[1]

Cell line	Relevant T-DNA genes	IAM (pmol/g fresh weight)	IAA (pmol/g fresh weight)
SR1 callus	-	31	10
SR1 C58	1^+ 2^+ 4^+	120	225
SR1 3132	1^+ 2^- 4^+	17160	107
SR1 3845	1^- 2^+ 4^+	11	90
rG1	1^+	1410	18
rG6a/6b	$6a^+$ $6b^+$	4	8
W38	-	< 1	14

[1]The transposed cell clones SR1 3132 and
SR1 3845 were obtained by cocultivation of mesophyll protoplasts of N. tabacum cv. Petit Havana
SR1 with agrobacteria as described [38]. The
T-DNA transposon mutants pGV3132 and pGV3845 are
described [13]. The N. tabacum cv. W38 plants,
rG1 and rG6a/6b contain respectively an active
gene 1 and active genes 6a/6b [20]. The extraction, purification and quantitative analysis of
IAM and IAA was as described [20]. The SR1 callus
and tumor tissues were analyzed 12 days after
subculture.

These data lead to the conclusion that the T-DNA gene 1
is responsible for the accumulation of IAM in transformed
plants. Either gene 1 encodes an enzyme that synthesizes
IAM or less likely encodes a regulator protein that activates a pre-existing plant gene.

Recently, we have been able to show that tobacco tissue
transformed with gene 1 contains a tryptophan-2-monooxy-
genase activity. Cell-free extracts of the SR1 3132 (2^-)
line were shown to convert ^{3}H-Trp into ^{3}H-IAM (Fig. 5). The
monooxygenase activity measured in SR1 3132 extracts taken
10 days after subculture corresponded to an IAM production
of about 1800 pmol/hr.g fresh weight. This activity is

certainly sufficient to account for the endogenous level of IAM which accumulates in the SR1 3132 tumor line. The very low Trp → IAM conversion observed in the SR1 callus and the SR1 3845 (1⁻) might be attributed to an aspecific peroxidase activity as was observed also in P. savastanoi [27].

In agreement with the above data, crude cell-free extracts prepared from three bacterial species harbouring pTiA6 gene 1 were shown to convert L-tryptophan into indole-3-acetamide [39]; control extracts lacking gene 1 could not carry out the reaction.

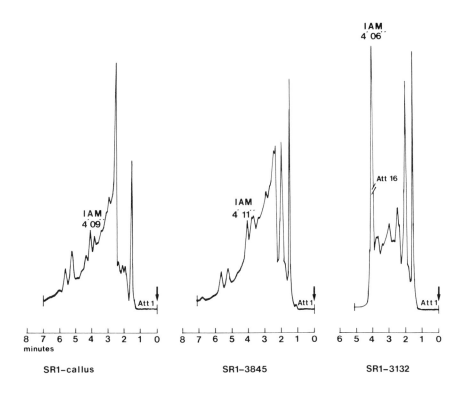

FIGURE 5. IS-HPLC spectrofluorimetric elution profiles of the tryptophan-2-monooxygenase reaction mixtures of SR1 callus, SR1 3132, and SR1 3845 extracts. ^{3}H-Trp elutes at 19.5 minutes (not shown). IS-HPLC, 5 μm Roslil (18 HL, 25 x 0.47 cm IS, MeOH-H$_2$O-HAc (40/60/0.05 v/v), 1.5 ml/min. Detection : on-line spectrofluorimeter (Shimadzu 530) λ_{ex} 285 nm; λ_{em} 360 nm.

Molecular Basis of the Limited Host Range of Pseudomonas savastanoi

Pseudomonas savastanoi is a gall-forming pathogen of olive (Olea europea) and oleander (Nerium oleander). The induction of callus on Nicotiana tabacum plants by a wild-type P. savastanoi strain (NCPPB639) is weak (Fig. 6). Interstingly, similar infections on rG1, but not on rG2 plants, resulted in large outgrowths that sprouted many roots. Since the rG1 plants contain high amounts of IAM, these observations indicate that the IAM production might be one of the limiting factors for the pathogenicity of P. savastanoi on tobacco plants. The "tumor-like" tissue, resulting from the infection of rG1 plants with NCPPB639 did not grow without plant hormones in axenic culture conditions. Up to now, there is no evidence for any kind of transfer of genetic material from P. savastanoi to plant cells.

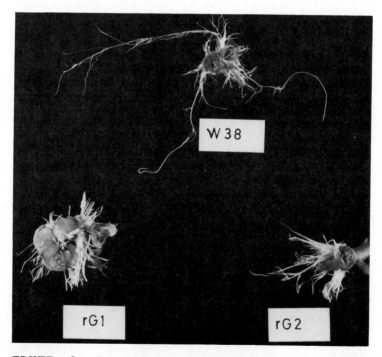

FIGURE 6. Infections of decapitated rG1, rG2, and control (W38) plants with Pseudomonas savastanoi NCPPB639. The photographs were taken 8 weeks after inoculation.

The Possible Use of Gene 1 as a Selectable Marker

The tryptophan-2-monooxygenase confers 5-methyl tryptophan resistance to P. savastanoi [34]. Therefore, we tested
different tissues transformed with the T-DNA gene 1 for
5-methyl tryptophan resistance. First, there is no significant difference in growth rate between untransformed N.
tabacum callus tissue, C58 crown gall tissue, and callus
tissue derived from rG1 plants on media containing 5-methyl
tryptophan in concentrations ranging from 1 to 10 mg/l. The
tissues were cultured on Linsmaier & Skoog medium [40],
supplemented with 2 mg/l naphthalene acetic acid and
0.3 mg/l benzyl-amino purine, except for the C58 crown gall
tissue which was grown without hormones. In general, the
5-methyl tryptophan slows down the growth rate at concentrations of 5 mg/l or higher. Second, seeds obtained from a
self-fertilization of rG1 plants germinated and developed in
the same way as seeds derived from untransformed plants on
media (1/2 of Linsmaier & Skoog [40] salts, 10 g/l sucrose,
8 g/l agar) with different concentrations of 5-methyl tryptophan. Both types of seedlings showed a reduction in
primary root formation and growth rate already on media with
1 mg/l 5-methyl tryptophan.

Recently, is has been reported [41] that tobacco cell
suspension transformed the Agrobacterium strain A6 are
approximately 3-fold more resistant to 5-methyl tryptophan
than untransformed tobacco cell suspensions.

ACKNOWLEDGMENTS

The authors wish to thank Dr. Allan Caplan for critical
reading of the manuscript; Ms Martine De Cock for typing,
and Karel Spruyt and Albert Verstraete for photographs and
drawings. HVO is a Senior Research Associate, and DI a
Senior Research Assistant of the National Fund for Scientific Research (Belgium); PR is a recipient of an IWONL fellowship.

REFERENCES

1.   De Cleene M, De Ley J (1976). The host range of crown-
      gall. Botan Rev 42:389.

2.   Hernalsteens J-P, Thia-Toong L, Schell J, Van Montagu M
     (1984). An Agrobacterium-transformed cell culture from
     the monocot Asparagus officinalis. EMBO J 3:3039.
3.   Hooykaas-Van Slogteren GMS, Hooykaas PJJ, Schilperoort
     RA (1984). Expression of Ti plasmid genes in monoco-
     tyledonous plants infected with Agrobacterium tumefa-
     ciens. Nature (London) 311:763.
4.   Petit A, Tempé J (1985). The function of T-DNA in
     nature. In Van Vloten-Doting L, Groot, GSP, Hall TC
     (eds): "Molecular form and function of the plant
     genome," New York: Plenum Press, p 625.
5.   Braun AC (1956). The activation of a growth-substance
     system accompanying the conversion of normal to tumor
     cells in crown gall. Cancer Res 16:53.
6.   Zaenen I, Van Larebeke N, Teuchy H, Van Montagu M,
     Schell J (1974). Supercoiled circular DNA in crown gall
     inducing Agrobacterium strains. J Mol Biol 86:109.
7.   Van Larebeke N, Genetello C, Schell J, Schilperoort RA,
     Hermans AK, Hernalsteens J-P, Van Montagu M (1975).
     Acquisition of tumour-inducing ability by non-oncogenic
     agrobacteria as a result of plasmid transfer. Nature
     (London) 255:742.
8.   Watson B, Currier TC, Gordon MP, Chilton M-D, Nester EW
     (1975). Plasmid required for virulence of Agrobacterium
     tumefaciens. J Bacteriol 123:255.
9.   Gheysen G, Dhaese P, Van Montagu M, Schell J (1985).
     DNA flux across genetic barriers : the crown gall
     phenomenon. In Hohn B, Dennis ES (eds): "Genetic flux
     in plants," (Advances in Plant Gene Research, Vol. 2),
     Wien: Springer Verlag, p 11.
10. Ooms G, Hooykaas PJ, Moleman G, Schilperoort RA (1981).
     Crown gall plant tumors of abnormal morphology, induced
     by Agrobacterium tumefaciens carrying mutated octopine
     Ti plasmids; analysis of T-DNA functions. Gene 14:33.
11. Garfinkel DJ, Simpson RB, Ream LW, White FF, Gordon MP,
     Nester EW (1981). Genetic analysis of crown gall : fine
     structure map of the T-DNA by site-directed mutagenesis.
     Cell 27:143.
12. Leemans J, Deblaere R, Willmitzer L, De Greve H,
     Hernalsteens J-P, Van Montagu M, Schell J (1982).
     Genetic identification of functions of TL-DNA tran-
     scripts in octopine crown galls. EMBO J 1:147.
13. Joos H, Inzé D, Caplan A, Sormann M, Van Montagu M,
     Schell J (1983). Genetic analysis of T-DNA transcripts
     in nopaline crown galls. Cell 32:1057.

14. Morris RO, Akiyoshi DE, MacDonald EMS, Morris JW, Regier DA, Zaerr JB (1982). Cytokinin metabolism in relation to tumor induction by Agrobacterium tumefaciens. In Wareing PF (ed), "Plant growth substances 1982," London: Academic press, p 175.

15. Akiyoshi DE, Morris RO, Hinz R, Mischke BS, Kosuge T, Garfinkel DJ, Gordon MP, Nester EW (1983). Cytokinin/ auxin balance in crown gall tumors is regulated by specific loci in the T-DNA. Proc Natl Acad Sci USA 80:407.

16. Van Onckelen H, Rüdelsheim P, Hermans R, Horemans S, Messens E, Hernalsteens J-P, Van Montagu M, De Greef J (1984). Kinetics of endogenous cytokinin, IAA and ABA levels in relation to the growth and morphology of tobacco crown gall tissue. Plant & Cell Physiol 25: 1017.

17. Amasino RM, Miller CO (1982). Hormonal control of tobacco crown gall tumor morphology. Plant Physiol 69:389.

18. Inzé D, Follin A, Van Lijsebettens M, Simoens C, Genetello C, Van Montagu M, Schell J (1984). Genetic analysis of the individual T-DNA genes of Agrobacterium tumefaciens; further evidence that two genes are involved in indole-3-acetic acid synthesis. Mol Gen Genet 194:265.

19. Follin A, Inzé D, Budar F, Genetello C, Van Montagu M, Schell J (1985). Genetic evidence that the tryptophan 2-mono-oxygenase gene of Pseudomonas savastanoi is functionally equivalent to one of the T-DNA genes involved in plant tumour formation by Agrobacterium tumefaciens. Mol Gen Genet 201:178.

20. Van Onckelen H, Rüdelsheim P, Inzé D, Follin A, Messens E, Horemans S, Schell J, Van Montagu M, De Greef J (1985). Tobacco plants transformed with the Agrobacterium T-DNA gene 1 contain high amounts of indole-3-acetamide. FEBS Lett 181:373.

21. Engler G, Depicker A, Maenhaut R, Villarroel-Mandiola R, Van Montagu M, Schell J (1981). Physical mapping of DNA base sequence homologies between an octopine and a nopaline Ti-plasmid of Agrobacterium tumefaciens. J Mol Biol 152:183.

22. Willmitzer L, Simons G, Schell J (1982). The TL-DNA in octopine crown gall tumours codes for seven well-defined polyadenylated transcripts. EMBO J 1:139.

23. Willmitzer L, Dhaese P, Schreier PH, Schmalenbach W, Van Montagu M, Schell J (1983). Size, location, and polarity of T-DNA-encoded transcripts in nopaline crown gall tumors; evidence for common transcripts present in both octopine and nopaline tumors. Cell 32:1045.

24. Zambryski P, Joos H, Genetello C, Leemans J, Van Montagu M, Schell J (1983). Ti plasmid vector for the introduction of DNA into plant cells without alteration of their normal regeneration capacity. EMBO J 2:2143.

25. Depicker A, De Wilde M, De Vos G, De Vos R, Van Montagu M, Schell J (1980). Molecular cloning of overlapping segments of the nopaline Ti-plasmid pTiC58 as a means to restriction endonuclease mapping. Plasmid 3:193.

26. Akiyoshi DE, Klee H, Amasino RM, Nester EW, Gordon MP (1984). T-DNA of Agrobacterium tumefaciens encodes an enzyme of cytokinin biosynthesis. Proc Natl Acad Sci USA 81:5994.

27. Barry GF, Rogers SG, Fraley RT, Brand L (1984). Identification of a cloned cytokinin biosynthetic gene. Proc Natl Acad Sci USA 81:4776.

28. Schröder G, Waffenschmidt S, Weiler EW, Schröder J (1984). The T-region of Ti plasmids codes for an enzyme synthesizing indole-3-acetic acid. Eur J Biochem 138: 387.

29. Thomashow LS, Reeves S, Thomashow MF (1984). Crown gall oncogenesis : evidence that a T-DNA gene from the Agrobacterium Ti plasmid pTiA6 encodes an enzyme that catalyzes synthesis of indoleacetic acid. Proc Natl Acad Sci USA 81:5071.

30. Kemper E, Waffenschmidt S, Weiler EW, Rausch T, Schröder J (1985). T-DNA encoded auxin formation in crown gall cells. Planta 163:257.

31. Sembdner G, Gross D, Liebisch H-W, Schneider G (1980). Biosynthesis and metabolism of plant hormones. In MacMillan J (ed), "Hormonal regulation of development I," (Encyclopedia of Plant Physiology, New Series, Vol. 9), Berlin: Springer-Verlag, p 281.

32. Kosuge T, Heskett MG, Wilson EE (1966). Microbial synthesis and degradation of indole-3-acetic acid. I. The conversion of L-tryptophan. J Biol Chem 241:3738.

33. Smidt M, Kosuge T (1978). The role of indole-3-acetic acid accumulation by alfa-methyl trypthophan resistant mutants of Pseudomonas savastanoi in gall formation on oleanders. Physiol Plant Pathol 13:203.

34. Comai L, Kosuge T (1980). Involvement of plasmid deoxyribonucleic acid in indoleacetic acid synthesis in Pseudomonas savastanoi. J Bacteriol 143:950.
35. Comai L, Kosuge T (1982). Cloning and characterization of iaaM, a virulence determinant of Pseudomonas savastanoi. J Bacteriol 149:40.
36. Miller JH (1972). "Experiments in Molecular Genetics." New York: Cold Spring Harbor Laboratory, p 466.
37. Yamada T, Palm CJ, Brooks B, Kosuge T (1985). Nucleotide sequences of the Pseudomonas savastanoi indoleacetic acid genes show homology with Agrobacterium tumefaciens T-DNA. Proc Natl Acad Sci USA 82:6522.
38. Van Lijsebettens M, Inzé D, Van Montagu M, Schell J (1986). Transformed cell clones as a tool to study T-DNA integration mediated by Agrobacterium tumefaciens. J Mol Biol 188: in press.
39. Thomashow MF, Hugly S, Buchholz WG, Thomashow LS (1986). Molecular basis for the auxin-independent phenotype of crown gall tumor tissues. Science 231:616.
40. Linsmaier EM, Skoog F (1965). Organic growth factor requirements of tobacco tissue cultures. Physiol Plantarum 18:100.
41. Sanger M, Kosuge T (1984). Evidence that indoleacetamide is an intermediate in IAA biosynthesis of crown gall tumor tissue. Plant Physiol Suppl 75:42.

Molecular Biology of Plant Growth Control, pages 197–207
© 1987 Alan R. Liss, Inc.

AN ENZYME-IMMUNOASSAY FOR QUANTITATIVE ANALYSIS OF
ABSCISIC ACID IN WHEAT[1]

Natasha V. Raikhel, D. Wayne Hughes, and Glenn A. Galau

Department of Botany, University of Georgia
Athens, Georgia 30602

ABSTRACT. Two antibody preparations against abscisic acid (ABA) are commercially available. The polyclonal from Miles was raised against (±)ABA(C-1)HSA and should preferentially recognize only free and conjugated (-)ABA [Walton et al. (1979) Planta 146:139; Weiler (1980) Planta 148: 262-272]. The monoclonal antibody from Idetek was raised against (±)ABA(C-4')BSA. It is clone C5 of Mertens et al. (1983, FEBS Let. 160:269-272), reported to recognize only free (+)ABA. Using ELISAs based on competitive binding between free and enzyme-linked ABA, we show that these antibodies recognize the predicted enantiomers. Consequently, only the monoclonal antibody is useful for measurements of ABA in biological samples. The ELISA with the monoclonal antibody gives estimates of free (+)ABA in extracts of wheat tissues identical to those achieved with gas chromatography which measures both enantiomers. Free (+)ABA was measured in wheat embryos during embryogenesis and in normal and Fluridone-treated wheat seedlings.

[1]This work was supported by National Science Foundation Grant DMB83-14374 to NAR and by a grant GM 29495 to GAG from the United States National Institute of Health.

INTRODUCTION

Wheat germ agglutinin (WGA) has been identified as a wheat embryo-specific protein, and its synthesis and localization during grain development appears to be under abscisic acid (ABA) control (1). It has been demonstrated that in dry, mature embryos WGA is localized in particular organs (2) and that several of these organs accumulate WGA as early as 10 days post-anthesis (DPA) (3). However, only after 15 DPA has ABA been detected by gas chromatography in the whole wheat grain (4). We wished to corroborate prior evidence for ABA involvement in WGA accumulation by asking if ABA is in fact present in the embryo during the time WGA first accumulates. Additionally, there is WGA in wheat seedlings, and its concentration is increased by ABA and decreased by Fluridone (5), an herbicide that reduces endogenous ABA levels in maize embryos *in situ* (6). Consequently we wished to determine if Fluridone also reduces the level of endogenous ABA in wheat, and possibly thereby, reduces WGA accumulation.

The concentration of other specific polypeptides and mRNAs in excised embryos have been shown to be modulated by ABA (7). There is thus a general need for a method that permits quick, precise determination of the endogenous level of this hormone. Immunological methods have recently been adapted for quantitation of plant growth regulators, including ABA. These provide several advantages over the physicochemical methods, most importantly allowing the use of crude or only partially purified samples without expensive instrumentation (8, 9, 10, 11, 12).

Two antibodies to ABA are commercially available. We have examined their usefulness in the enzyme-immunoassay method of Weiler (10) and Daie and Wyse (11), based on the competitive binding of sample and alkaline phosphatase-labeled ABA to immobilized anti-ABA antibody. As has been extensively reported elsewhere (8, 9, 10), the stereoisomer specificity of anti-ABA antibodies is highly dependent on the immunogen. As can be predicted from these observations and knowledge of the immunogens used to prepare the commercial antibodies, we report here that only the monoclonal antibody recognizes (+)ABA with sufficient specificity to be useful.

MATERIALS AND METHODS

Plant Material.

Wheat (*Triticum aestivum* L. cv. Marshall) was grown in soil in a growth chamber with a period of 18 h light at 24°C and 6 h dark at 20°C. Individual plants were tagged on the day of anthesis, defined by the first appearance of bright yellow anthers outside of the glumes. Grains were collected at 10, 25, and 40 DPA. One hundred to two hundred embryos at each stage were isolated and pooled. Growth and treatment of seedlings were performed as described (5).

Preparation of Plant Extracts for ELISA.

The entire procedure was carried out in dim green light and extracts stored at -80°C. Fresh material was ground to a fine powder in liquid nitrogen with a mortar and pestle. Using a glass homogenizer or Polytron homogenizer (Brinkmann, Westbury NY), the powder was homogenized in (per g powder) 20 ml freshly prepared 80% MeOH (pH adjusted to 7.0 with 1 N NaOH) containing 10 mg/l 2,6-Di-*tert*-butyl-4-methylphenol (BHT; Aldrich, Milwaukee WI) and 10 μCi/l [^{3}H](±)ABA (22 Ci/mmole; Amersham, Arlington Heights IL). The homogenate was shaken in the dark for 48 h at 4°C, after which it was centrifuged at 15,000 x g for 15 min. The supernatant was evaporated to dryness under nitrogen and the residue dissolved in (per g starting powder) 0.5-1.0 ml TBS (25 mM Tris-HCl, 100 mM NaCl, 1 mM MgCl$_2$, 0.1% sodium azide, pH 8.5). Measurement of radioactivity in a small aliquot was used to estimate, and correct for recovery of endogenous (+)ABA. Recovery of [^{3}H](±)ABA was 72-96%.

ELISA Procedures.

The lyophilized monoclonal antibody was purchased from Idetek, San Bruno CA (#P-B02-002, Lot 2-121084), dissolved in TBS to 100 μg/ml, and stored at 4°C. The alkaline phosphatase-labeled ABA was also from Idetek (#PA03-001, Lot 88512). It was prepared according to their instructions and stored in 25 mM Tris-HCl, 1 mM MgCl$_2$, 0.1% sodium azide, 0.1% gelatin, pH 7.5, at 4°C. The ELISA was performed essentially as described by Weiler (10) and Daie and Wyse (11), with some modifications. Polystyrene flat-bottom Nunc-Immuno or Micro Well Module plates (Scientific Resource Assoc., Mellevue WA) were used. 1) Wells were coated with 200 μl 5 μg/ml monospecific rabbit anti-mouse IgG in bicarbonate buffer (50 mM sodium bicarbonate, 0.1% sodium

azide, pH 9.6) for 24-48 h at 4°C, and the wells were rinsed
three times with WB (0.85% NaCl, 0.05% Tween-20, 0.1% sodium
azide, pH adjusted to 7.0 with NaOH). 2) The wells were
coated with 200 µl monoclonal antibody for 24-48 h at 4°C,
and rinsed three times with WB. 3) Remaining nonspecific
binding sites were covered by adding 200 µl 1% rabbit
albumin (Sigma) in TBS for 1 h at room temperature and
rinsing three times with WB. 4) Under dim green light,
100 µl ABA in TBS and 100 µl alkaline phosphatase-labeled
ABA were added and the plate incubated 3 h at 4°C and rinsed
three times with WB. 5) Under green light, the activity of
the bound enzyme was assayed by the addition of 200 µl
1 mg/ml p-nitrophenyl phosphate in 9.6% (v:v) diethanol-
amine, 0.5 mM MgCl$_2$ (pH adjusted to 9.6 with HCl), followed
by incubation for 1-1.5 h at 37°C. The reaction was stopped
with the addition of 100 µl 1 N NaOH. 6) The absorbance of
the solution was measured between 400 and 500 nm with a
custom-built, automated spectrophotometer provided by Dr.
Lee Pratt of this department.

Kits are available from Idetek (#P-02D-096) with plates
already coated with the monoclonal antibody. If these were
used, the procedure was carried out beginning with step 4
above.

Lyophilized polyclonal anti-ABA antibodies were
purchased from Miles, Naperville IL (#61-392, Lot AB1),
reconstituted as recommended by the supplier, and stored at
-20°C. It was used at a 1:10 dilution in an ELISA as
described by Weiler (10) with the modifications, where
appropriate, listed above. The alkaline phosphatase-labeled
ABA was prepared by coupling (±)ABA (#A 1012, Sigma, St.
Louis MO) through C-1 (10).

Two standards, namely (±)ABA (Sigma #A 1012) and (+)ABA
(Sigma #A 4906) were used in each assay. Non-specific binding of the alkaline phosphatase-labeled ABA conjugate was
defined as the absorbance (0.00-0.01) achieved when clearly
saturating amounts of unlabeled (±)ABA was present in the
well. Maximum binding of the conjugate was defined as the
absorbance (1.8-2.1) obtained when no unlabeled ABA was
present.

Measurement of ABA by Gas Chromatography.

All preparative steps were performed under dim green
light. Plant tissue was homogenized in 80% methanol
containing BHT and [^{3}H](±)ABA and clarified by
centrifugation as described above. Methanol was removed by
flash-evaporation, and the remaining aqueous fraction
adjusted to pH 8.0 with NaOH and extracted with

dichloromethane. The pH was then adjusted to 2.5 with HCl
and the free ABA was partitioned into dicloromethane. The
organic phase was dried with sodium sulfate, flash
evaporated to dryness, and the residue dissolved in
methanol:diethylether (1:9; v:v). Esterification of ABA and
its purification on florisil were performed as described
(13). The pooled eluates were dried under nitrogen and
dissolved in diethylether. Recovery of the [^{3}H]($\pm$)ABA in
its methyl ester form at this point was 80-90%. The analysis
of methyl-ABA was carried out on a Tracor 550 gas chromato-
graph equipped with a ^{63}Ni electron capture detector. The
column was 0.6 X 180 cm glass, packed with 3% OV-1 on 80-100
mesh Anakrom ABS, with nitrogen as carrier gas. The injection port and detector were at 225°C while the column
temperature was maintained at 200°C. Peak height was used
as the measure of detector response to ABA. Standard
methyl-ABA ester was prepared from ($\pm$)ABA and used without
purification, assuming that 100% of the input ABA was in
fact esterified.

RESULTS AND DISCUSSION

ELISA using the Polyclonal Antibody.

     The immunogen used in the preparation of the polyclonal
antibody available from Miles was ($\pm$)ABA(C-1)HSA (Miles
Technical Representatives, personal communications). This
antibody should recognize the ring region of the molecule
and thus both free ABA and the biological conjugates of ABA,
which are through C-1 (Fig. 1). However, both Walton (8) and
Weiler (9) report that such an immunogen results in antibodies reacting preferentially with (−)ABA. (+)ABA (C-1)
conjugates are required as immunogen if resulting antibodies
are to preferentially recognize (+)ABA, either its free or
conjugated forms (9, 10).
     The experimental results in Fig. 2 confirm that the
polyclonal sees (−)ABA preferentially. While a direct test
of (−)ABA *vs* (+)ABA was not possible due to the unavail-
ability of (−)ABA, the greater than 5000-fold difference in
the apparent molar cross-reactivities of (+)ABA and the
equal molar mixture of (+)ABA and (−)ABA [the ($\pm$)ABA
product] conclusively demonstrates the preference for
(−)ABA. The apparent reaction with (+)ABA may in fact be
with trace (−)ABA contaminants. Consequently, using ($\pm$)ABA
as the standard without correction, the polyclonal antibody-
based ELISA would underestimate the concentration of pure
(+)ABA by at least 200-fold. It is impossible with this

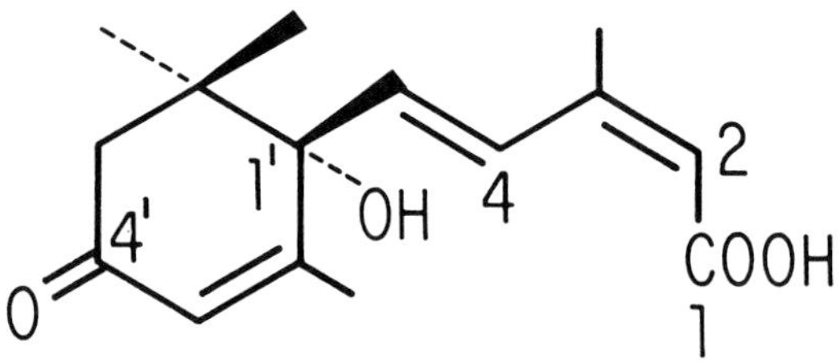

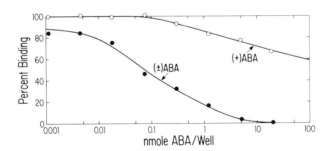

FIGURE 1.   Structure of (+)ABA (Milborrow 1978).

antibody   to   estimate   the   total   amount   or   relative
composition of   ABA in mixtures   of (-)ABA and   (+)ABA, even
where   (-)ABA   comprises   as   little   as 10^{-3} of the (±)ABA.
Plant   ABA   is   reported   to   be   solely   (+)ABA (14). These
studies do not exclude (-)ABA at this level, however. A significant fraction of   the ABA extracted from   plants exposed
to (±)ABA will of course be (-)ABA. These several uncertain-
ties,   along with   its apparent   poor sensitivity to (+)ABA,
render   this   ELISA   virtually   useless   for   its   intended
purpose.   It would   be applicable, however, as   an assay for
measuring (-)ABA   in the presence of   (+)ABA,   as in the case
of ABA determinations in plants exposed to (±)ABA.

FIGURE 2.   Reactions of (+)ABA and (±)ABA standards in
the polyclonal antibody-based ELISA.

ELISA with Monoclonal Antibody.

The monoclonal anti-ABA antibody sold by Idetek is clone C5 described by Mertens *et al*. (15) (Idetek Technical Representatives, personal communications). The immunogen was (±)ABA conjugated to BSA through C-4', resulting in the monoclone, which like polyclonals directed to this immunogen (9), react preferentially with free (+)ABA (15). When used as anti-ABA antibody in the ELISA (Fig. 3), the monoclonal antibody preferentially recognizes (+)ABA. The molar cross-reactivity of (±)ABA relative to (+)ABA is 0.54 ± 0.06 (N=3), in agreement with prior measurements, by radioimmuno-assay, of zero cross-reaction with pure (-)ABA (15). Since (+)ABA is the endogenous plant ABA (14), and will comprise at least 50% of ABA recovered from tissues exposed to exogenous (±)ABA, the monoclonal antibody-based ELISA should specifically measure free (+)ABA in all biological samples. The detection limit is about 50 fmole (+)ABA (13 pg) in 100 μl per well. The useful range is 50-2000 fmole per well.

The validity of the immunoassay was confirmed by show-ing that plant extract dilutions give measurements that decrease in parallel with those of similarily diluted ABA standards (Fig. 3; additional data not shown). Standard addition experiments also showed that known quantities of (+)ABA, when added to appropriately diluted aliquots of plant extracts, were subsequently correctly detected (Table 1; additional data not shown). These results indicate the absence, in wheat extracts, of non-specific inhibitors of antigen-antibody binding, or of interfering materials cross-reacting with the antibodies.

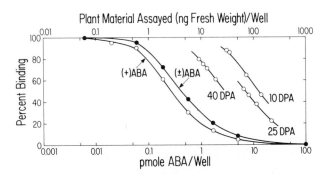

FIGURE 3. Reactions of (+)ABA, (±)ABA, and wheat extracts in the monoclonal antibody-based ELISA. Extracts are of wheat embryos at 10, 25, and 40 DPA.

TABLE 1
(+)ABA STANDARD ADDITION TO 10 DPA WHEAT EMBRYO EXTRACT
IN THE MONOCLONAL ANTIBODY-BASED ELISA

Sample	Free (+)ABA Content (pmoles/Well)		
	No Addition	+0.25 pmole (+)ABA	Difference
1	0.30	0.62	0.32
2	0.16	0.41	0.25
3	0.07	0.32	0.25
4	0.07	0.37	0.30

Levels of Free (+)ABA in Wheat Embryos.

Free (+)ABA was detected in wheat embryos as early as 10 DPA (Table 2). However, there is a substantial increase in the content of free (+)ABA by 25 DPA, with little further increase at 40 DPA. It is already known that WGA can be detected first in the radicle and coleorhiza at 10 DPA (3). Since ABA is also present at this time, perhaps in locally higher concentration, it is possible the WGA in the radicle and coleorhiza accumulates under the control of ABA as it most likely does later in the coleoptile and epiblast (3).

TABLE 2
FREE (+)ABA CONTENT IN WHEAT EMBRYOS

Embryo Age (DPA)	Free (+)ABA Content[a] (ng/Embryo ± SD)
10	0.25 ± 0.05
25	2.54 ± 0.04
40	3.07 ± 0.03

[a]Measurements are the average and standard deviations of three independent determinations using the monoclonal antibody-based ELISA.

Quantitation of Free (+)ABA in Wheat Seedlings.

It is known that exogenously applied ABA enhances the accumulation of WGA-like lectin in wheat seedlings, and that the lectin is decreased if plants are treated with Fluridone (5). We thus wished to know if Fluridone also reduces the level of endogenous ABA in wheat as it does in maize (6). Seedlings grown on nutritional medium normally have about 300 ng free (+)ABA/g fresh weight in the shoot-base region (Table 3). Treatment with 10 mg/L Fluridone reduces free (+)ABA to one-half this level. The estimates of free (+)ABA obtained with ELISA were virtually identical with those for free (±)ABA obtained with gas chromatography (Table 3). These results demonstrate a partial effect of Fluridone on free (+)ABA levels. However, the observed coincident reduction in ABA and WGA is consistent with the notion that ABA levels determine WGA accumulation in seedlings. Thus the reason the herbicide inhibits only half the normal accumulation of WGA lectin (5) may be due to its inhibiting only half the normal ABA accumulation. A similar effect of Fluridone on ABA levels, and on the level of an ABA-inducible mRNA, has also been observed in excised soybean cotyledons (16).

TABLE 3

ESTIMATION OF FREE ABA IN WHEAT SEEDLING SHOOT-BASE BY MONOCLONAL ANTIBODY-BASED ELISA AND GAS CHROMATOGRAPHY

Treatment	Free ABA Content (ng/g Fresh Weight)	
	ELISA[a]	Gas Chromatography
Untreated	297 ± 13	310
+10 mg/L Fluridone	146 ± 15	150

[a]Measurements are the average and standard deviations of two independent determinations.

SUMMARY

An ELISA using the commercially available monoclonal anti-free (+)ABA antibody has the required availability, specificity, and ease of use for routine measurements of free (+)ABA, the entantiomer of ABA which is believed to be biologically important. Using this assay, free (+)ABA is present at 10 DPA in wheat embryos at about one-tenth the level reached at 25-40 DPA. In wheat seedlings treated with Fluridone, free (+)ABA is reduced two-fold. In both systems there is a strong correlation between endogenous levels of free (+)ABA and WGA, suggesting that (+)ABA may be the endogenous regulator of lectin accumulation in wheat.

ACKNOWLEDGEMENTS

We thank Mrs. Kim Pearson and Ms. Barbara Stewart for technical assistance.

REFERENCES

1. Triplett BA, Quatrano RS (1982). Timing, localization, and control of wheat germ agglutinin synthesis in developing wheat embryos. Dev Biol 90:491-496.
2. Mishkind M, Raikhel NV, Palevitz BA, Keegstra K (1982). Immunocytochemical localization of wheat germ agglutinin. J Cell Biol 92:753-764.
3. Raikhel NV, Quatrano RS (1986). Localization of wheat germ agglutinin in developing wheat embryos and those cultured in abscisic acid. Planta, in press.
4. King RW (1976). Abscisic acid in developing wheat grains and its relationship to grain growth and maturation. Planta 132:43-51.
5. Raikhel NV, Palevitz BA, Haigler CH (1986). Abscisic acid control of lectin accumulation in wheat seedlings and callus cultures. Plant Physiol 80:167-171.
6. Fong F, Smith JD, Koehler DE (1983). Early events in maize seed development. Plant Phsiol 73:899-901.
7. Quatrano RS (1986). Regulation of gene expression by abscisic acid during angiosperm embryo development. In Miflin B (ed): "Oxford Surveys of Plant Molecular and Cell Biology," Oxford U.K.: Oxford Press, in press.
8. Walton D, Dashek W, Galson E (1979). A radioimmunoassay for abscisic acid. Planta 146:139-145.

9. Weiler EW (1980). Radioimmunoassays for the differential and direct analysis of free and conjugated abscisic acid in plant extracts. Planta 148:262-272.
10. Weiler EW (1982). An enzyme-immunoassay for *cis*-(+)-abscisic acid. Physiol Plant 54:510-514.
11. Daie JD, Wyse R (1982). Adaptation of the enzyme-linked immunosorbent assay (ELISA) to the quantitative analysis of abscisic acid. Anal Biochem 119:365-371.
12. Rosher PH, Jones HG, Hedden P (1985). Validation of a radioimmunoassay for (+)-abscisic acid in extracts of apple and sweet-pepper tissue using high-pressure liquid chromatography and combined gas chromatography-mass spectrometry. Planta 165:91-99.
13. Powell LE (1964). Preparation of indole extracts from plants for gas chromatography and spectrophotofluoro-metry. Plant Physiol 39:836-842.
14. Milborrow BR (1978). Abscisic acid. In Letham, Goodwin, Higgins (eds): "Phytohormones and Related Compounds," New York: Elsevier/North-Holland Biomedical Press, p 295-347.
15. Mertens R, Deus-Neumann B, Weiler EW (1983). Monoclonal antibodies for the detection and quantitation of the endogenous plant growth regulator, abscisic acid. FEBS Let 160:269-272.
16. Bray EA, Beachy RN (1985). Regulation by ABA of β-conglycinin expression in cultured developing soybean cotyledons. Plant Physiol 79:746-750.

Molecular Biology of Plant Growth Control, pages 209–217
© 1987 Alan R. Liss, Inc.

MONOCLONAL ANTIBODY-BASED APPROACH TO DISTINGUISH
CLASSES OF CYTOKININS[1]

David L. Brandon and Joseph W. Corse

USDA Western Regional Research Center
Albany, CA 94710

Arieh Maoz

ARO Volcani Center
Bet Dagan 50250, Israel

ABSTRACT   Derivatives of the naturally occurring cytokinins were synthesized for use as haptens in eliciting
antibodies which can differentiate between classes of
cytokinins (the purines, ribosides, ribotides, and glucosides). The 9-carboxyethyl derivative of E-zeatin was
synthesized and coupled to proteins for use as immunogens
and ligands. Monoclonal antibodies raised with either
9-carboxyethylzeatin or zeatin riboside as haptens were
analyzed, and found to fall into two patterns of binding.
In one, the purine and riboside were bound equivalently,
the side chain appearing immunodominant. In the second
pattern, the riboside was bound preferentially, and
only bulky side chains dramatically impaired binding.
High resolution immunoassays using specific antibodies
will allow analysis of the role of cytokinin conjugates
in plant senescence and responses to stress.

INTRODUCTION

Immunoassay of Cytokinins

Immunoassays offer a quick and potentially simple approach
to the measurement of biologically important compounds, and

[1]Supported, in part, by BARD grant US-711-83.

they have been applied to the quantitation of plant growth
regulators since 1969 (1). For a recent review, see (2).
Immunoassays, in addition to their speed, avoid the problems
of differential uptake of test substances and sensitivity to
physiological inhibitors.

Antibodies directed against purines and pyrimidines were
described by Erlanger and Beiser (3). Hacker et al. (4) first
prepared antibodies against a cytokinin by these procedures.
Khan et al. (5) demonstrated the use of anti-cytokinin antibodies in a radioimmunoassay, and other workers have extended
this approach with assays (6), immunocytochemical techniques
(7), and affinity purification (8,9).

The most widely used radiolabels have been tritiated ligands produced by oxidative cleavage and reduction (5) or by
exchange (8); radioiodinated ligands have also been exploited
(10). More recently, enzyme labels and monoclonal antibodies
have been used in immunoassays of cytokinins (e.g., 11,12).

Naturally Occurring Phytohormones

The transport, binding and degradation of endogenous
phytohormones, in addition to their synthesis, may be essential features of regulation by plant growth substances.
Many naturally occurring growth regulators can exist in
several forms. For example, cytokinins can exist as free
bases, ribosides, ribotides, glucosides (glucosylated in
either of two positions), and as riboside-glucoside. These
forms exist in vivo and their activities may depend on the
physiological system studied (e.g., 13)

It is possible that the translocation and degradation of
phytohormones depend on their derivatization. Thus, the
quantitation of individual forms of the cytokinins could contribute to the understanding of important physiological processes such as senescence and responses to stress. It would
be useful to distinguish between related hormones, and this
paper describes our approach to this objective.

Resolution of Immunoassays

To distinguish between closely related cytokinins, possibly no method can compare with HPLC. Immunoassay of cytokinins for quantitation has been combined with the analytical
power of HPLC to estimate individual cytokinins (14,15). Since
immunoassays are faster and less expensive than physicochemical
methods, we wish to know whether they can be used to quantitate
the individual forms of cytokinins in plant tissues.

METHODS

Chemicals

The cytokinins were synthesized (16-18) or purified from commercial products. The 9-carboxyethyl derivatives of the cytokinins (Figure 1) were synthesized by a Michael addition, essentially according to Baker and Tanna (19) and Letham et al. (13). Purity and structures were verified by HPLC and NMR spectroscopy.

Protein Conjugates

Conjugation of the cytokinin ribosides to carrier proteins was performed essentially according to Erlanger and Beiser (3). The various conjugates contained 10 to 20 molecules of cytokinin per 100,000 daltons of protein. Conjugates of 9-carboxyethyl-E-zeatin (CEZ) were prepared by derivatization of amino groups of protein carriers and enzymes.

Solid-phase Immunoassays

Assays to screen and titer sera and tissue culture fluids were performed on flexible polyvinylchloride microtitration plates, by procedures adapted from Tsu and Herzenberg (20). Assays were performed by direct binding of mouse antibodies to plates coated with cytokinin conjugates. In inhibition assays, antibodies were preincubated with cytokinins prior to assay. Mouse antibodies bound to the assay plate were visualized with horseradish peroxidase-labeled rabbit anti-mouse IgG, using 2,2'-azinobis-3-ethylbenzthiazoline as chromogenic substrate.

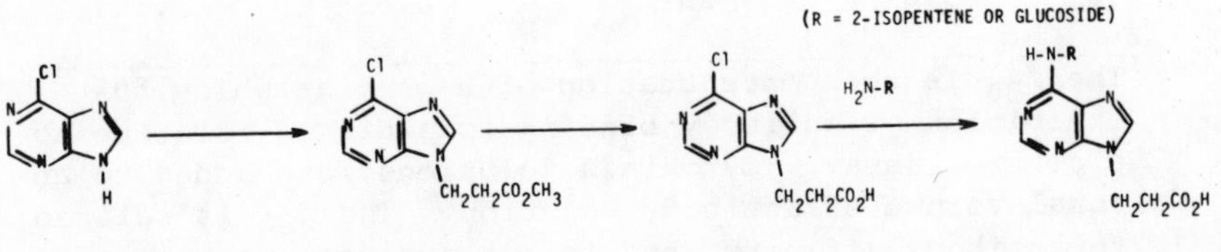

FIGURE 1.   Synthesis of 9-carboxyethyl-E-zeatin

Assays conducted in the ELISA format employed conjugates of zeatin riboside with peroxidase or alkaline phosphatase. Antibody was coated directly onto assay plates or bound via a primary coating of rabbit-anti-mouse IgG.

Antibody Production

Hybridomas were derived from Balb/c mice immunized with cytokinin conjugates. The myeloma "NS-1" was used as the fusion partner, using procedures described by Oi and Herzenberg (21)

RESULTS

Antibodies

Several clones were obtained from independent cell fusion

TABLE 1

COMPARISON OF ZR- AND CEZ-DERIVED MONOCLONAL ANTIBODIES[a]

Clone	Immunizing hapten	I_{50} (ZR) ($\underline{M}$, x 10^8)	$\dfrac{I_{50} \text{ (ZR)}}{I_{50} \text{ (Z)}}$	$\dfrac{I_{50} \text{ (ZR)}}{I_{50} \text{ (ZROG)}}$
94	ZR	7.5	2.5	.015
103	ZR	6.5	1.1	.014
98	ZR	18	2.8	.023
105	ZR	15	1.4	.016
70	CEZ	15	0.19	.017
112	CEZ	45	0.50	.009
93	CEZ	25	1.6	.010
118	CEZ	25	0.83	.007

[a]The I_{50} is the concentration of hapten at which 50% inhibition of antibody binding is achieved relative to a control assay. Cytokinin solutions were added to an equal volume of antibody solution. The I_{50} is related to antibody affinity, but is not a direct measure of affinity. (Z, zeatin; ZR, zeatin riboside; ZROG, zeatin riboside-O-glucoside)

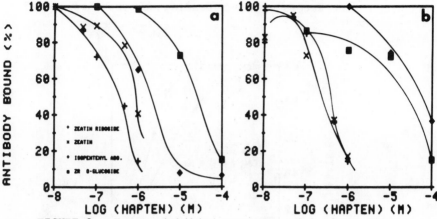

FIGURE 2.   Hapten inhibition assay.   (a) clone 112, and (b) clone 118.

experiments in which the immunizing antigens were conjugates of either E-zeatin riboside (ZR) or carboxyethylzeatin.   Table 1 summarizes the specificities as determined by inhibition of solid-phase binding (see Figure 2).

Antibodies derived from CEZ-immunized mice and antibodies derived from ZR-immunized mice generally have I_{50} values for Z and ZR within one order of magnitude of each other.   However, antibodies derived from CEZ-immunized mice (e.g., clones 112 and 118) appear less able to accommodate bulky side chains (e.g., the O-glucoside) compared to antibodies derived from ZR-immunized mice.

Other Influences on Assay Specificity

The specificity and sensitivity of immunoassays depend not only on the antibody but on the assay format   Some of the clones described above were compared in the ELISA format, which yielded I_{50} values one order of magnitude lower than the inhibition assays (Figure 3).   In addition to the format of the assay, other parameters such as ionic strength can influence the specificity of solid-phase immunoassays.   The results of using the chaotropic agent sodium thiocyanate in the assay are shown in Figure 4.   Binding of clone 93 antibody to iso-pentenyladenosine-coated wells was suppressed.   However, sup-pression of crossreactivity is not a general result with high ionic strength, detergents, and chaotropic agents.   We have found that different monoclonal antibodies vary in the influ-

ence of these factors on their binding to cytokinin-coated
assay plates.

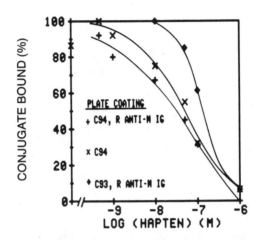

FIGURE 3.   ELISA assay.   Ligand was zeatin riboside-
ovalbumin-alkaline phosphatase conjugate, with p-nitrophenylphosphate as substrate.   Analyte was zeatin riboside.

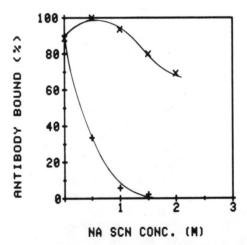

FIGURE 4.   A chaotropic agent enhances the selectivity of
binding by antibody.   The binding of clone 93 to isopentenyladenosine (+) was suppressed relative to the binding to a
CEZ-conjugate (x).   On each coating, maximal binding was
considered 100%.

DISCUSSION

We have defined at least two groups of zeatin-specific antibodies (idiotypes, in immunochemical terms). Type I, represented by clone 118, appears specific for the zeatin side chain. Type II, represented by clone 112, appears to recognize other features of the cytokinin, and only a bulky group in the side chain severely alters binding. Type I antibodies appear indifferent to substituents in the 9-position. We hypothesize that the Type I binding site is too small to accommodate such substituents. It should be noted that purines lacking the isopentenyl side chain bind with much lower affinity than the cytokinins considered in this paper. Further definition of binding patterns and their correlation with the physiological activity of different cytokinins would allow detection of an appropriate idiotype for developing anti-idiotypic antibodies as probes for cytokinin receptors (22).

The relative specificity of the antibodies for Z and ZR, measured by hapten inhibition of solid-phase antibody binding, does not correlate strongly with the linkage of the cytokinin to protein. This seems reasonable in view of the linkage in both cases through the 9-position, and the cleavage of the ribose ring in the ZR immunogen. However, attachment through the 9-carboxyethyl group will facilitate eliciting antibodies specific for zeatin-O-glucoside. We will use these reagents to study the role of cytokinin conjugates in senescence and responses to stress.

Assay specificity and sensitivity depend on factors such as assay format and conditions, in addition to the specificity of the antibody. Variations in solvent conditions (such as the use of chaotropic salts) may effectively suppress unwanted crossreactivities and should have applications in both immuno-assay and immunoaffinity chromatography.

In summary, the possibility of high-resolution immunoassay will depend on judicious selection of immunogens, extensive screening and characterization of antibodies, and modulation of specificity by choice of assay conditions.

ACKNOWLEDGMENTS

We appreciated the contributions of M. Axelrod, S. Prohaska, J. Tweedie, W. Zing, and L. Rabin to this project.

REFERENCES

1. Fuchs, S, Fuchs, Y (1969). Immunological assay for plant hormones using specific antibodies to indoleacetic acid and gibberellic acid. Biochim Biophys Acta 192:528-530.
2. Weiler, EW (1984). Immunoassay of plant growth regulators. Ann Rev Plant Physiol 35:85-95.
3. Erlanger, BF, Beiser, SM (1964). Antibodies specific for ribonucleosides and ribonucleotides and their reaction with DNA. Proc Natl Acad Sci USA 52:68-74.
4. Hacker, B, Van Vunakis, H, Levine, L (1972). Formation of an antibody with serologic specificity for N^6-(Δ^2-iso-pentenyl)adenosine. J Immunol 108:1726-1728.
5. Khan, SA, Humayun, MZ, Jacob, TM (1977). A sensitive radioimmunoassay for isopentenyladenosine. Anal Biochem 83:632-635.
6. Vold, BS, Leonard, NJ (1981). Production and characteriza-tion of antibodies and establishment of a radioimmunoassay for ribosylzeatin. Plant Physiol 67:401-403.
7. Zavala, ME, Brandon, DL (1983). Localization of a phyto-hormone using immunocytochemistry. J Cell Biol 97:1235-1239.
8. Constantinidou, HA, Steele, JA, Kozlowski, TT, Upper, CD (1978). Binding specificity and possible analytical appli-cations of the cytokinin-binding antibody, anti-N^6-benzyl-adenosine. Plant Physiol 62:968-974.
9. MacDonald, EMS, Morris, RO (1985). Isolation of cytokinins by immunoaffinity chromatography and analysis by high-per-formance liquid chromatography radioimmunoassay. Meth. Enzymol. 110:347-358.
10. Weiler, EW (1980). Radioimmunoassays for trans-zeatin and related cytokinins. Planta 149:155-162.
11. Brandon, DL (1984). Monoclonal antibodies for phytohormone research. In Stern, NJ, Gamble, NR (eds): "Hybridoma Tech-nology in Agriculural and Veterinary Research," Totowa, NJ: Rowman and Allanheld, pp 246-250.
12. Hansen, CE, Wenzler, H, Meins, F, jr (1984). Concentration gradients of trans-zeatin riboside and trans-zeatin in the maize stem. Measurement by a specific enzyme immunoassay. Plant Physiol 75:959-963.
13. Letham, DS, Palni, LMS, Tao, G-Q, Gollnow, BI, and Bates, CM (1983). Regulators of cell division in plant tissues XXIX. The activities of cytokinin glucosides and alanine conjugates in cytokinin bioassays. J Plant Growth Regul 2:103-115.

14. MacDonald, EMS, Akiyoshi, DE, Morris, RO (1981). Combined high-performance liquid chromatography-radioimmunoassay for cytokinins. J Chromatog 214:101-109.
15. Regier, DA, Morris, RO (1982) Secretion of trans-zeatin by Agrobacterium tumefaciens: a function determined by the nopaline TI plasmid. Biochem Biophys Res Commun 104: 1560-1566.
16. Hall, RH, Robins, MJ (1968). N-(3-methyl-2-butenyl)adenosine. In Zorbach, WW, Tipson, RS (eds): "Synthetic Procedures in Nucleic Acid Chemistry," vol 1, New York: Wiley-Interscience, pp 210-211.
17. Corse, JW, Kuhnle, J (1972). An improved synthesis of trans-zeatin. Synthesis 1972:618-619.
18. Corse, J, Gaffield, W, Lundin, RE (1983) Dihydrozeatin: an improved synthesis and resolution of both isomers. J Plant Growth Regul 2:47-57.
19. Baker, BR, Tanna, PM (1965) A Michael addition with 6-chloropurine. J Am Chem Soc 87:2857-2858.
20. Tsu, TT, Herzenberg, LA (1979) Solid-phase radioimmune assays. In Mishell, BB, Shiigi, SM (eds): "Selected Methods in Cellular Immunology," San Francisco: WH Freeman, pp 373-397.
21. Oi, VT, Herzenberg, LA (1979) Immunoglobulin-producing hybrid cell lines. In Mishell, BB, Shiigi, SM (eds): "Selected Methods in Cellular Immunology," San Francisco: WH Freeman, pp 351-372.
22. Schreiber, AB, Couraud, PO, Andre, C, Vray, B, Strosberg, AD (1980). Anti-alprenolol anti-idiotypic antibodies bind to beta-adrenergic receptors and modulate catecholamine-sensitive adenylate cyclase. Proc Natl Acad Sci USA 77:7385-7389.

Molecular Biology of Plant Growth Control, pages 219–228
© **1987 Alan R. Liss, Inc.**

AUXIN-BINDING PROTEINS IN MAIZE:
PURIFICATION AND RECEPTOR FUNCTION

Michael A. Venis

East Malling Research Station, Maidstone, Kent ME19 6BJ
England

ABSTRACT   High affinity binding sites for auxins are
found in membrane preparations from maize coleoptiles.
The sites can be solubilised without detergent and
purified by various means, including FPLC. The native
binding proteins are 40–45 kDa in size. Affinity
labelling studies have been carried out both in the
membrane-bound and solubilised states, using
radioactive and non-radioactive probes. Attempts are
currently being made to raise antibodies to the binding
proteins. The various lines of evidence pointing to an
auxin receptor function for the maize membrane binding
sites are described and discussed.

INTRODUCTION

Auxins can elicit very rapid responses (within minutes)
as well as longer-term responses requiring new macromolecule
synthesis. Research has therefore been directed both to
membrane-bound receptors presumed to be involved in the fast
responses and to soluble receptors that might mediate
transcriptional changes. Membranous binding sites for auxins
have been best studied in maize coleoptiles, a system
pioneered by Hertel and coworkers (1) and subsequently
studied in several laboratories, including our own. This
report summarises recent work on these sites in the
following areas:-   1. methods of purification, 2. molecular
size and other properties,   3. affinity labelling,   4.
evidence of receptor function.

METHODS

Membrane Preparation and Binding Assays

Membrane fractions were prepared from coleoptiles of
4-5 day maize seedlings (cvs. Kelvedon 33 or Beaupré) and
binding determined by a centrifugation assay (2). Binding
proteins were solubilised from the membranes using an
acetone procedure (3) and assayed by equilibrium dialysis
(3) or by centrifugation to completion through YMT
ultrafilter membranes in MPS-1 units (Amicon). Naphthaleneacetic acid (NAA) $-^{14}C$ (Amersham, 61 mCi/mmol) was used as
the radioactive ligand throughout in binding assays.

Purification

Solubilised binding proteins were partly purified
either by conventional column procedures (3, 4) or by Fast
Protein Liquid Chromatography (FPLC) using a Pharmacia
system and anion exchange (Mono Q) or gel filtration
(Superose 12) chromatography.

Electrophoresis

Fractions were separated on 12% SDS gels (5) and either
stained with Coomassie Brilliant Blue or silver, or else
blotted onto a 0.22 μ nitrocellulose membrane (100-120 ma,
16 h at 4°C). Blots were stained for total protein by
iodine-starch (6) or for glycoproteins by periodate
oxidation and alkaline phosphatase hydrazide coupling (7).

Affinity labelling

1. Solubilised preparations were purified on DEAE Bio
Gel and Sephadex G100 (3) and photolysed with 1.2μM (ca.
4μCi) 5-azido-IAA-^{3}H (8) by Dr. A.M. Jones in the presence
or absence of 0.1 μM IAA or NAA. They were then dialysed
overnight against distilled water, lyophilized and run on
12% SDS gels. Labelled polypeptides were visualised
fluorographically after impregnation of gels with Amplify
(Amersham) and drying.

2. Similar preparations were coupled with $30\,\mu M$
(ca. $25\,\mu$ Ci) diazotised Chloramben $-^3H$ (2.5-dichloro-3-
aminobenzoic acid, tritiated at Amersham by catalytic
exchange, purified, and used at 1.5 Ci/mmol). Coupling was
carried out with or without 0.2 mM NAA, IAA or benzoic acid,
at pH 5.6, 25°C for 15 min (9), followed by dialysis,
lyophilization and fluorography as above.

RESULTS

Purification methods

These have been facilitated by an effective
solubilisation technique for the binding sites that avoids
the use of detergent (3). Original chromatographic
procedures with standard citrate-acetate binding buffer
appeared to give effective small-scale purification (3, 4).
Problems were subsequently encountered on changing to
different seed batches, in that erratic and often large
losses were encountered at the DEAE step. These were
circumvented by changing from an anionic to a zwitterionic
buffer (MES) under conditions such that active eluates could
be taken directly from column to column without intermediate
dialysis or concentration. In this way, over 90% of crude
extract activity could be retained after three column steps.
On an analytical scale, it was possible to resolve two
peaks of binding activity either by Sephadex G15 isoelectric
focusing (IEF) or by shallow gradient elution of anion
exchange columns (4). However, IEF over the pH range
required for resolution (pH 3.5-5) gave only 10-15% recovery
of activity, and the gradient fractionation proved
inconsistent, perhaps because the anionic buffer system had
been used. More recently we have been exploring FPLC for
purification. When the post-DEAE binding fraction was
applied to a Mono Q column in MES buffer, elution in a
linear NaCl gradient (0-0.35 M) produced two peaks of auxin-
binding activity. The nature of the early part of the A_{280}
elution profile suggested that protein aggregation was
occurring, and to avoid this it was found necessary to
introduce a preceding FPLC gel filtration step. Even then,
the resolved binding peaks failed to rechromatograph in
their original positions. Consistent FPLC profiles on Mono Q
have now been achieved by changing from MES to a cationic

buffer, either histidine or piperazine at pH 5.5. These
reveal a single major peak of auxin binding activity, with
perhaps a low activity area eluting close by, and this major
peak re-runs in the same position as a single discrete A_{280}
peak.

Molecular Properties

The apparent molecular weight of the native solubilised
binding proteins (by gel permeation) was originally reported
to be in the 40-45 kDa range (3), but Cross and Briggs (10)
subsequently claimed a molecular size of 80 kDa, with
aggregation to 200 kDa in the absence of 0.1 M NaCl.
However, all later investigations (4, 11, 12) have confirmed
the original assignment in the lower size range irrespective
of the presence or absence of protease inhibitors, as have
our recent high resolution separations on Superose 12. The
findings of Cross and Briggs are therefore unique, and it
seems likely that for some reason they obtained fractions
that were particularly prone to aggregation.

In addition, maize mesocotyl membranes appear to
contain somewhat similar auxin-binding proteins.
Interestingly, these would appear to be in a different
membrane environment, since they cannot be extracted by the
acetone method. They can, however, be solubilised with
Triton X-100 and exhibit similar ion exchange behaviour and
molecular size to detergent-solubilised binding proteins
from coleoptiles, again in the 40-45 kDa range (Fig. 1).

From native and SDS gel electrophoresis of purified
fractions solubilised from coleoptile membranes we had
considered that the 40-45 kDa binding protein(s) were
monomeric (4). However, the report by Löbler and Klämbt
(13) of a single band at 20 kDa on SDS gels after
immunoaffinity purification of maize auxin-binding proteins
led us to re-examine this question using FPLC-purified
fractions. In partial agreement with these authors we now
find that on SDS gels, band enrichment does indeed occur in
the lower molecular weight region. We consistently find a
major polypeptide at 22 kDa, often associated with a second
band at 21 kDa. A dimeric structure therefore seems
probable, though it is not yet clear whether there is more
than one native binding species of slightly different
molecular size. The discrepancy between our current and
earlier findings is not clear.

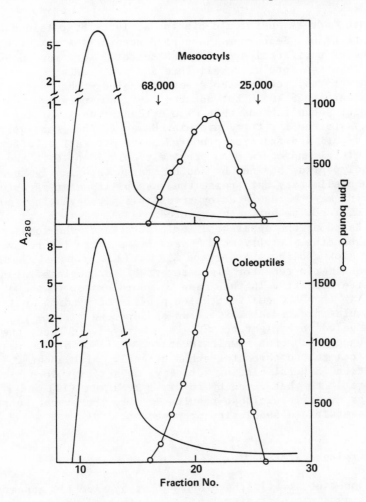

FIGURE 1. Sephadex G100 elution profiles of auxin-
binding proteins solubilised with Triton X-100 from either
coleoptile or mesocotyl membranes of maize.

The tendency to aggregation seen on FPLC ion exchange
is particularly characteristic of glycoproteins, though
inclusion of zwitterions or glycerol in buffers (14) did not

improve matters, nor could binding activity be retained on
concanavalin A-Sepharose. When binding proteins at different
stages of purification (using MES buffers) were run on SDS
gels, blotted onto nitrocellulose and stained for
glycoproteins, several bands were revealed – the most
prominent at 29 kDa – but not the 21-22 kDa putative
receptor subunit bands that were evidenced on adjoining
tracks stained for total protein. However, in a subsequent
highly purified fraction prepared using piperazine buffers,
a single glycoprotein band precisely coincident with the 22
kDa polypeptide stained by iodine-starch was clearly seen.
These preliminary data raise the possibility that receptor
activity may be regulated by reversible glycosylation, and
this will now be actively explored.

Based on the apparent dimeric structure, the loss of
auxin binding activity at 25°C and its substantial recovery
after slow cooling, Löbler and Klämbt (13) proposed that an
active dimeric receptor might reversibly dissociate into
inactive subunits. In this case we would expect to see a
shift in the FPLC gel filtration profile of binding activity
in samples treated at 25°C , run on Superose 12,
'reannealed' overnight to 4°C, then assayed. In fact, the
binding profile of a sample incubated at 25°C for 8 h was
identical to one chromatographed directly, with about 20%
reduction in total binding activity, an activity loss
comparable to that noted by Löbler and Klämbt (13) under
similar conditions. These results do not, therefore, support
the reversible dissociation proposed.

Affinity Labelling

Jones et al. (15) have shown that 5-azido-IAA appears
to act as a photoaffinity label of auxin binding sites in
maize membranes and that irreversible attachment of the
tritiated compound is reduced by IAA though not by NAA (8).
Since the NAA-binding sites are in fact inactivated by the
unlabelled compound (15), labelling of the binding proteins
by 5-azido-IAA-^{3}H must presumably represent only a small
proportion of the total. We reasoned that selectivity of
labelling might be improved in solubilised and partly
purified fractions and have carried out some collaborative
work with Dr. Alan Jones to this end. Binding fractions
purified on DEAE and Sephadex G100 were prepared,
lyophilised and despatched to Dr. Jones for photolabelling.

The re-lyophilised samples were then returned and labelling patterns compared on SDS gels from control and 'protected' irradiations. These showed that: 1. labelling was selective in that intensity was not proportional to polypeptide abundance, 2. the putative receptor subunits, (21-22 kDa) were labelled, but only faintly, 3. a polypeptide at 26 kDa was most heavily labelled, 4. IAA, but not NAA, clearly afforded protection against azido-IAA photolabelling, agreeing therefore with results obtained in the membrane-band state (8), where it was suggested that this might reflect preferential labelling of proteins concerned with auxin transport. Further collaborative experiments are planned.

From earlier studies with diazotised auxin analogues, diazo-Chloramben was found to act as an affinity label for the membrane-bound sites (9) and subsequently, tritiated material was obtained. Experiments using azo-coupling conditions, but otherwise similar to those described above, have shown that whereas the detailed labelling pattern differs from that obtained with azido-IAA, once again the receptor subunit polypeptides are only lightly labelled and clear protection is afforded by IAA, but not by NAA.

Physiological Relevance

As well as having appropriate binding affinities and selectivity between auxin analogues that is generally consistent with relative physiological activities (16), several pieces of correlative evidence support a receptor function for the membranous binding sites:
1. Red light reduces auxin responsiveness and auxin binding site concentration in parallel in maize mesocotyls. Auxin binding and the elongation response also decrease down the length of the mesocotyl (17).
2. As maize coleoptiles increase in size, auxin responsiveness and binding both reduce in concert (18).
3. Apical sections of oat coleoptiles are far more auxin-sensitive and show higher auxin binding than basal sections (18).
4. In maize coleoptiles the situation is reversed, i.e. basal sections are, somewhat surprisingly, more sensitive, and also have greater auxin binding than apical sections (18).

5.   The effectiveness with which naturally-occurring
benzoxazinones and their derivatives inhibit auxin
binding to maize membranes is in accord with their
ability to inhibit auxin-induced growth (19).

In addition, there are now two more direct pieces of
evidence of a receptor role for the binding sites. First,
maize membrane auxin-binding proteins (containing associated
ATPase) have been incorporated into a synthetic bilayer
lipid membrane, across which a fixed potential was applied.
Addition of auxin followed by ATP produced an immediate rise
in transmembrane current (20). All three components
(protein, auxin, ATP) were required for the response to
occur, but the order of addition was not critical. The pH
dependence of the response, together with the known pH
dependence of auxin binding, was consistent with the
operation of a proton-translocating ATPase stimulated by
auxin plus receptor, thus providing direct support for the
proton pump hypothesis of auxin action (21).

The other direct evidence comes from Löbler and Klämbt
(13), who prepared an antiserum using a heterogeneous
antigen containing maize membrane auxin-binding proteins
(ABP). By a circuitous route involving affinity and
immunoaffinity chromatography, an apparently monospecific
anti-ABP antiserum was derived and shown to be capable of
blocking auxin-induced growth responses (23). As well as
reinforcing the suggestion that these binding proteins are
receptors, these results indicate that auxin receptors
involved in cell expansion are located at the outer face of
the plasma membrane. This does not of course mean that
similar receptors, perhaps mediating other auxin responses,
do not also occur intracellularly - indeed there is general
agreement that there are auxin binding sites located on
endomembranes (16).

DISCUSSION

Auxin-binding proteins in maize membranes have been
characterised better than any other plant hormone binding
system, and considerable direct and indirect evidence exists
for their receptor function. Improvements in purification
procedures and the development of immunochemical approaches
should help to clarify a number of unresolved questions such
as the number of discrete binding species, their cellular
localisation and relationship to auxin binding proteins in

other species, both membrane-bound and soluble, and the nature of stimulus-response coupling. We have begun attempts to produce monoclonal antibodies using partly-purified antigen preparations, and with the aid of FPLC we hope to be able to obtain enough homogeneous antigen to raise polyclonal antisera also. To do this we will need to improve on existing low recoveries which may be associated with variable glycosylation. This aspect, and the possibility of receptor phosphorylation are currently being examined. In addition, we have confirmed the observation of soluble auxin binding proteins in maize (11) and will endeavour to establish their relationship, if any, to the membrane-bound species.

ACKNOWLEDGEMENTS

The azido-IAA experiments were carried out in collaboration with Dr. Alan Jones (University of Wisconsin). I am grateful to Mike Bolton for valuable technical assistance.

REFERENCES

1.   Hertel R, Thomson K-St, Russo VEA (1972). In vitro auxin binding to particulate cell fractions from corn coleoptiles. Planta 107:325.
2.   Batt S, Wilkins MB, Venis MA (1976). Auxin binding to corn coleoptile membranes: kinetics and specificity. Planta 130:7.
3.   Venis MA (1977). Solubilisation and partial purification of auxin-binding sites of corn membranes. Nature (London) 66:268.
4.   Venis MA (1980). Purification and properties of membrane-bound auxin receptors in corn. In Skoog F(ed): "Plant Growth Substances 1979". Berlin - Heidelberg - New York: Springer-Verlag: 61.
5.   Laemmli UK (1970). Cleavage of structural proteins during the assembly of the head of bacteriophage T4. Nature (London) 227:680.
6.   Kumar BV, Lakshmi MJ, Atkinson JP (1985). Fast and efficient method for detection and estimation of proteins. Biochem Biophys Res Commun 131:883.

7. Gershoni JM, Bayer EA, Wilchek M (1985). Blot analyses of glycoconjugates: enzyme hydrazide – a novel reagent for the detection of aldehydes. Anal Biochem 146:59.

8. Jones AM, Melhado LL, Ho THD, Pearce CJ, Leonard NJ (1984). Azido auxins: photaffinity labelling of auxin-binding proteins in maize coleoptile with tritiated 5-azidoindole-3-acetic acid. Plant Physiol 75:1111.

9. Venis MA (1977). Affinity labels for auxin binding sites in corn coleoptile membranes. Planta 134:145.

10. Cross JW, Briggs WR (1978). Properties of a solubilized auxin-binding protein from coleoptiles and primary leaves of Zea mays. Plant Physiol 62:152.

11. Murphy GJP (1980). Naphthaleneacetic acid binding by membrane-free preparations of cytosol from the maize coleoptile. Plant Sci Lett 19:157.

12. Tappeser B, Wellnitz D, Klämbt D (1981). Auxin affinity proteins prepared by affinity chromatography. Z Pflanzenphysiol 101:295.

13. Löbler M, Klämbt D (1985). Auxin-binding protein from coleoptile membranes of corn. I. Purification by immunological methods. J Biol Chem 260:9848.

14. Pharmacia booklet (1985). FPLC ion exchange and chromatofocusing.

15. Jones AM, Melhado, Ho THD, Leonard NJ (1984). Azido auxins. Quantitative binding data in maize. Plant Physiol 74:295.

16. Venis MA (1985). "Hormone Binding Sites in Plants." New York – London: Longman.

17. Walton JD, Ray PM (1981). Evidence for receptor function of auxin binding sites in maize. Plant Physiol 68:1334.

18. Kearns AW (1982). The search for the auxin receptor. D Phil thesis, University of York.

19. Venis MA, Watson PJ (1978). Naturally occurring modifiers of auxin-receptor interaction in corn: identification as benzoxazolinones. Planta 142:103.

20. Thompson M, Krull UL, Venis MA (1983). A chemoreceptive bilayer lipid membrane based on an auxin-receptor ATPase electrogenic pump. Biochem Biophys Res Commun 110:300.

21. Hager A, Menzel H, Krauss A (1971). Versuche und Hypothese zur Primärwirkung des Auxin beim Streckungswachstum. Planta 100:47.

22. Löbler M, Klämbt D (1985). Auxin-binding protein from coleoptile membranes of corn. II. Localisation of a putative auxin receptor. J Biol Chem 260:9854.

Molecular Biology of Plant Growth Control, pages 229–243
© 1987 Alan R. Liss, Inc.

CHARACTERIZATION AND FUNCTION ANALYSIS OF A
HIGH-AFFINITY CYTOPLASMIC AUXIN-BINDING PROTEIN

K.R. Libbenga, H.J. van Telgen, A.M. Mennes,
P.C.G. van der Linde,[1] and E.J. van der Zaal

Department of Plant Molecular Biology, University of Leiden
2311 VJ-Leiden, The Netherlands

ABSTRACT   Cytosol and high-salt nuclear extracts from
tobacco cells contain a high-affinity auxin-binding
protein ($Ka,IAA:10^8-10^9M^{-1}$). This protein could be
purified at a small scale by using an affinity matrix
prepared by coupling 5-hydroxy-indole-3-acetic acid to
Epoxy-activated Sepharose-6B. The protein profile of
purified receptor preparations indicates that the
binding protein consists of slightly different subunits
of ca 50,000 D. A protein fraction of about this molecular weight present in the purified preparations could
be phosphorylated in vitro, suggesting that subunits
might be phosphoproteins. Addition of auxin to auxin-
-starved stationary-phase cells in batch culture brought
about a rapid increase in the level of binding proteins
in the nuclei. This was accompanied by activation of
overall nuclear transcription and an increase in at
least three translatable mRNA-species. Overall transcription in isolated nuclei could be stimulated by purified preparations of the binding protein in the presence
of indole-3-acetic acid.

INTRODUCTION

A growing body of evidence suggests that auxin regulates
the expression of specific genes at the transcriptional level
(1 and references therein). To this date, however, very
little is understood of the mechanism by which plant cells
detect and transduce hormonal signals (2). It seems very
likely that plant cells too are equipped with receptor pro-

[1]Present address: Bulb Research Centre, 2160 AB-Lisse

teins that act as primary detectors and transducers of hormonal signals. In case of auxin, which is rapidly taken up by
cells, these receptors might be regulatory proteins which are
directly involved in auxin-regulated gene expression. In
fact, such a principle of detection and transduction has already been demonstrated for animal steroid hormones (3). In
order to explore this possibility we started with an investigation aimed at the identification of specific high-affinity auxin-binding proteins, i.e. putative receptors, in
(cultured) cells and tissues from Nicotiana tabacum. We were
able to identify two distinct classes of auxin-binding proteins: a binding protein present in cytosols and in high-salt
extracts from isolated nuclei (4,5), and a binding protein
present in microsome fractions (6,7). In addition to these
auxin-binding proteins, the cells contain a membrane-bound
protein with a high affinity towards the auxin-transport inhibitor Naphthylphthalamic acid (NPA). This protein is presumably an auxin-transport carrier (8).

We assume that the first-mentioned auxin-binding protein
(henceforth tentatively called cytoplasmic receptor) is the
most likely candidate to have a receptor function in auxin-
regulated gene expression. Some of the properties of this
binding protein have been summarized in table 1.

TABLE 1
PROPERTIES OF THE CYTOPLASMIC AUXIN RECEPTOR

$K_d(M)$ (IAA)	pH opt.	MW^a (KD)	Temp. of binding assay	Max. conc. in pmol mg^{-1} protein	Occur-rence in cell	Present in
$10^{-9} - 10^{-8}$	7.5	150-200	26°C	0.2	Cyto-plasm Nucleus	Callus Cell suspen-sions Shoot tips

[a]Determined by filtration over Sephadex-G200.

In subsequent studies we could demonstrate that partially purified receptor preparations are able to significantly

enhance overall transcription in isolated nuclei in an auxin-
-dependent way (5). This could be confirmed by Malcolm
Elliott's group at Leicester Polytechnic (9).

The purification and further characterization of this
auxin receptor were hampered by considerable variations of
receptor levels that were often even below the detection
level of the binding assay. Finally we discovered that
apparent receptor levels in crude preparations could be substantially increased under conditions which favour ATP-dependent protein kinase activity and which (partly) protect phosphoproteins against alkaline phosphatase activity in the
extracts (10). This suggests that the receptors might be
liable to enzymatic phosphorylation and dephosphorylation,
transforming them into high- or low-affinity form respectively. Similar affinity modulation has been described for some
steroid receptors (11). This emerging knowledge of (in vitro)
receptor modification has made possible further progress in
our laboratory. Most results which will be described below
are from current experiments and thus are more or less preliminary. Nevertheless, they will show the progress we are
making in the further characterization and function analysis
of the cytoplasmic auxin receptor.

METHODS

Tobacco plants (Nicotiana tabacum L.var. White Burley)
were grown from surface-sterilized seeds in a climate room at
25°C, 70-80% relative humidity and 16 h of illumination
(Philips TLF/65 W/55) a day. An established cell line derived
from tobacco leaves was grown in batch culture as described
in (7). This cell line requires 2,4-dichlorophenoxyacetic
acid (2,4-D) as growth factor.

About 4 mm-long shoot tips were excised from 8-weeks-old
non-flowering tobacco plants. All leaves over 3 mm long were
removed. Trimmed tips from 30 plants were grinded in a mortar
in 20 ml of homogenization buffer (50 mM Tris/HCl pH 7.5,
0.75 M KCl, 2 mM Na_2 EDTA, 2 mM dithiothreitol (DTT)). The
brei was centrifuged at 140,000 x g for 1 h. The supernatant
was concentrated to ca 10 ml in a Amicon ultrafiltration cell
(PM-10 membrane) at 2.75 bar pressure. After adding 100 ml of
binding buffer (see Fig. 2), the mixture was concentrated to
ca 10-20 ml. This receptor preparation was either used directly in the binding assay or was first purified.

Nuclei from cultured cells were isolated essentially according to the method described in (12). The isolated nuclei

were processed for isolation of receptors as shown in Fig. 2.
Auxin-Binding activity was determined as described in (4),
except that in a number of experiments a rapid filtration
method (13) was used instead of dextran-coated charcoal to
separate bound from free ligand. Moreover, 5 mM Mg^{2+} ATP and
5 mM p-Nitrophenylphosphate (PNPP) were routinely added to
protect the receptor against inactivation. Binding parameters
were computed from Scatchard plots by means of non-linear
regression analysis.

    Crude receptor preparations were purified using a
ligand-affinity matrix prepared by coupling 5-hydroxyindole-3
-acetic acid to epoxy-activated Sepharose-6B (Fig. 4). A few
preparations obtained by affinity chromatography were analysed by SDS-Polyacryl amide gelelectrophoresis as described
in (14).

    Overall transcription in isolated nuclei from cultured
cells was determined as described in (5).

    RNA was isolated from cultured cells by a LiCl method as
described in (15). Total RNA was translated in a rabbit reticulocyte lysate system (New England Nuclear) according to the
manufacturer's instructions. Isoelectrophocussing of translation products was performed according to O'Farrell and
O'Farrell (16). Protein samples containing approximately
800,000 CPM of incorporated ^{35}S-methionine were applied per
gel. The IEF gels were subjected to discontinuous SDS-gel-
electrophoresis according to Laemmli (14). After electrophoresis the gels were processed for autoradiography.

    For ^{32}P-labelling of receptors samples of 0.5 ml were
taken from crude receptor preparations from shoot tips. The
samples were incubated with 10 mM PNPP and 100-500 µCi
[γ-[32]P-] ATP, while circulating over the affinity column for
15 min at room temperature. After subsequent purification
steps (Fig. 4), the receptor preparations were loaded on a
SDS-polyacrylamide rod gel. After electrophoresis the gels
were sliced into 2 mm-long segments which were counted for
radioactivity.

RESULTS

    A number of experiments were performed in order to de-
sign an appropriate experimental system to be used for further
investigations on the detection and transduction mechanism
involved in auxin-controlled cell-division activity. For that
purpose we selected a well established batch-cultured cell
line from tobacco which requires only 2,4-D as growth factor.

The cells are equipped with the auxin-transport carrier and the cytoplasmic auxin receptor, but they apparently miss the membrane-bound auxin-binding protein (7). A typical growth curve of this culture is shown in Fig. 1.

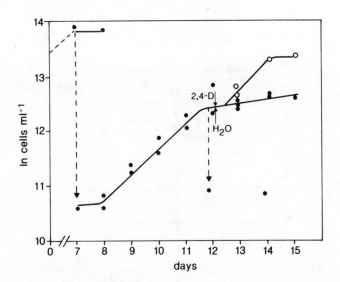

FIGURE 1.   Growth of tobacco cells after transfer to 2,4-D-free medium.

For experimental purposes the cells were depleted from auxin by subculturing inocula from early stationary-phase cells in 2,4-D lacking medium. In this medium the cells grow to a certain density and cell division can be restored by adding 2,4-D (final conc. 4.4 µM) to the culture medium. After this treatment the culture reaches a new stationary--phase when another factor, probably sucrose, becomes limiting. This cell culture provides us with a rather well defined and handsome auxin-target system: auxin-starved early stationary-phase cells (henceforth called stationary-phase cells) respond relatively rapidly to auxin with cell division; the only experimental intervention is injection of a small amount (0.3 ml) of a concentrated 2,4-D-solution into the culture medium or injection of the same amount of water as control; prior to the onset of cell division there is a lag time of ca 10 h, which enables us to extend early response times over time intervals of minutes and hours.

The first experiments with this system were aimed at the determination of receptor levels in nuclei from auxin--activated stationary-phase cells. A flow scheme of the experimental procedure is shown in Fig. 2.

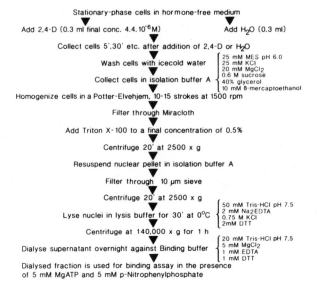

Flow scheme for the determination of receptor levels in nuclei from tobacco cells.

Stationary-phase cells in hormone-free medium

Add 2,4-D (0.3 ml final conc. 4.4.10⁻⁶M)   Add H₂O (0.3 ml)

Collect cells 5',30' etc. after addition of 2,4-D or H₂O

Wash cells with icecold water

Collect cells in isolation buffer A

$\left\{\begin{array}{l}\text{25 mM MES pH 6.0}\\\text{25 mM KCl}\\\text{20 mM MgCl}_2\\\text{0.6 M sucrose}\\\text{40\% glycerol}\\\text{10 mM ß-mercaptoethanol}\end{array}\right.$

Homogenize cells in a Potter-Elvehjem, 10-15 strokes at 1500 rpm

Filter through Miracloth

Add Triton X-100 to a final concentration of 0.5%

Centrifuge 20' at 2500 x g

Resuspend nuclear pellet in isolation buffer A

Filter through 10 µm sieve

Centrifuge 20' at 2500 x g

Lyse nuclei in lysis buffer for 30' at 0°C $\left\{\begin{array}{l}\text{50 mM Tris-HCl pH 7.5}\\\text{2 mM Na}_2\text{EDTA}\\\text{0.75 M KCl}\\\text{2mM DTT}\end{array}\right.$

Centrifuge at 140,000 x g for 1 h

Dialyse supernatant overnight against Binding buffer $\left\{\begin{array}{l}\text{20 mM Tris-HCl pH 7.5}\\\text{5 mM MgCl}_2\\\text{1 mM EDTA}\\\text{1 mM DTT}\end{array}\right.$

Dialysed fraction is used for binding assay in the presence of 5 mM MgATP and 5 mM p-Nitrophenylphosphate

FIGURE 2.   Determination of receptor levels in nuclei from cultured tobacco cells.

We first determined nuclear-receptor levels at hours after the addition of 2,4-D. We then found that receptor levels in nuclei from 2,4-D treated cells were substantially higher than those in nuclei from the water control. An example is shown by table 2.

TABLE 2

NUCLEAR RECEPTOR LEVELS IN STATIONARY-PHASE CELLS
AT 90 MIN AFTER ADDITION OF EITHER WATER OR 2,4-D

Treatment	Affinity constant (M^{-1}) towards IAA (26°C)	Concentration of binding sites (pmol mg^{-1} protein)
H_2O	8.5×10^8	0.09
2,4-D	1.5×10^9	0.20

The next trials were aimed at the determination of nuclear receptor levels at even shorter response times. To our surprise, we reproducibly found that receptor levels were already significantly increased over the water control at ca 5 min after the addition of 2,4-D (table 3).

TABLE 3

SPECIFIC BINDING OF IAA IN NUCLEAR-RECEPTOR PREPARATIONS AT
5 MIN AFTER ADDITION OF EITHER WATER OR 2,4-D TO
STATIONARY-PHASE CELLS

Treatment	DPM ^{3}H-IAA bound per mg of protein	pmoles ^{3}H-IAA bound per mg of protein
H_2O	3400	0.052
2,4-D	7800	0.118

The affinity constants for IAA varied between 10^9 and 10^8 M^{-1} and they showed a tendency to be higher in preparations from 2,4-D treated cells than in those from water controls. Experiments are in progress to analyse this phenomenon more rigorously.

The next question was whether the increase in nuclear

receptor levels is 2,4-D specific or that it is a general auxin effect. In order to examine this question nuclear--receptor levels were determined in parallel trials at 30 min after the addition of a few auxin analogues. It turned out that 2,4-D, 1-NAA and IAA were effective, whereas 2-NAA and NPA had no significant effect.

TABLE 4

EFFECT OF AUXIN ANALOGUES ON NUCLEAR-RECEPTOR
LEVELS IN STATIONARY-PHASE CELLS

Treatment (Final conc. of analogue in medium is 4.4 µM)	Receptor levels relative to the level in 2,4-D treated cells at 30 min after addition of analogue
2,4-D	100%
1-NAA	136%
2-NAA	0%
IAA	67%
NPA	6%

Among the auxins, IAA appeared to be the least effective. This might be due to the fact that this auxin is rapidly metabolized.

From earlier experiments we know that auxin-occupied receptors significantly stimulate overall transcription in isolated nuclei (5). Since the present results indicate that nuclear receptor levels rapidly increase after the addition of auxin to stationary-phase cells, we wondered whether this is accompanied by an increase in overall nuclear transcription.

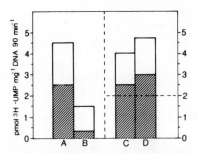

FIGURE 3.   Overall transcription in isolated nuclei from
cultured tobacco cells. A: log-phase; B: stationary-phase;
C and D: stationary-phase at 20 min and 120 min after
addition of 2,4-D respectively. Horizontal dashed line shows
level of transcription in water controls. Hatched areas give
RNA-polymerase-I activity; rest is RNA-polymerase-II activity.

Fig. 3 shows that the overall in vitro transcription in
nuclei isolated from stationary-phase cells was much lower
than in nuclei from log-phase cells. Addition of 2,4-D to the
stationary-phase cells brought about a rapid increase in both
RNA-polymerase-I and polymerase-II activity. Within 20 min
(the shortest time we have measured thus far), transcription
had reached a level which was similar to that in nuclei from
log-phase cells.
As a first attempt to examine the possibility of more
specific effects of auxin on the expression of genes, we
isolated RNA from auxin-activated stationary-phase cells.
This RNA was translated in vitro. We found that within 60 to
120 min after the addition of 2,4-D three clear protein spots
could be detected in autoradiographs of 2-D gels of the
translation products which had substantially increased in
intensity. These spots, which could hardly be detected in the
water controls, represent proteins with Mr 35,000, 24,500 and
27,800 and pI of 7.1, 6.3 and 5.6 respectively. Experiments
are in progress to construct cDNA clones from these mRNA
species. These clones will be used in more detailed studies
on the role of auxin and its cytoplasmic receptor in specific
gene expression in our experimental system. Of course, such
studies require purified receptors as well. In our laboratory
we are now trying to purify crude receptor extracts by affinity chromatography. The affinity matrix consists of

5-hydroxyindole-3-aceticacid coupled to Epoxy-activated
Sepharose-6B. A flow scheme of the purification procedure is
shown in Fig. 4.

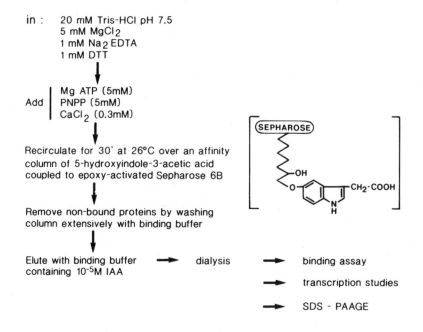

**Dialized nuclear extract, or concentrated receptor preparation
from shoot tips,**

in :   20 mM Tris-HCl pH 7.5
       5 mM MgCl2
       1 mM Na2 EDTA
       1 mM DTT

Add | Mg ATP (5mM)
    | PNPP (5mM)
    | CaCl2 (0.3mM)

Recirculate for 30' at 26°C over an affinity
column of 5-hydroxyindole-3-acetic acid
coupled to epoxy-activated Sepharose 6B

Remove non-bound proteins by washing
column extensively with binding buffer

Elute with binding buffer ➝ dialysis ➝ binding assay
containing 10⁻⁵M IAA
                              ➝ transcription studies
                              ➝ SDS - PAAGE

FIGURE 4.   Purification of receptor preparations by
affinity chromatography.

Preparations obtained by this procedure exhibited high-
-affinity IAA binding with Ka = 10^8-10^9 M^{-1}. The overall
protein concentration in the purified preparations was below
the detection level of the Bradford assay (<10 µg protein
ml^{-1}). As expressed per unit of volume, concentrations of
binding sites in the purified preparations as high as
10^{-10} mol ml^{-1} were found. This indicates a high degree of
enrichment in receptor. Since each purification results in
approximately 3 ml of purified preparation and the preparations cannot be stored, scaling up of the procedure is
required.

The fact that by this purification method preparations can be obtained that still exhibit high-affinity IAA binding indicates that the receptor is still in a functional conformation. This is supported by experiments with cytosolic extracts from tobacco shoot tips. The purification method was similar to that shown in Fig. 4. Receptors which were eluted from the affinity column by IAA (10^{-5}M), were kept saturated with the hormone and tested for their effect on overall in vitro transcription. Nuclei isolated from cells in either late log-phase or stationary-phase of the cell culture were used in the transcription assay. In most trials a significant stimulation of transcription was observed (table 5).

TABLE 5

EFFECT OF PURIFIED RECEPTOR PREPARATIONS
ON OVERALL IN VITRO TRANSCRIPTION

Experiment	Incorporation of ^{3}H-UMP (DPM)		
	nuclei + elution buffer + IAA(10^{-5}M)	nuclei + eluted fraction + IAA(10^{-5}M)	stimulation (%)
A	1728	2256	31
B	1494	1704	14
C	1440	2448	70
D	360	576	60
E	312	1398	346
F	300	570	90
G	216	660	206

In two experiments no stimulation was observed.

In order to examine their protein profiles, concentrated purified receptor preparations from shoot tips were loaded on SDS-polyacrylamide slab gels. After electrophoresis the gels were developed by Silver staining. We found that two proteins with almost equal molecular weight (ca 50,000 D) could be specifically eluted from the affinity column by IAA. This suggests that these proteins might represent two slightly different subunits of the receptor. The molecular weight of 150,000 - 200,000 D as determined by gel filtration (5) indicates that in its native form the receptor might consist of 3 to 4 of these subunits.

As we have explained in the introduction, addition of Mg^{2+} ATP and PNPP increases the number of binding sites in crude receptor preparations. The idea is that excess PNPP as substrate for alkaline phosphatases partly protects the phosphorylated receptors from being attacked by these enzymes, whereas at the same time dephosphorylated receptors are being phosphorylated by protein kinases using Mg^{2+} ATP as cosubstrate. Indeed, crude extracts from shoot tips and from isolated nuclei contain ATP-dependent protein kinase and alkaline phosphatase activity. If this idea is correct, it must in principle be possible to label receptors with ^{32}P by adding [γ-^{32}P] ATP to crude receptor preparations.

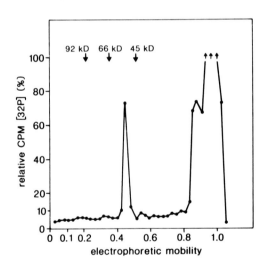

FIGURE 5.   Electropherograph of a ^{32}P-labelled receptor preparation purified by affinity chromatography. Arrows indicate electrophoretic mobility of marker proteins.

Figure 5 shows the distribution of radioactivity in a SDS-polyacrylamide rod gel after electrophoresis of a receptor preparation from shoot tips which was treated with [γ-^{32}P] ATP and then purified by affinity chromatography.

Three peaks of radioactivity were present. Two coincided with the ion front and the buffer front representing $[\gamma-^{32}P]$ ATP and/or ^{32}Pi. The third peak represents a macromolecule with a relative molecular weight of 51, 677 D (mean of 6 determinations, range 47,500 - 55,000 D). This molecular weight agrees with that determined from protein profiles of purified receptor preparations in other experiments (see above). Experiments, using non-denaturing gelelectrophoresis are in progress to show that the ^{32}P-labelled macromolecule specifically binds ^3H-IAA.

DISCUSSION

The batch-cultured tobacco cell line as described in the preceding section turns out to be an appropriate experimental system for further investigations on detection and transduction of auxin by the cytoplasmic receptor. The results thus far obtained in our laboratory show that:
- Addition of auxin brings about a rapid increase ($\leqq$ 5 min) in the level of nucleus-bound receptors; a rapid activation ($\leqq$ 20 min) of overall transcription in nuclei and a rapid increase ($\leqq$ 60 min) in the level of at least 3 translatable mRNA species.
- Receptor preparations purified by affinity chromatography stimulate overall transcription in nuclei isolated from early log-phase or stationary-phase cells.
These effects are quite similar to those described in pioneering studies on steroid hormone receptors, in particular studies on estrogen-stimulated rat uterus (3,11).
Our experimental system will be used to study in more detail affinity modulation by enzymatic phosphorylation and dephosphorylation, transformation of receptor into DNA- or nuclear-binding form and characterization of nuclear binding sites for transformed receptors. In this connection we realize that even in steroid-receptor research quite a number of problems concerning these processes are still unsolved. Our primary goals, however, are clear: scaling up of the present receptor purification method and, if necessary, further purification of the receptor preparations to homogeneity; preparing (monoclonal) antibodies against purified receptors; construction of cDNA clones and clones of auxin-regulated genes from our experimental system.
Since all evidence suggests that the cytoplasmic binding protein is an auxin receptor, a basic question will be how it is involved in the regulation of overall transcription and

specific gene expression. It might be that among others sequence-specific receptor-DNA interactions are involved, which is, of course, mere speculation at the present moment.

REFERENCES

1. Hagen G, Guilfoyle TJ (1985). Rapid induction of selective transcription by auxins. Mol Cell Biol 5:1197.
2. American Society of Plant Physiologists summaries-II. The molecular biology of plant hormone action: Research Directions. In Vanderhoef LN, Kosuge T (eds). Waverly press, Baltimore, Maryland 21202.
3. Anderson JE (1984). The effect of steroid hormones on gene transcription.   Golderger RF, Yamamoto KR (eds): "Biological Regulation and Development, Vol. 3B: Hormone Action," New York: Plenum Publishing Corp., p 169.
4. Oostrom H, Kulescha Z, van Vliet ThB, Libbenga KR (1980). Characterization of a cytoplasmic auxin receptor from tobacco pith callus. Planta 149:44.
5. Van der Linde PCG, Bouman H, Mennes AM, Libbenga KR (1984). A soluble auxin-binding protein from cultured tobacco tissues stimulates RNA synthesis in vitro. Planta 160:102.
6. Maan AC, Vreugdenhil D, Bogers RJ, Libbenga KR (1983). The complex kinetics of auxin-binding to a particulate fraction from tobacco-pith callus. Planta 158:10.
7. Maan AC, van der Linde PCG, Harkes PAA, Libbenga KR (1985). Correlation between the presence of membrane-bound auxin binding and root regeneration in cultured tobacco cells. Planta 164:376.
8. Maan AC, Kühnel B, Beukers JJB, Libbenga KR (1985). Naphthylphthalamic acid-binding sites in cultured cells from Nicotiana tabacum. Planta 164:69.
9. Bailey HM, Barker RDJ, Libbenga KR, van der Linde PCG, Mennes AM, Elliott MC (1985). Auxin binding site in tobacco cells. Biol Plant 27:105.
10. Van der Linde PCG, Maan AC, Mennes AM, Libbenga KR (1985). Auxin receptors in tobacco. In "Proc. 16th FEBS Congress Part C", VNU Science press, p 397.
11. Grody WW, Schrader WT, O'Malley BW (1982). Activation, transformation, and subunit structure of steroid hormone receptors. Endocrine Reviews 3:141.
12. Mennes AM, Bouman H, van der Burg MPM, Libbenga KR (1978). RNA synthesis in isolated tobacco callus nuclei and the influence of phytohormones. Plant Sci Lett 13:329.

13. Bruns R, Lawson-Wendling K, Pugsley T (1983). A rapid filtration assay for soluble receptors using polyethylenimine-treated filters. Anal Biochem 132:74.

14. Laemmli UK (1970). Cleavage of structural proteins during the assembly of the head of bacteriophage T4. Nature 227:680.

15. Van Slogteren GMS, Hoge JHC, Hooykaas PJJ, Schilperoort RA (1983). Clonal analysis of heterogeneous crown-gall tumor tissues induced by wild-type and shooter mutant strains of Agrobacterium tumefaciens. Expression of T-DNA genes. Mol Biol 2:321.

16. O'Farrell PH, O'Farrell PZ (1977). In Stein G, Stein J, Kleinsmith CJ (eds): "Methods in Cell Biology XVI," New York: Academic press, p 407.

Molecular Biology of Plant Growth Control, pages 245–255
© 1987 Alan R. Liss, Inc.

PLANT CELL DIVISION - THE ROLES OF
IAA AND IAA BINDING PROTEINS[1]

Malcolm C.Elliott, Angela M.O'Sullivan, J.F.Hall,
Gillian M.Robinson[2], Jane A.Lewis[3], D.A.Armitage,
Helen M.Bailey and R.D.J.Barker

School of Life Sciences, Leicester Polytechnic,
Scraptoft Campus, Scraptoft, Leicester,
LE7 9SU, U.K.

K.R.Libbenga and A.M.Mennes,

Department of Plant Molecular Biology,
University of Leiden, Nonnensteeg 3,
NL-2311 VJ Leiden, The Netherlands.

ABSTRACT   Intra-cellular IAA levels of Acer
pseudoplatanus, L. cells whose divisions were
synchronised in various ways, were compared with
those of cells maintained in steady state cultures.
Although some data were in accord with the simple
hypothesis that cells had to achieve a critical
peak ("trigger") concentration of endogenous IAA
before each division the overall conclusion was
that this hypothesis required modification. It
is clear that cellular concentrations of IAA
immediately before and throughout a sequence of
divisions are always such as to achieve 50%
saturation of an auxin binding protein which has
been isolated from the cells.   The importance
of compartmentalisation of the IAA is emphasised

[1]This work was supported by grants from the SERC which
enabled four of us (A.M.O'S., J.F.H., G.M.R. and J.A.L.) to
participate.
[2]Present address: Wyeth Laboratories, Taplow, Berks, U.K.
[3]Present address: Biotechnology Department, Glaxo Group
Research Ltd., Greenford Road, Greenford, Middlesex,
UB6 OHE, U.K.

as are the implications of our demonstration of an
auxin receptor from Nicotiana tabacum cells which
appears to cycle between the cytoplasm and the
nucleus.

INTRODUCTION

The hypothesis that there is a precise correlation
between auxin content and growth of plant parts arose from
classical experiments (1).   Critics of the classical view
(2,3,4) have drawn attention to other workers' deficiencies
in experimentation and interpretation.   We have argued
(5,6) that plant cell suspension cultures are particularly
appropriate for studies of hormonal control of cell division
and enlargement since they tend to be uniform and free of
differential patterns of distribution of nutrients and phytohormones between cells.   It was anticipated (6) that the
application of rigorous physico-chemical methods of phytohormone analysis to such cultures would permit identification
of the specific factors regulating the onset and termination
of cell division and cell enlargement.
     Gautheret (7) noted that cultures of plant cells often
required an auxin in their culture medium if cell division
and enlargement were to occur.   It has been assumed that
such cells had a growth limiting defect in their capacity to
synthesise IAA (8,9,10).   The Acer pseudoplatanus, L. cell
suspension cultures used in our work (5) were grown in a
culture medium which contained the auxin 2,4-D (11, minus
urea and kinetin) and had been assumed to synthesise no IAA
(12).   However, we were able to establish (by GC-MS) that
the cells did, in fact, synthesise IAA and to show that the
kinetics of changes in cellular IAA levels through the
culture cycle were in accord with a concentration dependent
role for IAA in the regulation of cell division (6).   Cells
of plant tumour tissues divide and enlarge on culture media
lacking added auxin.   Such cultures have been shown to
produce IAA but there is no simple relationship between
cellular IAA levels and rates of cell division (13,14).   Our
earlier work involved the use of batch cultures of A.
pseudoplatanus cells.   The interpretation of data obtained
from such cultures is complicated.   Here we report the
results of experiments with synchronously dividing and steady
state cultures and we discuss the new data on intra and extracellular IAA levels in the context of our recent studies on
IAA binding proteins in plant cells.   Full experimental

details have necessarily been omitted. They will be
published elsewhere.

RESULTS AND DISCUSSION

A. pseudoplatanus cells grown in our standard culture
medium quickly become nitrogen limited (15) and cell division
synchrony has been obtained using starvation/regrowth techniques in combination with sub-culture at low inoculum
densities (< 30,000 cells cm^{-3}) (16,17). Such low inoculum
density cultures were of no value for our present studies
but we noted that the first division after transfer of
stationary phase (23 day) cells to fresh medium was routinely
synchronised (up to 80% division in 2h) and this observation
has been exploited in our detailed studies of the relationship between cell numbers, mitotic indices and cellular IAA
levels (Figure 1). These results show that the IAA concentration peaked sharply between 81 and 85h after transfer to

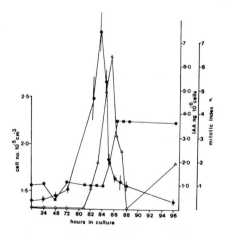

FIGURE 1.   The sequence of events through the first
division of A. pseudoplatanus cells transferred to fresh
culture medium.   IAA (●—●), cell number (■—■), mitotic
index (△—△).   Error bars represent the standard deviation
of mean.   Standard deviations for cell numbers (which did
not exceed 6%) have been omitted for clarity.

fresh medium, reaching a maximum at 83h. This preceded the maximum mitotic index measured (6.5% at 85h) and the partially synchronised division (in this case 44% from 85 to 87h). The biosynthetic capability of the cells for IAA production peaked sharply just before the rise in the concentration of IAA, suggesting that the higher concentration of IAA found was a result of de novo synthesis. Although the actual timing and degree of sychrony varied from one experiment to another it was found that the sequence of peaks, viz. IAA synthesising capability, IAA concentration, mitotic index, cell division was unaltered. This strict chronological sequence was in agreement with other workers' data for tobacco cells (18,19) and in accord with the notion that the endogenous IAA is a key factor in the control of division of plant cells in suspension culture, perhaps by induction of mitosis (20,21,22).

Work with asynchronously dividing batch cultures and with cells studied through a single synchronous cell division gave results in accord with the hypothesis that there was a concentration dependent relationship between intracellular IAA and cell division. It was anticipated that continuous culture techniques would facilitate a quantitative study of the role of IAA in plant cell division since growth takes place at a constant rate and in a constant environment. Thus cells should display constant cellular composition and constant cellular activity (23). An apparatus similar to that described previously (24) was used to grow A. pseudoplatanus cell cultures at four different dilution rates. A positive correlation was found between IAA levels in the cells and specific growth rate of the culture. Figure 2 shows the linear-log correlation between specific growth rate and IAA content per 10^6 cells. Studies of this type have given scant attention to the possible presence or role of IAA released into the culture medium by the cells (25). In fact our cultures have been shown (by GC-MS) to contain IAA at concentrations (determined by SPF, 6) vastly in excess of those found in the cells themselves. No correlation was observed between the amounts of IAA present per unit volume of culture medium and dilution rate. These data from studies of continuous cultures are in accord with the proposal that endogenous IAA levels are determinants of cell division.

Although, as we noted earlier, the first division of batch cultured cells is synchronous (Figure 1) divisions in these cultures tend quickly towards asynchrony. However, if cells which had undergone the first (synchronous) division

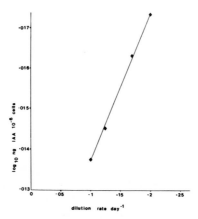

FIGURE 2. The relationship between the common
logarithm of the IAA levels of A. pseudoplatanus cells
maintained in continuous culture at different dilution
rates, and dilution rate.

were subjected to alternate periods at $10^{\circ}C$ (72h) and $25^{\circ}C$
(70-80h) a series of partially synchronous divisions
followed (Figure 3). The relationships between peaks of
intracellular IAA, mitotic index and cytokinesis observed
during the first synchronous division persisted but a second
peak in intracellular IAA was routinely detected during the
period of incubation at $10^{\circ}C$. This peak was not associated
with changes in any of the growth parameters monitored. It
is suspected that the second peak is a consequence of a
change in the biosynthesis: degradation ratio at the lower
temperature. Nevertheless it occurred to us that the variations in cellular IAA level may be less important than the
absolute concentrations. We have detected a specific high
affinity IAA binding protein in sycamore cells which has
characteristics similar to the "soluble receptor" described
in a series of papers from Libbenga's group (26,27,28,29,30,
31). The apparent K_a for IAA was approximately $1x10^8$ M^{-1}.
Figure 4 shows the variation of $\log_{10}$ endogenous IAA concentration with time for the culture whose data are shown in
Figure 3. This shows that the intracellular concentration
of IAA is below the K_a value of the binding protein on
transfer of cells to fresh medium from stationary phase
batch cultures (23 days). It quickly (prior to the onset of

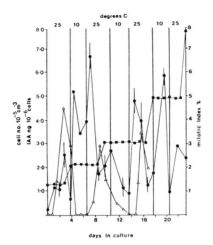

FIGURE 3. Variations in the level of "free endogenous IAA related to mitosis and cytokinesis through four cell divisions in cultures synchronised by repeated 25°C-10°C cycles. IAA (●—●) cell number (■—■) mitotic index (△—△). Mitotic indices are based on 1,000 nuclei scored per data point. Error bars represent the standard deviation of the mean. Standard deviations for cell numbers (which did not exceed 6%) have been omitted for clarity.

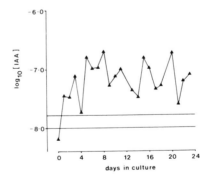

FIGURE 4. Intracellular concentrations of IAA, calculated on the basis of a single compartment cell, of the cells featured in Figure 3. Horizontal lines define the limits of the K_a values determined for the auxin binding protein (1.6×10^8 M^{-1} to 1×10^8 M^{-1}).

cell division) rises above the 50% saturation level and, despite considerable fluctuation, remains above this level for the remainder of the period covered by Figure 3.

Some fluctuations in IAA levels in the culture medium containing the cells analysed for Figures 3 and 4 occurred but the overall trend was one of accumulation of IAA in the medium at both $10^\circ C$ and $25^\circ C$. The intracellular IAA accounts for only a small proportion of the total IAA in the culture. Thus at day 9 the total IAA content of the cells was 202 ng while that of the culture medium was 2,324 ng. The extra-cellular IAA seemed to be stabilised by some product of the cells' metabolism since IAA added to fresh medium broke down rapidly but by day 9 the composition of the medium had changed in some way which rendered the IAA very stable in solution in it. The extracellular IAA is not resorbed and it seems unlikely to serve any purpose in our cultures. Stange (20) would see this release as compatible with her perception of IAA as both an intracellular regulator and an intercellular signal.

The cold synchronisation technique may well induce atypical changes in cell physiology and biochemistry. Accordingly Firby et al. (31) have defined a synchronisation technique which depends upon the restoration of nitrate and phosphate levels in the culture medium to their optimal values after each division. By this method a series of highly synchronised divisions (circa 80% in 2h) could be reliably and reproducibly produced. In these "semi-steady state" (in terms of nutrients) conditions the pattern of changes in IAA is dramatically different than in other experimental systems. Data from other systems were compat-ible with the view that before each division cells had to achieve a critical "trigger" concentration of endogenous IAA (between 5.0 and 7.5 ng/10^6 cells). In our nitrate/phosphate synchronised system (Figure 5) lag phase cells contained high (5.0 to 7.0 ng/10^6 cells) but the levels decreased rapidly to a relatively stable value of 0.5 ng/10^6 cells during the period of rapid synchronous divisions. Such fluctuations as did occur yielded "peaks" at quite different times relative to cytokinesis than hitherto.

Figure 6 shows that relationship between $\log_{10}$ intra-cellular IAA concentration and time for the culture shown in Figure 5 but for these calculations we have made corrections necessary because the cells are not single compartments. The cytoplasm/vacuole ratio for suspension cultured sycamore cells in approximately 1:9 (33) and at the reported values for cytoplasmic pH (7.3) and vacuolar pH (5.7) (34) the IAA

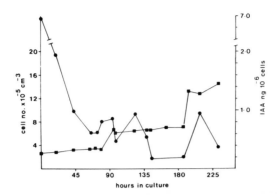

FIGURE 5. Intracellular levels of IAA in A. pseudoplatanus cell cultures synchronised by the nitrate/phosphate feeding technique. IAA (●—●), cell number (■—■). Standard deviations have been omitted for clarity.

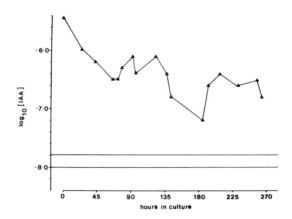

FIGURE 6. Intracellular concentrations of IAA, corrected for cytoplasm: vacuole ratios, of the cells featured in Figure 5. Horizontal lines define the limits of the K_a values determined for the auxin binding protein ($1.6 \times 10^8 M^{-1}$ to $1 \times 10^8 M^{-1}$).

will be almost exclusively in the cytoplasm, raising the effective concentration tenfold. Clearly the concentration of IAA in the cells is far above that necessary to achieve 50% saturation of the auxin binding protein.

We have previously reported (35) that a specific, high affinity IAA binding site has been demonstrated in both a cytosol fraction and in isolated nuclei from <u>Nicotiana</u> <u>tabacum</u> cv. Wisconsin No.38 cells grown in suspension culture. The amount of the binding site detected in both these fractions changed during the culture cycle according to a strict pattern. When partially purified by affinity chromatography and allowed to pre-incubate with IAA, the site had a significant stimulatory effect on total RNA synthesis, as measured by a cell-free assay system.

It was always naive to assume that hormone concentrations alone determined developmental changes (4). Our data suggest that approaches which take account of hormone concentration and distribution within the cell together with receptor availability and distribution will prove fruitful.

REFERENCES

1. Went FW (1928). Wuchstaff und Wachstum. Recl Trav Bot Néerl 25:1.
2. Dennis FG Jr (1977). Growth hormones: pool size, diffusion, or metabolism? Hortscience 12:217.
3. Mer CL (1969). Plant growth in relation to endogenous auxin, with special reference to cereal seedlings. New Phytol 68:275.
4. Trewavas A (1981). How do plant growth substances work? Plant, Cell and Environment 4:203.
5. Elliott MC, Farrimond JA, Hall JF, Clack DW (1977). Hormonal regulation of growth of higher plant cell suspension cultures. In Kudrev T, Ivanova I, Karanov E (eds): "Plant Growth Regulators", Publishing House of the Bulgarian Academy of Sciences, Sofia, p 459.
6. Moloney MM, Hall JF, Robinson GM, Elliott MC (1983). Auxin requirements of sycamore cells in suspension culture. Plant Physiol 71:927.
7. Gautheret RJ (1955). The nutrition of plant tissue cultures. Ann Rev Plant Phys 6:433.
8. Henderson JHM, Bonner J (1952). Auxin metabolism in normal and crown-gall tissue of sunflower. Am J Bot 39:444.
9. Gautheret RJ (1955). Sur la variabilité des propriétés

physiologiques des cultures de tissus végétaux. Rev Gen Botan 62:1.

10. Mousdale DM (1982). Endogenous IAA and the growth of auxin-dependent and auxin autotrophic (crown gall) plant tissue. Biochem Physiol Pflanzen 177:9.

11. Stuart R, Street HE (1969). Studies on the growth in culture of plant cells. IV. The initiation of division in suspensions of stationary phase cells of Acer pseudoplatanus L. J Exp Bot 22:735.

12. Street HE, Collin HA, Short KC, Simpkins I (1968). Hormonal control of cell division and expansion in suspension cultures of Acer pseudoplatanus L.: the action of kinetin. In Wightman F, Setterfield G (eds): "Biochemistry and Physiology of Plant Growth Substances", Runge Press, Ottawa, p 489.

13. Pengelly WL, Meins F Jr (1983). Growth, auxin requirement and indole-3-acetic acid content of cultured crown gall and habituated tissues of tobacco. Differentiation 2:101.

14 Weiler EW, Spanier K (1981). Phytohormones in the formation of crown gall tumors. Planta 153:326.

15. Young M (1973). Studies on the growth in culture of plant cells. XVI. Nitrogen assimilation during nitrogen-limited growth of Acer pseudoplatanus L. cells in chemostat culture. J Exp Bot 23:1172.

16. King PJ, Mansfield KJ, Street HE (1973). Control of growth and cell division in plant cell suspension cultures. Can J Bot 51:1807.

17. King PJ, Cox BJ, Fowler MW, Street HE (1974). Metabolic events in synchronized cell cultures of Acer pseudoplatanus L. Planta 117:109.

18. Nishinari N, Yamaki T (1976). Relationships between cell division and endogenous auxin in synchronously-cultured tobacco cells. Bot Mag Tokyo 89:73.

19. Nishinari N, Yamaki T (1978). Increase in activity of IAA-synthesising enzymes prior to cell division. Bot Mag Tokyo 91:93.

20. Stange L (1977). Meristem differentiation in Riella helicophylla (Bory et Mont.) Mont. under the influence of auxin or anti-auxin. Planta 135:389.

21. Van't Hof J (1968). The action of IAA and kinetin on the mitotic cycle of proliferative and stationary phase excised root meristems. Exp Cell Res 51:167.

22. Patau K, Das NK, Skoog F (1957). Induction of DNA synthesis by kinetin and indoleacetic acid in excised tobacco pith tissue. Physiol Plant 10:949.

23. King PJ, Street HE (1973). Growth patterns in cell cultures. In Street HE (ed): "Plant Tissue and Cell Culture", Blackwell Scientific Publications, Oxford, p 269.

24. Wilson SB, King PJ, Street HE (1971). Studies on the growth in culture of plant cells. XII. A versatile system for the large scale batch or continuous culture of plant cell suspensions. J Exp Bot 21:177.

25. Weiler EW (1981). Dynamics of endogenous growth regulators during the growth cycle of a hormone – autotrophic plant cell culture. Naturwiss 68:377.

26. Libbenga KR (1978). Hormone receptors in plants. In Thorpe TA (ed): "Frontiers of Plant Tissue Culture", IAPTC, Calgary, p 325.

27. Mennes AM, Bouman H, Van der Berg MPM, Libbenga KR (1978). RNA synthesis in isolated tobacco callus nuclei, and the influence of phytohormones, Plant Sci Lett 13:229.

28. Bouman H, Mennes AM, Libbenga KR (1979). Transcription in nuclei isolated from tobacco tissues. FEBS Lett 101: 369.

29. Bogers RJ, Kulescha Z, Quint A, Van Vliet ThB (1980). The presence of a soluble auxin receptor and the metabolism of 3-indole-acetic acid in tobacco pith explants. Plant Sci Lett 19:311.

30. Oostrom H, Kulescha Z, Van Vliet ThB, Libbenga KR (1980). Characterization of a cytoplasmic auxin receptor from tobacco pith callus. Planta 149:44.

31. Van der Linde PCG, Bouman H, Mennes AM, Libbenga KR (1984). A soluble auxin-binding protein from cultured tobacco tissues stimulates RNA synthesis in vitro. Planta 160:102.

32. Firby DJ, O'Sullivan AM, Leach CK, Elliott MC (in preparation). Partial synchronization of Acer pseudoplatanus L. cell suspension cultures induced by addition of limiting nutrients.

33. Bligny R, Doucé R (1976). Les mitochondries des cellules végétales isolées (Acer pseudoplatanus L.) I. Etude comparee des propriétés oxydatives des mitochondries extraites et des mitochondries placées dan leur contexte cellulaire.

34. Rebeillé F, Bligny R, Martin J-B, Douce R (1985). Effects of sucrose starvation on sycamore (Acer pseudoplatanus) cell carbohydrate and phosphate status. Biochem J 226:679.

35. Bailey HM, Barker RDJ, Libbenga KR, Van der Linde PCG, Mennes AM, Elliott MC (1985). Auxin binding site in tobacco cells. Biol Plant 27:105.

Molecular Biology of Plant Growth Control, pages 257-266
© 1987 Alan R. Liss, Inc.

GROWTH-STAGE DEPENDENT
OCCURRENCE OF SOLUBLE AUXIN-BINDING
PROTEINS IN PEA

Hans-Jörg Jacobsen and Karin Hajek

Institut für Genetik
Universität Bonn
D 53oo Bonn - 1
FRG

ABSTRACT   One of the fundamental laws in biology
says that a signal without recognition is not a
signal. When we understand plant hormones as sig-
nalling compounds, we have to consider specific
recognition sites, since we observe specific, obviously signal-depending responses. The responses to
the application of phytohormones, however, are
multiple ones: They depend on hormone concentration and differ with respect to the type of cells
or tissues, physiological age, or even genotype.
Consequently, we should expect different specific
recognition sites for the same class of hormones.
This contribution gives evidence for the presence
of biochemically different soluble auxin-binding
sites in different tissues of pea seedlings.

INTRODUCTION

Evidence has accumulated in the past 15 years that
plants possess specific proteins, which are able to bind with
a high affinity and a satisfying specificity the various
groups of plant hormones, thus fulfilling two main requirements for a possible receptor function (1-8). In the case of
auxins, we presently have evidence for the existence of two
classes of such binding sites, namely the membrane-located
ABP and the soluble cytoplasmic ABP (sABP). Membrane located
ABP are characterized by a pH-optimum for binding in the
range of about pH 5.o and dissociation constants (K_d) between

$10^{-6} - 10^{-7}$M in crude preparations(1, 3-4) or 5.7 x 10^{-8}M as purified protein (9), whereas $sABP_7$bind auxins with an opti- mal pH between pH 7-8 and $K_d < 10^{-7}$M in crude and $< 10^{-8}$M in partially purified preparations (2, 5-8). Generally, the auxin binding in the membrane fractions are associated with rapid auxin effects on cell elongation (1o) or auxin trans- port (1,3-4,12), while sABP are understood as mediators bet- ween auxin application and long-term morphogenetic effects (2,5-7, 13-14). The present state of our knowledge on long- term auxin effects at the molecular level may be characte- rized as follows:

a) auxins induce multiple morphogenetic effects,
b) auxins induce the formation of specific mRNA and proteins,
c) auxins bind to cytoplasmatic proteins with a specificity
   and affinity required for a receptor function, (2,5-8,13-14),
d) auxins + sABP translocate from the cytosol to the nucleus
   (2o)
e) auxin + sABP induce a higher transcriptional activity, and
   neither auxin nor sABP alone give the same effect (21).

   All these data fit quite well into an analogy to the steroid action model of animal or fungal systems. Since auxins are a unique signal causing various responses in different tissues, one can expect more than one receptor pro- tein, because it seems rather unlikely that the very same re- cognition site shall exert various specific effects. We aim to present experimental evidence for this hypothesis.

METHODS

   Most experiments reported here were carried out with etiolated pea seedlings (cv. "Dippes gelbe Viktoria", a com- mercial german pea variety), grown in moist vermiculite at room temperature. Epicotyls were harvested in the cold under dim light, and for some experiments apical hooks were sepa- rated from the epicotyls. In some yet preliminary experiments the epicotyls were further subdivided into nodal and inter- nodal tissues under the same conditions. The material was either directly used for preparation of cytosol or kept frozen at - 2o°C until use.
   Preparation of cytosol was as previously described (8,14), further separation of cytosol proteins using a preparative chromatofocusing system was according to (14) with modifica- tions (22-23).
   High- affinity auxin-binding was determined with various

ligand assays (8, 22-23), calculation of kinetic data was performed according to (24), using a computer program in BASIC,
kindly provided by A.Maan(25).

RESULTS

   First evidence for growth-stage depending differences
in the occurrence of sABP was obtained when chromatofocused
cytosol fractions of seedlings with different age were
screened for the presence of high-affinity auxin binding (14):
Epicotyls of yound seedlings (7-9 days) exhibited only one
binding site, while in older seedlings two sites could be
detected (> 9-1o days).

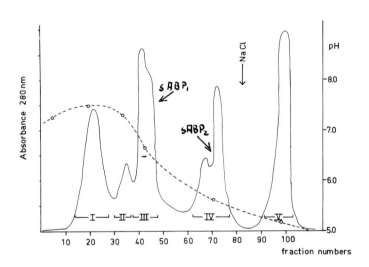

Fig.1: Elution profile of chromatofocused cytosol from etiolated pea seedling epicotyls (1o days)

   The chromatofocusing column designed for the separation
of cytosol proteins enables the separation of 25o - 3oo mg
protein within 6 - 7 hours, each of teh fractions obtained
is composed of proteins with approximately similar pIThis system gives reproducible results without the application

of ampholyte-containing buffers (14, 22-23), thus avoiding
the necessity to remove these compounds, which interfere with
the binding assays (unpublished observation).

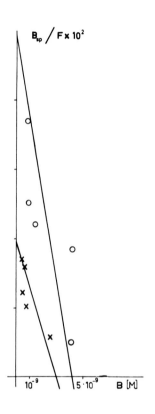

Fig.2:Scatchard analysis of
soluble auxin-binding sites
from epcotyls of 11 days
old seedlings (o———o, sABP$_1$,
present in all seedlings
from days 7-11, K_d:5.5 x lo^{-8}M;
x———x; sABP$_2$, only present
in seedlings > 9 -lo days,
K_d: 2.24 x lo^{-7}M; kinetic
data corrected for the presence of low affinity binding
according to Chamness and
Mc Guire (26, for details
see also ref. 8)

    When the same mode of analysis was applied to cytosol
from etiolated apical hooks of seedlings > lo days, a slightly differing elution pattern was obtained (Fig. 3), and high
affibity auxin binding was found not only in the same fractions as obtained from epicotyl tissue, but also in another
fraction, characterized by its more neutral isoelectric point.
    It was an interesting observation that fractions IVa
and IVb show only minor differences in their polypeptide
patterns: this obviously is due to diferences in the phosphorylation of these proteins, since fraction IVb contained
more phosphate groups than fraction IVa (16o nM/mg protein,

S.D. 6o nM vs. 26o nM/mg protein, S.D. 3o nM, assayed according to Hedrick and Fischer (26).

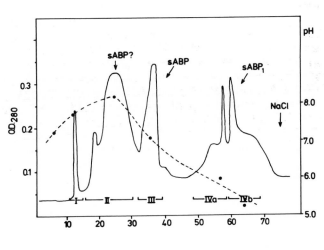

fractions

Fig. 3: Elution profile of chromatofocused cytosol proteins
from etiolated apical hooks of 12 days old pea seedlings

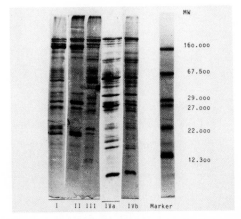

Fig. 4: Polypeptide composition of cytosol fractions I-IVb
from Fig.3(SDS-gradient gel, 1o-15%, Coomassie stain)

The data on soluble high-affinity auxin-binding in pea
epicotyls and apical hooks are compiled in table 1.

TABLE   1
High-affinity auxin-binding in
apical hooks and epicotyls   of
11 days old etiolated   peas

Fractions	I	II	III	IVa	IVb
Apical hooks	-	+	-	-	+
Epicotyls	-	-	-	+	+

In recent experiments we obtained preliminary evidence
for the distribution of $sABP_1$ and $sABP_2$ in the epicotyls.
When nodal and internodal tissues were separated, $sABP_1$ was
found only in the internodes, whereas $sABP_2$ and $sABP_1$ both
could be identified in nodal tissues. However, these data
have to be confirmed by immunological methods.

In another experiment, we compared the pattern of high
affinity auxin-binding in epicotyls of the pea line used in
most of our studies reproted here with that of a fasciated
mutant derived from this line (489 C) and a recombinant,
coming from a cross of these two genotypes (R 4111, 28).
This recombinant differs from the initial line by its almost doubled internode length, a character, which is expressed
already at the seedlings stage (table 2).

TABLE   2
Number of nodes and length of epicotyls in lines DGV, 489 C and R4111
at days 7 and 9 of germination

Genotype	DGV		489 C		R 4111	
Days	7	9	7	9	7	9
No.of nodes	2	3	2	3	2	3
Epicot.length	5.1	6.1	6.2	1o.1	7.5	11.7

When we compare the pattern of high-affinity auxin-binding in these three genotypes at day 12, we observe a complex binding kinetics in the parent lines (apparent cooperative binding ?), while in the recombinant only one type of binding sites is present (Fig. 5).

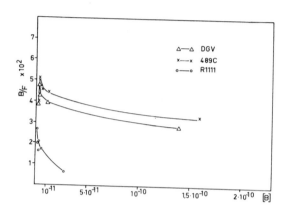

Fig. 5: Scatchard plot analysis of high-affinity auxin-binding in etiolated epicotyls of three pea genotypes (No. of recombinant was changed from R1111 to R 4111 in order to avoid confusion with a mutant having the same No.)

Obviously the recombinant, which conserves the juvenile 2-nodes-stage up to day 12 of germination, expresses the "juvenile" high-affinity auxin-binding found in the other genotypes at that stage: It seems rather likely that this stage is characterized by the sole presence of $sABP_1$, while in later stages $sABP_2$ is synthesized and the binding pattern becomes more complex. However, further investigations, requiring more seed material of the recombinant, are necessary to prove a causal relationship between the deviant growth characteristics and the sABP-pattern.-

DISCUSSION

It was argued by Trewavas (29) for numerous reasons that plant development is rather regulated by changes in the sensitivity to growth substances than by changes in the concentration of growth substances. We believe that the data presented here give some evidence for assuming more than just

one class of soluble auxin binding sites with putative receptor functions. Since various independent binding assay procedures have been applied to prove the general existence of
high-affinity auxin-binding (ammonium-sulfate precipitation,
nitrocellulose filters, PEI-treated glass fiber filters (3o),
and equilibrium dialysis for more details on reliability of
binding assay see 22,23), we have to consider the high-affinity auxin-binding as a fact, although the recognition function of receptors is only part of the truth: Still an open
question are the subsequent steps in the chain of transducing
the morphogenetic information carried by a hormone molecule.
First promising evidence that these subsequent events are on
the level of transcriptional control has been presented
(2o,21), abd data are available showing specific transcription upon auxin application (15-19). However, these reports
so far give no evidence that hormone receptors are involved
in mediating these specific responses to auxins. Here apparently a gap has to be closed.

REFERENCES

1.   Hertel R, Thomson K-St, Russo VEA(1972). In vitro auxin
     binding to particulate cell fractions from corn coleop-
     tiles. Planta 1o7: 325-34o.

2.   Oostrom H, van Loopik-Detmers MA, Libbenga K (1975)
     A high-affinity receptor for indoleacetic acid in cul-
     tured tobacco pith explants. FEBS-Lett. 59:194-197.

3.   Batt S, Venis MA (1976). Separation and localization of
     two classes of auxin binding sites in corn coleoptiles.
     Planta 13o:15-21.

4.   Ray PM, Dohrmann U, Hertel R (1977). Characterization of
     naphtalenacetic acid binding to receptor sites on cellu-
     lar membranes of maize coleoptile tissue. Plant Physiol.
     59: 357-364.

5.   BogersRJ, Kulescha Z, Quint A, van Vliet TB, Libbenga K
     (198o). The presence of a soluble auxin receptor and the
     metabolism of 3-indoleacetic acid in tobacco pith ex-
     plants. Plant Sci.Lett. 19:311-317.

6.   Oostrom H,Kulescha Z, van Vliet,TB, Libbenga K (198o).
     Characterization of a cytoplasmic auxin receptor from
     tobacco pith callus. Planta 149: 44-47.

7.   Jacobsen HJ (1981). Soluble auxin-binding proteins in
     pea. Cell Biol.Intern.Rep. 5:768.

8.   Jacobsen HJ (1982). Soluble auxin binding proteins in
     pea epicotyls. Physiol. Plant.56: 161-167.

9.   Löbler M, Klämbt D (1985). Auxin-binding protein from
     coleoptile membranes of corn. J.Biol. Chem. 26o(17):9848-53

1o. Walton JD, Ray PM (1981). Evidence for receptor function of auxin binding sites in maize. Plant Physiol.68:1334-1338.

11. Venis MA (1977). Solubilzation and partial characterization of auxin-binding sites of corn. Nature 266:268-269

12. Hertel R (1983) The mechanism of auxin transport as a model for auxin action. Z.Pflanzenphysiol.112:53-67.

13. Libbenga K (1978) Hormone receptors in plants. In Thorpe TA (ed): Frontiers of plant tissue culture 1978:Calgary: IAPTC, p. 325-333.

14. Jacobsen HJ (1984). Two different soluble cytoplasmic auxin-binding sites in etiolated pea epicotyls. Plant & Cell Physiol. 25(6): 867-873.

15. Theologis A, Ray PM (1982). Early auxin-regulated polyadenylated mRA sequences in pea stem tissue. PNAS(USA) 79:418-421.

16. Melanson D, Trewavas AJ (1982). Changes in tissue protein patterns in relation to auxin induction of DNA synthesis. Plant,Cell &Environm. 5:53-64.

17. Zurfluh LL, Guilfoyle TJ (198o) Auxin-induced changes in the patterns of protein synthesis in soybean hypocotyls-PNAS(USA) 77(1): 357-361.

18. Hagen G, Kleinschmidt A. Guilfoyle TJ (1984). Auxin-regulated gene expression in intact soybaen hyocotyl and excised hypocotyl sections. Planta 162: 147-153.

19. Hagen G., Guilfoyle TJ (1985). Rapid induction of selective transcription by auxin. Molec.Cell.Biol. June 1985: 1197-12o3.

2o. Bailey HM,Barker RD,Libbenga K, van der Linde PCG, Mennes AM, Elliott MC (1985) An auxin receptor in plant cells. Biol. Plant. 27: 1o5-112.

21. van der Linde PCG, Bouman H, Mennes AM, Libbenga K (1984). A soluble auxin binding protein from tobacco tissues stimulates RNA-synthesis in vitro. Planta 16o:1o2-1o8.

22. Hajek K, Jacobsen HJ(1985). genotype-specific soluble auxin-binding in etiolated pea epicotyls. Biol. Plant. 27(2-3): 11o-113.

23. Jacobsen HJ, Hajek K (1986). A critical reassessment of binding assay procedures. (submitted).

24. Scatchard G (1949). The attraction of proteins for small molecules and ions. Ann.N.Y.Acad.Sci. 51: 66o-672.

25. Maan A. (1983) pers. comm.

26. Chamness GC, Mc Guire WL (1975). Scatchard plots:Common errors in correction and interpretation. Steroids 26(4): 538-542.

27. Hedrick JL, Fischer EH (1965).On the role of pyridoxal
     5phosphate in phosphorylase.I. Absence of classical vita-
     min B_6-dependent enzymatic activities in muscle glycogen
     phosphorylase. Biochem. 4: 1337-1342.
28. Lönnig WE (1985) Quantitative inheritance in pea: Lines
     with several mutant genes for the number of sterile
     nodes. Genetica 66:29-4ọ.
29. Trewavas AJ(1983). Sensitivity is the regulating factor.
     Discussion Forum, TIBS: 35$-357.
3o. Bruns R, Lawson-Wendling K, Pugsley T (1983).A rapid
     filtration assay for soluble receptors using polyethy-
     lenimine treated filters. Anal. Biochem. 132:74-81.

Molecular Biology of Plant Growth Control, pages 267–277
© 1987 Alan R. Liss, Inc.

AUXIN TRANSPORT: INHIBITION BY 1-N-NAPHTHYLPHTHALAMIC ACID
(NPA) AND 2,3,5-TRIIODOBENZOIC ACID (2,3,5-TIBA)

Hans Depta

Pflanzenphysiologisches Institut, Universität Göttingen,
Cytologische Abteilung, Untere Karspüle 2,
D - 3400 Göttingen. FRG.

SOME FEATURES OF POLAR AUXIN TRANSPORT

Auxin transport in coleoptile segments is known to be
polar, basipetal from cell to cell including the free space
between successive cells and is not due to simple diffusion
(1, 2, 3). Further characterization with the help of the
specific inhibitors of polar transport NPA and 2,3,5-TIBA has
led to the polar secretion hypothesis (4), which proposed a
passive auxin uptake into cells and an active secretion at
the plasmamembrane. Preferential distribution of these
"secretion sites" towards the basal end of each cell is
suggested to account for the polarity of transport. This
latter argument which does not concern the origin and
development of polarity as pointed out by Hertel (5) is
supported by substantial experimental evidence (6, 7).
In accordance with the chemiosmotic diffusion hypothesis
(8, for reviews see 9, 10) uptake kinetics of auxin by tissue
segments (11, 12) cultured cells (8, 13) and membrane
vesicles (14) indicate that at least two types of carrier may
catalyse transmembrane IAA transport, in addition to non-
carrier mediated diffusive fluxes of the lipid soluble
undissociated acid. Recent studies on membrane vesicles (15,
16, 17) provide evidence for an electrogenic rather than an
electroneutral uptake carrier, thus accomodating the
electrical potential dependence and the high accumulation
ratio of IAA uptake into vesicles. The efflux carrier is
suggested to act as a "saturable leak" for IAA efflux
out of the cell. Whether this carrier operates as a
electrogenic uniport for IAA anions (8) is not clear, since
it has recently been shown, that IAA pulse velocity and basal

secretion in coleoptile segments is not stimulated by decreased cell wall pH (18) thus indicating a non electrogenic transport component for polar auxin secretion.

POLAR AUXIN TRANSPORT INHIBITION BY NPA AND 2,3,5-TIBA

NPA and 2,3,5-TIBA inhibit totally polar auxin transport within minutes to a saturable level. With both substances half saturation of inhibition is reached at 0.5-1 µM concentrations (19, 20). Uptake studies provide evidence for a specific interference of these inhibitors with the efflux carrier, thus stimulating net IAA uptake (4, 8, 14, 21).

Although 2,3,5-TIBA has an action on polar transport almost identical to NPA the results from transport tests and in vitro binding studies (19, 22, 23, 24, 25) reveal a more complicated interaction between these compounds. 2,3,5-TIBA in contrast to NPA is polarly transported (19) and second unlike NPA it inhibits auxin binding sites found in particulate membrane fractions (25). From these findings it has been suggested that NPA and 2,3,5-TIBA act via separate sites on IAA transport inhibition (19, 22).

In Hertel's laboratory polar auxin transport was studied in coleoptile segments to find out whether NPA and 2,3,5-TIBA inhibit auxin transport by competitive or rather non competitive interaction (26). In transport tests with a high IAA concentration in the donor block and using increasing concentrations of 1-NAA or 2-NAA to block IAA transport, a shift of half-maximum inhibition towards higher inhibitor concentrations was found (Fig. 1A, B). Such a shift is expected in the case of a competitive interaction between IAA and inhibitor at the transport site. On the other hand it could be shown, that the inhibition of IAA transport by NPA, 2,3,5-TIBA and 3,4,5-TIBA is independent of the IAA concentration applied in the donor block (Fig. 1C), indicating a none-competitive interaction of these substances with auxin. Further analysis raised evidence for a "common target process" with respect to tranport inhibition by NPA, 2,3,5-TIBA and 3,4,5-TIBA. Increasing concentrations of 3,4,5-TIBA, which only partially inhibit polar IAA transport can de-block the total inhibition caused by NPA or 2,3,5-TIBA (Fig. 2). This relative stimulation of auxin transport is not due to an "unspecific leakage", since it can be depressed again when an additional excess concentration of NPA or

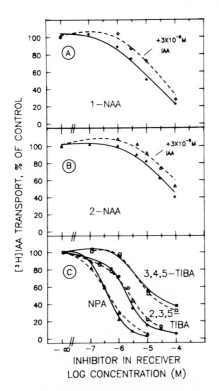

FIGURE 1.   Inhibition of IAA transport in maize
coleoptile segments. Increasing concentrations of 1-NAA (A),
2-NAA (B), NPA, 2,3,5-TIBA and 3,4,5-TIBA (C) were present
in basal receiver blocks during 1 h of transport. Closed
symbols: donors containing 0.1 μM ^{3}H IAA. Open symbols:
donors contained 0.1 μM ^{3}H IAA + 30 μM nonradioactive
IAA. Reprinted from (26).

2,3,5-TIBA is present in the receiver block. Similar effects
of the "soft" inhibitor 3,4,5-TIBA on IAA transport have been
observed in zucchini hypocotyl segments (27).

TANSMEMBRANE 2,3,5-TIBA TRANSPORT

The uptake of ^{14}C 2,3,5-TIBA and ^{14}C IAA into zucchini
hypocotyl segments has been studied in Rubery's laboratory
in order to answer the question whether 2,3,5-TIBA transport
involves carrier mediated components and whether or not these
are related to auxin transport sites (28). 2,3,5-TIBA uptake

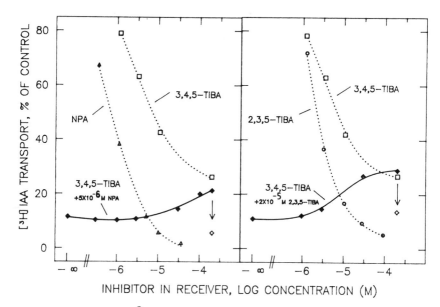

FIGURE 2.   ³H IAA (0.1 µM) transport inhibition by
NPA, 2,3,5-TIBA and 3,4,5-TIBA. The release of blocked
transport by increasing 3,4,5-TIBA concentrations. Closed
diamonds indicate transport tests in which receiver blocks
contained either 5 µM NPA (left) or 20 µM 2,3,5-TIBA (right),
plus the indicated 3,4,5-TIBA concentrations. In tests
marked by open diamonds receivers contained 10 µM NPA plus
200 µM 3,4,5-TIBA (left), or 100 µM 2,3,5-TIBA plus 200 µM
3,4,5-TIBA (right). Reprinted from (26).

is drastically increased when the pH of the external medium
is lowered. Similar pH dependencies have previously been
reported for IAA uptake into pea stem and maize coleoptile
segments (11, 12). In the case of 2,3,5-TIBA this finding may
entirely be attributed to a passive diffusion of undissociated
molecules into the cells due to the pH gradients between
external medium and cytoplasma respectively. Increased
radioactive 2,3,5-TIBA uptake with time was observed.
Radioactive TIBA uptake is increased by nonradioactive
2,3,5-TIBA or IAA after about 20 min of uptake throughout the
time course examinend (see Fig. 2A, B in 28). On the other
hand and in agreement with previous work (11, 12) the time
course of ^{14}C IAA uptake is affected differently by
competing "cold" IAA concentrations. In the presence of low
nonradioactive IAA ^{14}C IAA uptake is depressed regardless

of the uptake time measured, whereas high concentrations of "cold" IAA have opposing effects on the uptake course: an initial supression changes into a stimulation of uptake from about 40 min on (see Fig. 2C in 28).

The net uptake of ^{14}C -2,3,5-TIBA is significantly stimulated by increasing concentrations of nonradioactive 2,3,5-TIBA, NPA and auxins including IAA (Fig. 3A). A reciprocal specificity pattern is revealed for ^{14}C IAA uptake with respect to uptake stimulation by 2,3,5-TIBA, NPA and 1-NAA (Fig. 3B). But in contrast to 2,3,5-TIBA, radioactive IAA uptake is progressively depressed by increasing 2,4-D concentrations or by IAA concentrations up to 10 µM. With higher nonradioactive IAA concentrations ^{14}C-IAA uptake is again raised leading to a slight stimulation in the presence of 100 µM IAA (Fig. 3B). Such biphasic effects can be explained by simultaneously occuring inhibitory effects of nonradioactive IAA with both influx and efflux carriers. High cellular levels of nonradioactive IAA caused either by high concentrations in the external medium or by long incubation times are assumed to be responsible for a proportionally higher inhibition of ^{14}C-IAA efflux thus leading to an overall enhancement of uptake (11, 12). In our experiments no inhibitory effects on ^{14}C 2,3,5-TIBA uptake were detectable, which suggest that under these conditions saturable carrier mediated influx components do not participate in TIBA uptake. The reciprocal specificity pattern with regard to stimulation of radiolabelled 2,3,5-TIBA and IAA uptake suggest a common shared efflux carrier for both substances.

IN VITRO BINDING STUDIES

NPA binds with high affinity (K_d= 0.05 µM) to membrane associated sites (19). The binding affinities of NPA and other synthetic auxin transport inhibitors to these sites is in perfect harmony with the ability of these substances to block polar auxin transport (29). With respect to its inhibitory activity on transport 2,3,5-TIBA does not displace NPA from its high affinity binding sites except at much higher concentrations (19, 22, 23, 24).

To obtain further insight into the interaction between NPA and 2,3,5-TIBA, radiolabelled 2,3,5-TIBA was used for in vitro binding studies (26). These studies were complicated by different interactions of 2,3,5-TIBA with membrane fractions from zucchini hypocotyls and maize coleoptiles. First a heat

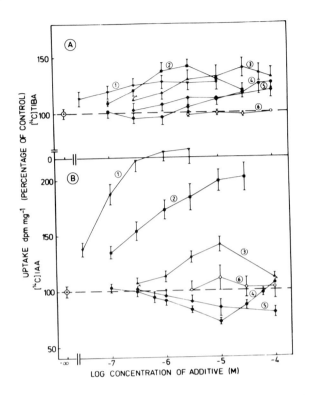

FIGURE 3.   The effect of nonradioactive NPA (1),
2,3,5-TIBA (2), NAA (3), IAA (4), 2,4-D (5) and benzoic acid
(6) on the uptake of A: ^{14}C 2,3,5-TIBA (0.06 µM) and B:
^{14}C IAA (0.1 µM) into zucchini hypocotyl segments.
Reprinted from (28).

stable binding was observed. The affinities of 2,3,5-TIBA,
other transport inhibitors and auxins to these sites do not
correlate with the biological activity of these ligands (for
further characterization see 30). This heat stable TIBA
binding also detectable in liver membrane fractions is not
considered relevant to specific auxin transport inhibition.
Second 2,3,5-TIBA accumulates into vesicles derived from
zucchini hypocotyls (31). TIBA uptake is sensitive to
artificially imposed pH gradients across the vesicle membrane
as judged from the effects of the proton ionophore carbonyl-
cyanide-p-trifluoro-methoxyphenyl-hydrazone (FCCP). Similar
vesicle preparations have recently been used to reconstruct

transmembrane auxin transport in vitro (14, 16).
    Nevertheless an in vitro binding of 2,3,5-TIBA could be
demonstrated which is influenced by NPA: In the presence of
3 µM NPA, radiolabelled 2,3,5-TIBA binding is less
efficiently inhibited by non-labelled 2,3,5-TIBA up to 5 µM
concentrations (Fig. 4). This NPA sensitive TIBA binding
represented by the cpm-difference in 2,3,5-TIBA binding
curves in the presence or absence of NPA was found to be
saturated by 2,3,5-TIBA with a K_d of approx. 0.3 µM (see
Fig. 3A in 26). In this context a small partial inhibition of
^{14}C 2,3,5-TIBA binding by less than micromolar NPA
concentrations was also detectable. The interaction of
micromolar NPA concentrations with 2,3,5-TIBA binding is
consistent with this type of higher affinity binding having a
physiological role. Since in agreement with previous work
(19, 22, 23) NPA binding was only slightly inhibited by
physiological concentrations (10 µM) of 2,3,5-TIBA we
approached the possibility of two classes of NPA binding
sites with different affinites for NPA, whereby one may have
a high affinity for 2,3,5-TIBA as well. However inhibition of
radiolabelled NPA binding by increasing 2,3,5-TIBA
concentrations measured at different NPA concentrations
revealed no competitive interaction between these two

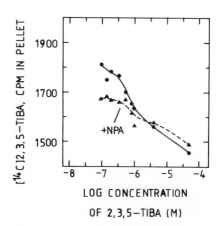

    FIGURE 4.   NPA-sensitive binding of 2,3,5-TIBA in
particulate fractions of maize coleoptiles. Inhibition of
^{14}C -2,3,5-TIBA binding in presence and absence of NPA
(3 µM). Reprinted from (26).

substances. Furthermore a separation by sucrose density
gradients of the bulk of NPA binding sites from TIBA
sensitive NPA binding was not achieved (26). Thus from the
detailed binding data currently available, it is not clear
whether inhibition of ^{3}H NPA binding by 2,3,5-TIBA is due
to a competitive binding of 2,3,5-TIBA to the NPA receptors
or due to a more complex inhibition of total NPA binding
involving allosteric interactions.

CONCLUSIONS

The well-documented ability of NPA and 2,3,5-TIBA to
inhibit auxin transport suggests specific interactions with
the IAA efflux carrier (8, 14, 22).

NPA has been found to bind with high affinity to plasmamembrane enriched membrane fractions (32). Since in such
preparations (see 23 for solubilized NPA receptors) no cross
competition between auxins and NPA could be demonstrated, NPA
does not seem to act on IAA transport by directly competing
for a common binding site (19, 22). 2,3,5-TIBA in contrast to
NPA is polarly transported probably using the same efflux
carrier as IAA (19, 28). These results fit easily into the
long-standing hypothesis that 2,3,5-TIBA acts directly at the
auxin transport site on the efflux carrier whereas NPA acts
via another discrete binding site (19, 22). On the other hand
in vivo transport inhibiton studies indicate that both
compounds act at the same site; albeit noncompetitively with
respect to auxin (26). Furthermore in vitro binding studies
with NPA and 2,3,5-TIBA show complex interactions which
cannot be explained by a simple competitive inhibition at the
appropriate sites for the substrate.

Thus a model for the IAA efflux carrier complex with
respect to NPA and 2,3,5-TIBA should be extended to a "three
site" model.

We have proposed such a model (28) in which one site can
bind auxin and a second 2,3,5-TIBA. Occupation of one of
these sites by its respective ligand should lead to
translocation across the membrane. Simultaneous occupation of
both catalytic transport sites on the other hand would cause
a translocation blockade. These proposed features take into
account firstly the non-competitive inhibition of polar
transport of IAA by 2,3,5-TIBA, secondly the polar transport
of IAA and 2,3,5-TIBA and thirdly the reciprocal specificity
pattern of both substances with respect to their stimulation

of uptake. However, a third binding site seems to be neccessary to explain the noncompetitive inhibitory effects of NPA on auxin transport. This site to which 2,3,5-TIBA putatively binds as well may operate as a pure regulatory site: occupied by ligands this subunit prevents efflux without having any catalytic translocation activity itself. Very recent work using protein synthesis inhibitors provide evidence that NPA binding sites are located on a separate polypeptide from the efflux carrier (Rubery personal communication). In these experiments membrane vesicles from cycloheximide pretreated tissue show reduced NPA and 2,3,5-TIBA stimulation of IAA uptake, whereas IAA efflux in the absence of NPA is not affected.

    Nevertheless alternative models (28) can be envisaged which account for partial binding inhibition. Generally the binding of a ligand to its specific site could induce conformational changes which inturn would decrease the binding affinity for a second ligand. Such models should predict a mixed rather than a pure noncompetitive inhibition of transport.

REFERENCES

1. Went FW (1928). Wuchsstoff und Wachstum. Rec Trav Bot Neerl 25: 1.
2. Weij HG van der (1934). Der Mechanismus des Wuchsstoff-transportes. II. Rec Trav Bot Neerl 31: 810.
3. Cande WZ, Ray PM (1976). Nature of cell-to-cell transfer of auxin in polar transport. Planta 129: 43.
4. Hertel R, Leopold AC (1963). Versuche zur Analyse des Auxintransports in der Koleoptile von Zea mays L. Planta 59: 535.
5. Hertel R (1985). Auxin transport, auxin action: some progress, many open questions. Jiorn Bot Ital. 119: in press.
6. Goldsmith MHM, Ray PM (1973). Intracellular localization of the active process in polar transport of auxin. Planta 111: 297.
7. Jacobs M, Gilbert SF (1983). Basal localization of the presumptive auxin transport carrier in pea stem cells. Science 220: 1297.
8. Rubery PH, Sheldrake AR (1974). Carrier-mediated auxin transport. Planta 118: 101.
9. Goldsmith MHM (1977). The polar transport of auxin. Ann Rev Plant Phys 28: 439.

10. Rubery PH (1980). The mechanism of transmembrane auxin transport and its relation to the chemiosmotic hypothesis of the polar transport of auxin. In Skoog F (ed): "Plant Growth Substances 1979," Heidelberg: Springer, p 50.

11. Davies PJ, Rubery PH (1978). Components of auxin transport in stem segments of Pisum sativum L. Planta 142: 211.

12. Edwards KL, Goldsmith MHM (1980). pH-Dependent accumulation of indole acetic acid by corn coleoptile segments. Planta 147: 457.

13. Rubery PH (1977). The specificity of carrier-mediated auxin transport by suspension-cultured crown gall cells. Planta 135: 275.

14. Hertel R, Lomax TL, Briggs WR (1983). Auxin transport in membrane vesicles from Cucurbita pepo L. Planta 157: 193.

15. Hertel R (1983). The mechanism of auxin transport as a model for auxin action. J Plant Phys 112: 53.

16. Lomax TL, Mehlhorn RJ, Briggs WR (1985). Active auxin uptake by zucchini membrane vesicles: quantitation using ESR volume and pH determinations. Proc Nat Acad Sci USA 82: 6541.

17. Benning C (1986). "Untersuchung an Membranvesikeln aus Cucurbita pepo: Der elektrochemische Potential-gradient als Triebkraft für den Auxintransport." Freiburg: Diplomarbeit. Fakultät für Biologie der Albert-Ludwigs-Universität.

18. Hasenstein K-H, Rayle D (1984). Cell wall pH and auxin transport velocity. Plant Phys 76: 65.

19. Thomson K-S, Hertel R, Müller S, Tavares JE (1973). N-1-Naphthylphthalamic acid and 2,3,5-triiodobenzoic acid. In vitro binding to particulate cell fractions and action on auxin transport in corn coleoptiles. Planta 109: 337.

20. Katekar GF, Geissler AE (1975). Auxin transport inhibitors. Plant Phys 56: 645.

21. Sussman MR, Goldsmith MHM (1981). Auxin uptake and action of N-1-naphthylphthalamic acid in corn coleoptiles. Planta 151: 15.

22. Sussman MR, Goldsmith MHM (1981). The action of specific inhibitors of auxin transport on uptake of auxin and binding of N-1-naphthylphthalamic acid to a membrane site in maize coleoptiles. Planta 152: 13.

23. Sussman MR, Gardner G (1980). Solubilization of the receptor for N-1-naphthylphthalamic acid. Plant Phys.

66: 1074.
24. Maan AC, Kühnel B, Beukers JJB, Libbenga KR (1985).
     Naphthylphthalamic acid-binding sites in cultured cells
     from Nicotiana tabacum. Planta 164: 69.
25. Ray RM, Dohrmann U, Hertel R (1977). Specificity of
     auxin-binding sites on maize coleoptile membranes as
     possible receptor sites for auxin action. Plant Phys
     60: 585.
26. Depta H, Eisele K-H, Hertel R (1983). Specific
     inhibitors of auxin transport: action on tissue
     segments and in vitro binding to membranes from maize
     coleoptiles. Plant Sci Lett 31: 181.
27. Jacobs M, Hertel R (1978). Auxin binding to subcellular
     fractions from Cucurbita hypocotyls: in vitro evidence
     for an auxin transport carrier. Planta 142: 1.
28. Depta H, Rubery PH (1984). A comparative study of
     carrier participation in the transport of 2,3,5-tri
     iodobenzoic acid, indole-3-acetic acid, and 2,4-di-
     chlorophenoxyacetic acid by Cucurbita pepo L. hypocotyl
     segments. J Plant Phys 115: 371.
29. Thomson K-S, Leopold AC (1974). In vitro binding of
     morphactins and 1-naphthylphthalamic acid in corn
     coleoptiles and their effects on auxin transport.
     Planta 115: 259.
30. Depta H (1981). "In vitro-Bindung des Auxintransport-
     hemmstoffs 2,3,5-Trijodbenzoesäure (2,3,5-TIBA) an
     Membranfraktionen aus Koleoptilen von Zea mays L. und
     Hypokotylen von Cucurbita pepo L." Freiburg: Inaugural-
     Dissertation. Albert-Ludwigs-Universität.
31. Depta H, Hertel R (1982). FCCP-sensitive association
     of weak organic acids to membrane vesicles. In Marmé
     D, Marre E, Hertel R (eds): "Plasmalemma and Tonoplast:
     Their Functions in the Plant Cell," Amsterdam:
     Elsevier, p 137.
32. Lembi CA, Morré DJ, Thomson K-S, Hertel R (1971).
     N-1-Naphthylphthalamic acid binding of a plasma
     membrane-rich fraction from maize coleoptiles. Planta
     99: 37.

Molecular Biology of Plant Growth Control, pages 279–288
© 1987 Alan R. Liss, Inc.

AUXIN RECEPTORS IN TARGET TISSUE

Marian Löbler, Karoline Simon,
Thomas Hesse, Dieter Klämbt

Botanical Institute, University of Bonn, Meckenheimer
Allee 170, D-5300 Bonn 1, FRG

ABSTRACT   A fast and reproducible method for the isolation of an auxin receptor was developed in order to obtain enough pure antigen for rabbit immunization. Solubilized membrane proteins from maize coleoptiles were
applied to ethanolamine-Sepharose. Fractions from the
flow through containing auxin binding activity were
further affinity purified on 2-OH-3,5-diiodobenzoic
acid-Sepharose. The 0.15 M NaCl eluate was subjected to
electrophoresis under denaturing conditions. The 20 kDa
auxin receptor was cut out of the gel, gel slices were
homogenized and used for rabbit immunization. The obtained IgG fraction was suitable for immunostaining of
the auxin receptor on Western blots. The receptor was
shown to be a glycoprotein by concanavalin A binding
with approximately 15-20 sugar residues. From in vitro
translation products of coleoptile mRNA in a wheat germ
system it can be calculated that the unprocessed receptor polypeptide contains a 20-30 amino acid leader sequence. The occurrence of the receptor within a variety
of maize tissues was followed by the Western blot technique and successive immunostaining.

INTRODUCTION

One of the multiple effects of auxin, a plant hormone,
is to induce cell elongation growth. The hormone signal has
to be recognized by a receptor which then triggers the
cell's response.
Very recently, an auxin binding protein with high affinity for 1-naphthylacetic acid was purified by immunological methods. The native auxin binding protein is a dimer

composed of two 20 kDa subunits (1). The pure protein was used to isolate monospecific antibodies against the auxin binding protein (IgG$_{anti\ ABP}$). Using these antibodies the auxin binding protein was localized within coleoptiles (2). In agreement with physiological data (3,4) the auxin binding protein was visible in the outer epidermal cells of the coleoptile by indirect immunofluorescence. Moreover, the monospecific IgG$_{anti\ ABP}$ blocked the auxin induced elongation growth of coleoptile tissue (2). These data justify the use of the term auxin receptor. In addition it was concluded that the auxin receptor resides at the outer surface of the plasmalemma (2).

The purification of monospecific antibodies yielded low amounts only (2). Higher amounts of antibodies are needed to screen maize tissues for the occurrence of auxin receptor protein. A pure receptor protein was necessary for rabbit immunization to obtain monospecific antisera thus circumventing the needs of additional affinity chromatography of the IgG fraction.

MATERIALS AND METHODS

Zea mays L. var. "Mutin" was grown at 27 C either in darkness or under summer daylight conditions in the greenhouse. Plants older than one week were taken from the fields.

Membrane proteins were prepared as described previously (1). Tissues with a high content of phenol oxidases were homogenized in the presence of insoluble polyvinyl pyrrolidone (1-1.5 g per g fresh tissue).

The auxin receptor was prepared from solubilized membrane proteins applied to ethanolamine-Sepharose equilibrated with binding buffer (10 mM Na-citrate pH 5.5, 5 mM MgSO$_4$, 250 mM sucrose). The flow through was tested by equilibrium dialysis (1). Fractions with auxin binding activity were pooled and chromatographed on 2-OH-3,5-diiodobenzoic acid-Sepharose (DIBA-Sepharose) which was eluted with 0.15 M NaCl.

Some samples of the salt eluate were digested with 0.01 units/ml of endoglucosidase H (Sigma) for 1h at 37 C. Proteins were then precipitated with 5 % trichloracetic acid (TCA) and processed for sodiumdodecylsulfate polyacrylamide gel electrophoresis (SDS-PAGE).

Most of the salt eluate was directly precipitated with TCA, prepared for and subjected to SDS-PAGE (1). Coomassie brilliant blue R 250 stained receptor protein bands were cut from the polyacrylamide gel, homogenized with Freund´s adjuvant, and used for rabbit immunization (5).

The IgG fraction was obtained from the antisera by three ammonium sulfate precipitations (33 % saturation) and affinity chromatography on protein A-Sepharose (Pharmacia) following manufacturer's instructions.

Blotted proteins were visible after amido black staining (6), immunostaining (2), or glycoprotein staining (7).

RNA was isolated according to the method of Chirgwin et al. (8), except that after RNA was dissolved in water a deproteination step with phenol:chloroform:isoamylalcohol (25: 24:1, v:v:v) was included.

Poly (A)$^+$ RNA was isolated from total RNA by affinity chromatography on oligo dT-Cellulose (Collaborated Research) according to Aviv and Leder (9) and translated in a wheat germ in vitro translation system (10) at an optimal Mg^{++} concentration of 2 mM.

The auxin receptor was immunoprecipitated from the translation products with anti receptor antibodies following the protocol of Anderson and Blobel (11).

RESULTS

The needs to obtain monospecific antisera forced us to develop a rapid preparation method for the auxin receptor. Figure 1 shows a flow diagram of auxin receptor preparation by two affinity chromatography steps. The protein absorbance profiles for both columns are shown together with auxin binding activity detected in single fractions. The salt eluate of DIBA-Sepharose was subjected to SDS-PAGE. Stained receptor protein bands were used for rabbit immunization and subsequently the IgG fraction was prepared from the antisera. IgG binds to the auxin receptor on Western blots, but a few cross reactivities still remain (Figs 1, 2).

From the conclusion that the auxin receptor resides at the outer surface of the plasmalemma (2) it is suggestive that it might be a glycoprotein.

The assay for glycoproteins with concanavalin A (con A) clearly shows the glycoprotein nature of the auxin receptor (Fig 2 lane 2). Endoglucosidase H digestion of the receptor preparation prior to electrophoresis shows a new protein band after immunostaining (Fig 2 lane 6) which is not visible in blots stained for glycoproteins (Fig 2 lane 7). From the observed difference in molecular weight of 3 kDa it can be calculated that approximately 15-20 sugar residues are attached to the auxin receptor monomer.

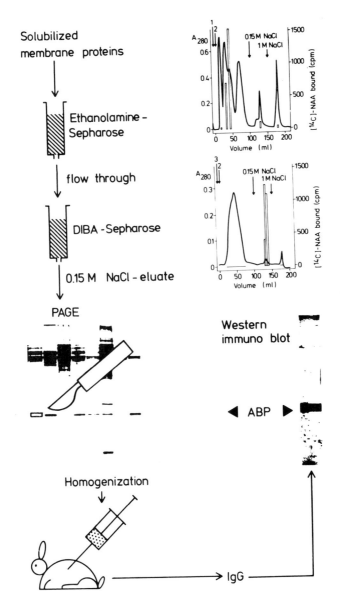

FIGURE 1. Flow diagram for the rapid preparation of auxin receptor protein. 1 – membrane proteins, 2 – binding buffer, 3 – flow through of ethanolamine-Sepharose.

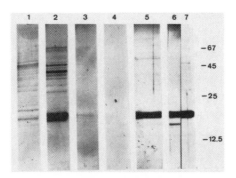

FIGURE 2. Western blot of receptor protein fractions after SDS-PAGE stained with amido black (lane 1), con A and horseradish peroxidase (lanes 2,7), con A in the presence of α-methylmannopyranoside and horseradish peroxidase (lane 3), horseradish peroxidase only (lane 4), anti receptor antibody and anti rabbit IgG-peroxidase conjugate (Sigma) (lanes 5,6). in lanes 6 and 7 the receptor protein fraction was treated with endoglucosidase prior to electrophoresis.

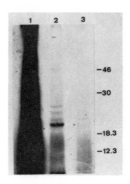

FIGURE 3. Fluorography of translation products from the wheat germ assay after SDS-PAGE. TCA precipitable proteins (lane 1), immunoprecipitated proteins (lane 2), competition of immunoprecipitation with unlabeled auxin receptor protein (lane 3).

Glycoproteins are usually synthesized at the rough endo-plasmic reticulum as precursors from which the signal peptide is split off after translocation into the ER lumen.

TABLE 1
DETECTION OF AUXIN RECEPTOR PROTEIN IN IMMUNOBLOTS AND
AUXIN BINDING ACTIVITY DURING DEVELOPMENT OF THE MAIZE
PLANT

Tissue	Age (d)	3	7	28	42	56	77	91
Root		+++						
		ooo						
Coleoptile (dark)		+++						
		ooo						
(light)		+++	+					
		oo_	oo					
Leaf (dark)		+++						
		⌀						
(light)		+++	++	+	+	+		
		o	⌀	⌀	⌀	⌀		
Internode[a]			++	++	++	++	++	-
			⌀	⌀	⌀	o	ooo	⌀
Tassel						+++	++	++
						ooo	ooo	o
Style[a]						++	++	++
						oo	ooᴐ	oo

+++ high, ++ moderate, + low amount of auxin receptor;
ooo high, oo moderate, o low auxin binding activity;
- no auxin receptor, ⌀ no auxin binding detected;
[a]Tissues from similar physiological stages.

As the wheat germ in vitro translation system does not
process newly synthesized proteins (12), we used this system
to translate poly (A)$^+$RNA from maize coleoptiles. The translation products were analysed by SDS-PAGE and fluorography.
In an immunoprecipitate one main protein band at a Mr of ca
20 kDa was detected (Fig 3 lane 2). In a competition assay
with unlabeled auxin receptor the newly synthesized receptor
protein is not precipitated (Fig 3 lane 3).

It can be calculated that the signal peptide of the precursor polypeptide should be around 20-30 amino acids in
length.

To investigate whether other growing tissues of the
maize plant may be auxin regulated by the same receptor different tissues were screened for receptor protein.

After SDS-PAGE membrane proteins were blotted onto nitrocellulose and then immunostaining was performed.

In roots from three day old etiolated maize seedlings
the content of auxin receptor increased from undetectable
levels within the root tip to high amounts within the root
hair zone, correlating with auxin binding activity (Table 1).

In coleoptiles from etiolated and green three day old
plants the auxin receptor is found, but auxin binding activity
was not detected in coleoptiles from green plants (Table 1).

Auxin receptor protein as well as auxin binding activity
can be found in etiolated and green leaves from three day old
seedlings, whereas both decline to undetectable levels in
green leaves from older plants (Table 1).

Within the upper three internodes of growing maize
plants auxin receptor was always found and was correlated
with binding activity. The auxin receptor was no longer detectable in internodes from plants not growing any longer
(Table 1).

In the tassel, again, high auxin binding activity was observed and auxin receptor was present in high amounts (Tab. 1).

Within female flowers auxin binding and receptor protein
was found in styles (Tab. 1), the cob was not tested.

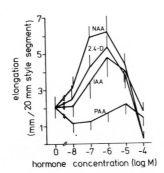

FIGURE 4. Elongation growth of 2 cm sections cut from
3-4 cm long styles at different hormone concentrations after
20 h in destilled water. Standard deviations are given by
vertical bars.

In elongation tests the styles showed a high rate of elongation growth upon auxin application. The effectiveness of inducing elongation growth is in the order NAA > 2,4-D > IAA ≫ PAA (Fig 4). Cell number per style remains constant from 0.5 to 20 cm long styles in vivo.

DISCUSSION

Rabbit antisera against the auxin receptor were produced according to the method of Boulard and Lecroisey (5). Initially we had to deal with considerable cross rectivities of the antibodies to other than auxin receptor proteins (data not shown). These cross reactivities could be highly reduced although not be eliminated completely, by adding SDS denatured bovine serum albumine at 1 % to the IgG solution. It is known that SDS denatured proteins are highly loaded with SDS. It seems that some of the antibodies are directed against SDS and SDS-protein complexes. In immuno double diffusion tests where SDS was the only antigen present precipitation lines were found with the antisera. However, these precipitation lines disappeared when the agar gels were washed in physiological salt (data not shown). A few cross reactivities that still remain (Fig 2 lane 5) seem to be due to sugar residues of glycoproteins. After sugar residues had been partially removed by endoglucosidase H digestion almost no cross reactivities were left, in parallel con A binding to glycoproteins is reduced (Fig 2 lanes 6,7).

The auxin receptor was shown to be glycosylated (Fig 2 lane 2) with approximately 15-20 sugar residues. This indicates that one or two glycosylation sites are present. On the other hand the unglycosylated precursor of the auxin receptor shows a Mr of approximately 20 kDa (Fig 3 lane 2). This points to a signal peptide of 20-30 amino acids in length which is in the usually observed range (13).

The findings that the auxin receptor is synthesized as a precursor and is glycosylated show that it resides in an extracytoplasmic compartment and supports our conclusion of a plasmalemma localization (2).

After the auxin receptor had been found in maize coleoptiles we were interested whether other fast growing tissues also contained auxin receptor and showed auxin binding activity.

Our results (Table 1) should be interpreted with caution. This is the more the case as in green tissues and in female flowers the high amount of phenole oxidases makes it very difficult to isolate membrane proteins. However, it seems that the auxin receptor is mainly expressed in fast growing non green tissues : coleoptile, root, tassel, style. Although young green leaves contain the auxin receptor it seems to disappear in older still growing leaves (Table 1).

We are now developing an enzyme linked immuno sorbent assay to quantify the amount of auxin receptor protein even in small tissue samples.

ACKNOWLEDGEMENTS

We thank U. Tillmann for critical reading the manuscript and H. Geithmann for preparing the figures.

REFERENCES

1. Löbler M,Klämbt D (1985a). Auxin-binding Protein from Coleoptile Membranes of Corn (Zea mays L.) I. Purification by Immunological Methods and Characterization. J Biol Chem 260:9848.

2. ----------------- (1985b). II. Localization of a Putative Auxin Receptor. J Biol Chem 260:9854.

3. Thimann KV, Schneider CL (1938). Differential Growth in Plant Tissues. Am J Bot 25:627.

4. Pope DG (1982). Effect of Peeling on IAA-induced Growth in Avena Coleoptiles. Ann Bot 49:493.

5. Boulard C, Lecroisey A (1982). Specific Antisera Produced by Direct Immunization with Slices of Polyacrylamide Gel Containing Small Amounts of Protein. J Immunol Meth 50:221.

6. Khyse-Andersen J (1984). Electroblotting of Multiple Gels: A Simple Apparatus Without Buffer Tank for Rapid Transfer of Proteins from Polyacrylamide to Nitrocellulose. J Biochem Biophys Meth 10:203.

7. Hawkes R (1982). Identification of Concanavalin A-Binding Proteins after Sodium Dodecyl Sulfate-Gel Electrophoresis and Protein Blotting. Anal Biochem 123:146.

8. Chirgwin JM, Przbyla AE,MacDonald RJ, Rutter WJ (1979). Isolation of Biologically Active Ribonucleic Acid from Sources Enriched in Ribonuclease. Biochemistry 18:5294.

9. Aviv H, Leder P (1972). Purification of Biologically Active Globin Messenger RNA by Chromatography on Oligothymidylic acid-Cellulose. Proc Nat Acad Sci USA 69:5294.

10. Anderson CW, Straus JW, Dudock BS (1983). Preparation of a Cell-Free Protein-Synthesizing System from Wheat Germ. In Wu R. Grossman L, Moldave K (eds):"Methods in Enzymology" Vol 101, Recombinant DNA Part C, New York: Academic Press, p 635.

11. Anderson DJ, Blobel G (1983). Immunoprecipitation of Proteins from Cell-Free Translations. In Fleischer S, Fleischer B (eds): "Methods in Enzymology" Vol 96, Biomembranes Part J, New York: Academic Press, p 111.

12. Burr FA, Burr B (1983). In Vitro Rocessing of Plant Preproteins. In Fleischer S, Fleischer B (eds): "Methods in Enzymology" Vol 96, Biomembranes Part J, New York: Academic Press, p 716.

13. Messing J, Geraghty D, Heidecker G, Hu NT, Kridl J, Rubenstein I (1983). Plant Gene Structure. In Kosuge T, Meredith CP, Hollaender A (eds): "Genetic Engineering of Plants," New York: Plenum Press, p 211.

Molecular Biology of Plant Growth Control, pages 289–298
© 1987 Alan R. Liss, Inc.

IN VITRO [^{3}H]GIBBERELLIN A$_1$ BINDING
TO SOLUBLE PROTEINS FROM GA-SENSITIVE
AND GA-INSENSITIVE DWARF MAIZE MUTANTS[1]

Brian Keith and Lawrence Rappaport

Plant Growth Laboratory/Department of Vegetable Crops
University of California at Davis
Davis, California 95616

ABSTRACT   A 100,000 x g supernatant fraction was
prepared from the first and second leaf sheaths of
light-grown Zea mays L. cv. Golden Jubilee. [^{3}H]GA$_1$
binding to a high molecular weight (HMW) fraction
(<500 Kdaltons) was demonstrated at 4°C using
Sephadex G-200 chromatography. The HMW component
was shown to be a protein and the [^{3}H] activity
bound to this protein was [^{3}H]GA$_1$ and not a metabolite. [^{3}H]GA$_1$ binding was pH-sensitive but largely
non-exchangeable on addition of unlabelled GA$_1$.
Both biologically active and inactive GAs were able
to inhibit [^{3}H]GA$_1$ binding. [^{3}H]GA$_1$ binding to an
intermediate molecular weight (IMW) fraction (40-100
Kdaltons) was also detected, provided cytosol was
first desalted using Sephadex chromatography.
[^{3}H]GA$_1$ binding to HMW and IMW fractions was also
detected in some dwarf, single-gene mutants of corn
that are either GA-sensitive (d$_1$, d$_2$, d$_3$, and d$_5$) or
GA-insensitive (D$_8$). HMW and IMW binding components
exhibit similar elution profiles after DEAE-
Sepharose chromatography. Additional evidence
suggests that the HMW-binding components are aggre-
gates derived from the IMW fraction. Neither the
HMW nor IMW binding components exhibit 2β-hydroxy-
lase activity.

[1]Supported by National Science Foundation Grant
NSF-PCM80-14942.

INTRODUCTION

Recent studies with cucumber hypocotyls have indicated that a soluble binding protein is capable of binding [^{3}H]GA$_4$ in vitro with many of the characteristics expected for a putative gibberellin receptor (1, 2, 10). Binding is saturable, exchangeable, pH-sensitive, and of high affinity; only biologically active GAs, but not inactive GAs, are capable of competing for these [^{3}H]GA$_4$ binding sites.

Attempts to demonstrate such in vitro binding properties with other plants have failed (3, 4, 9). In this present paper, we report on some [^{3}H]GA$_1$ binding characteristics in normal corn and five single-gene dwarf mutants of corn.

MATERIALS AND METHODS

Caryopses of maize (Zea mays L. cv. Golden Jubilee; Lagomarsino Seeds, Inc., Sacramento, CA) were grown in vermiculite for 7 days at 30°C on a 16-h photoperiod.

Caryopses of dwarf-1, dwarf-2, dwarf-3, dwarf-5, and Dwarf-8 (developed from lines originally provided by Dr. B. O. Phinney, Department of Biology, University of California, Los Angeles, CA 90024, and Dr. R. J. Lambert, Maize Cooperative, Department of Agronomy, University of Illinois, 1102 South Goodwin Avenue, Urbana, IL 61801) were also sown in vermiculite. After the first week of growth, they were watered with Hoagland solution for 6 weeks.

One- to two-cm lengths of the basal portion of the first leaf sheath, and the enclosed leaves and sheaths were excised, pooled, and weighed. To each g fresh weight of leaf sheaths was added 2 ml of 30-mM Tris-maleate buffer (pH 6.2) containing 0.1 g ml^{-1} PVP, 5 mM DTT, and 70 μM PMSF and the mixture homogenized in a Wareing blender. The extract was passed through 4 layers of cheesecloth and centrifuged at 100,000 x g for 1 h at 0-4°C.

The dwarf mutant plants were extracted when 6 weeks old. Plants at this stage possessed 13 leaves, and only the internode tissue and basal portions of the leaf sheaths above the fifth eldest leaf were extracted.

[^{3}H]GA$_1$ binding to soluble macromolecules was measured following the separation of bound from free [^{3}H]GA$_1$, using either Sephadex G-200 (2.5 cm x 35 cm gel bed) or G-100 (1 cm x 30 cm gel bed) chromatography. For anion exchange chromatography studies, leaf sheaths were extracted in 30-mM Tris-HCl (pH 7.0), instead of 30-mM Tris-maleate (pH 6.2). PVP, DTT, and PMSF were used at concentrations cited earlier.

The 2β-hydroxylase assay of Patterson <u>et al</u>. (1975), and Smith and MacMillan (1984) was used. The specific activity of [^{3}H]GA$_1$ was 41.5 Ci mmol^{-1}.

RESULTS

After 100,000 x g cytosol from 'Golden Jubilee' was incubated with [^{3}H]GA$_1$ (10^{-7} M concentration), the bound activity was separated from unbound activity using Sephadex G-200 gel filtration (Fig. 1). The bulk of the binding was of high molecular weight (HMW) -- occurring in the void volume of the gel bed which possessed an exclusion limit of 500 Kdaltons. There was very little binding in the intermediate molecular weight (IMW) range (fraction numbers 16-26; Fig. 1).

We used reversed phase HPLC to demonstrate that all the bound activity was in the form of [^{3}H]GA$_1$ (data not shown). Maximal binding levels were achieved in less than one hour and remained stable for up to 3 days. Only 20% of specific binding levels of [^{3}H]GA$_1$

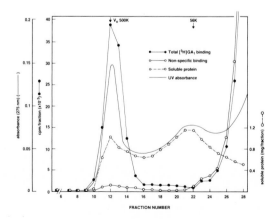

FIGURE 1.   Sephadex G-200 chromatography of [^{3}H]GA$_1$ binding components in 'Golden Jubilee' cytosol.

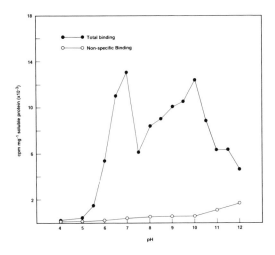

FIGURE 2. The effect of pH on $[^{3}H]GA_1$ binding levels by 'Golden Jubilee' cytosol.

TABLE 1
The effect of heat and degradative enzymes on $[^{3}H]GA_1$ binding levels from 'Golden Jubilee' cytosol.

Treatment	$[^{3}H]GA_1$ bound (% of control)
Control	100
Heat (85°C for 1 h)	76
Protease	103
Pepsin	97
Trypsin	1
Papain	2
RNase A	106
DNase 1	108
Phospholipase C	107

were exchangeable on addition of a 100-fold excess of
unlabelled GA_1 (data not shown).

[^{3}H]GA_1 binding was strongly pH-dependent (Fig. 2)
with two optima -- one at pH 7.0 and another at pH 10.0.
The physiological significance of the binding at pH 10.0
is unclear at the present time.

When 100,000 x $\underline{g}$ cytosol was heated for up to 1 h at
85°C or treated with a variety of degradative enzymes (1
mg ml^{-1} final concentration) for 5 h and then incubated
with [^{3}H]GA_1 (10^{-7} M concentration), only trypsin and
papain completely abolished binding (Table 1). These
results suggest that [^{3}H]GA_1 binds to a protein, although
it should be noted that protease and pepsin were relatively ineffective in abolishing binding, as was heat
treatment.

TABLE 2

Effect of different GAs and non-GAs on [^{3}H]GA_1 binding to
100,000 x $\underline{g}$ cytosol of 'Golden Jubilee' at pH 6.5 and
10.0.

Treatment	% inhibition		Biological Activity[d]
	pH 6.5	pH 10.0	
[^{3}H]GA_1 (control) 10^{-7} M[c]	0[a]	0[b]	
+ GA_1	98	98	+++
+ GA_3	88	85	++++
+ GA_4	98	98	+++
+ GA_{13}	63	95	0
+ GA_{17}	77	86	0
+ GA_{19}	45	77	++
+ GA_{20}	93	63	+++
+ IAA	48	87	0
+ ABA	62	92	0

[a] Control=0% inhibition=8,400 cpm mg^{-1} protein at pH 6.5.
[b] Control=0% inhibition=9,100 cpm mg^{-1} protein at pH 10.0.
[c] Unlabeled GAs and non-GAs used at 20-fold concen-
    tration.
[d] Relative activities: ++++ very high, +++ high, ++
    moderate, + low, 0 very low, inactive.

At pH 6.5, [^{3}H]GA$_1$ binding levels were inhibited by a variety of GAs and even non-GAs, such as ABA and IAA (Table 2). It has not been ascertained whether this inhibition is of a competitive nature. GA$_{17}$ and GA$_{13}$, which are biologically inactive in corn, significantly inhibited [^{3}H]GA$_1$ binding while GA$_{19}$ and GA$_{20}$, which are thought to be biologically active only by virtue of metabolic conversion to the active GA$_1$ (7), were also able to inhibit [^{3}H]GA$_1$ binding. At pH 10.0, all GAs and non-GAs tested were able to inhibit [^{3}H]GA$_1$ binding.

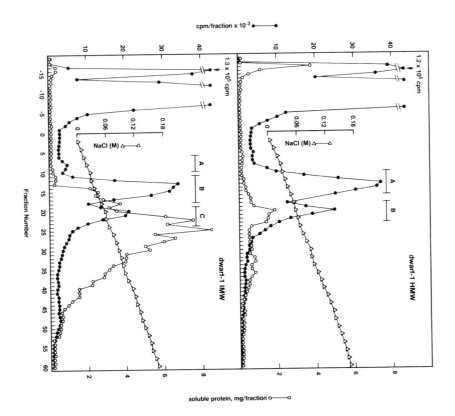

FIGURE 3. Separation by anion-exchange chromatography of HMW and IMW binding components from extracts of dwarf-1 corn.

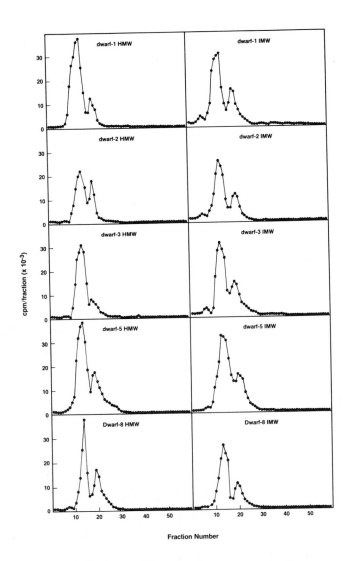

FIGURE 4. Separation by anion exchange chromatography of HMW and IMW binding components from extracts of some single-gene dwarf mutants of corn.

As stated earlier, $[^3H]GA_1$ binding could not be observed in the IMW fraction (fraction numbers 16-26, Fig. 1) when 100,000 x $\underline{g}$ cytosol was incubated with $[^3H]GA_1$. However, if cytosol was first passed through Sephadex G-200, followed by concentration (by ultrafiltration) of the IMW fraction and incubation with $[^3H]GA_1$, then binding was observed after rechromatography using G-200. All the binding, however, was in the void volume, suggesting that IMW binding components had aggregated to HMW binding components.

Both 100,000 x $\underline{g}$ cytosol and the HMW and IMW fractions, prepared after G-200 chromatography, failed to exhibit any 2β-hydroxylase activity (data not shown).

Similar $[^3H]GA_1$ binding properties were observed in extracts of the different dwarf mutants. An HMW fraction from d_1 was fractionated using Sephadex G-200 chromatography. After incubation for 1 h with $[^3H]GA_1$ (10^{-7} M concentration), HMW binding components were separated by anion exchange chromatography (DEAE-Sepharose; 2.5 cm diam. x 2.5 cm ht.), using a salt gradient (0 to 0.175 M) to achieve separation (Fig. 3). An initial peak of binding eluted from the column and represented $[^3H]GA_1$-bound activity that failed to adsorb to the anion exchanger. This was followed by a large peak representing free $[^3H]GA_1$, and the salt gradient was not begun until this peak had eluted. Two HMW binding components (A and B, see Fig. 3) were eluted at about 30 and 50 mM NaCl concentration, respectively.

Using anion-exchange chromatography, at least 3 IMW binding components were separable from d_1 (see Fig. 3). Two of the binding peaks (B and C) coincided exactly with the two binding peaks seen in the HMW fraction. A third and minor binding peak was seen in the IMW fraction eluting early in the salt gradient (see peak A).

Figure 4 shows the $[^3H]GA_1$ binding profiles of d_1, d_2, d_3, d_5, and D_8 for the HMW and IMW fractions, using anion-exchange chromatography. No consistent differences were noted between the different mutants, irrespective of whether the mutants were GA-sensitive (d_1, d_2, d_3, and d_5) or GA-insensitive (D_8) (see ref. 6).

DISCUSSION

Although we have demonstrated pH-sensitive binding of $[^3H]GA_1$ to a soluble protein(s) present in corn cytosol, this binding is largely non-exchangeable and exhibits too poor a specificity for the binding to be of any apparent functional significance. However, the observation that the IMW fractions can aggregate to give HMW-$[^3H]GA_1$ binding complexes may be indicative of extraction artifacts: different binding characteristics might be observed with the same HMW-binding protein extracted under different conditions.

At least one test for a functional role of GA binding proteins in GA-induced responses would be the discovery of binding mutants (4). Dwarf-8, a GA-insensitive mutant of corn, binds $[^3H]GA_1$ as well as the GA-sensitive mutants (d_1, d_2, d_3, and d_5). That D_8 can bind $[^3H]GA_1$ need not imply that this binding lacks functional significance, since the gene conferring GA-insensitivity may be involved in some process beyond the level of GA perception.

Stoddart (1982) has recently proposed that some $[^3H]GA_1$-binding proteins may be enzymes involved in GA_1 metabolism such as the 2β-hydroxylase that converts GA_1 to GA_8. It is of interest, therefore, that under conditions in which we have isolated $[^3H]GA_1$-binding fractions, these same fractions fail to hydroxylate GA_1. The possibility that other enzymes of metabolism in the GA pathway are responsible for the $[^3H]GA_1$ binding observed here receive little support from the evidence obtained from the mutants d_1, d_2, d_3, and d_5. These mutants are known to have specific genetic lesions in the GA biosynthetic pathway (7) and yet they all yielded similar binding profiles (see Fig. 4). The similarity of the HMW- and IMW-binding profiles obtained with the dwarf mutants may be partly explained on the basis that the IMW-binding components appear to aggregate to HMW binding components. The significance of this, if any, is unclear at the present time.

ACKNOWLEDGEMENTS

We thank Robert H. Thompson for technical assistance and Professors J. MacMillan and N. Takahashi for GAs used in this study.

REFERENCES

1. Keith B, Brown S, Srivastava LM (1982). In vitro binding of gibberellin A_4 in extracts of cucumber using DEAE-cellulose filters. Proc Nat Acad Sci 79:1515-1519.

2. Keith B, Foster NA, Bonettemaker M, Srivastava LM (1981). In vitro gibberellin A_4 binding to extracts of cucumber hypocotyls. Plant Physiol 68:344-348.

3. Keith B, Srivastava LM (1980). In vivo binding of gibberellin A_1 in dwarf pea epicotyls. Plant Physiol 66:962-967.

4. Kende H, Gardner G (1976). Hormone binding in plants. Ann Rev Plant Physiol 27:267-290.

5. Patterson R, Rappaport L, Breidenbach RW (1975). Characterization of an enzyme from Phaseolus vulgaris seeds which hydroxylates GA_1 to GA_8. Phytochemistry 14:363-368.

6. Phinney BO (1956). Growth response of single-gene dwarf mutants in maize to gibberellic acid. Proc Nat Acad Sci 42:185-189.

7. Phinney BO, Spray C (1982). Chemical genetics and the gibberellin pathway in Zea mays L. In Wareing PF (ed): "Growth Substances," Academic Press, London, pp 101-110.

8. Smith VA, MacMillan J (1984). Purification and partial characterization of a gibberellin 2β-hydroxylase from Phaseolus vulgaris. J Plant Growth Regul 2:251-264.

9. Stoddart JL (1982). Gibberellin perception and its primary consequences: the current status. In Wareing PF (ed): "Plant Growth Substances 1982," Academic Press, London, pp 131-140.

10. Yalpani N, Srivastava LM (1985). Competition for in vitro [^{3}H]gibberellin A_4 binding in cucumber by gibberellins and their derivatives. Plant Physiol 79:963-967.

Molecular Biology of Plant Growth Control, pages 299–308
© 1987 Alan R. Liss, Inc.

IN-VITRO GIBBERELLIN A_1 BINDING
TO A SOLUBLE FRACTION FROM DWARF PEA EPICOTYLS[1]

Coralie C. Lashbrook, Brian Keith, and Lawrence Rappaport

Plant Growth Laboratory/Department of Vegetable Crops
University of California, Davis, California 95616

ABSTRACT   A 100,000 x $\underline{g}$ cytosol was prepared from
the epicotyls of 4-day-old dark grown peas. Incu-
bation of crude cytosol in 10^{-7} $\underline{M}$ [^{3}H]GA$_1$ ± 10^{-5} $\underline{M}$
unlabelled GA$_1$ resulted in low levels of specific
binding to two macromolecular components. Further
purification of the cytosol by G-200 chromatography
produced an intermediate molecular weight fraction
greatly enriched in the ability to bind [^{3}H]GA$_1$.
Binding of [^{3}H]GA$_1$ to the G-200 fraction was rapid,
saturable, and exchangeable with unlabelled GA$_1$. A
biologically inactive GA, GA$_{17}$, did not reduce
[^{3}H]GA$_1$ binding levels. Two pH optima for binding
were observed, one at pH 6.75-7.0 and the other at
pH 7.75-8.0.
    Binding specificity was negligible in vitro.
In contrast, an in vivo pea binding system retained
specificity for biologically active GAs. Evidence
for a common [^{3}H]GA$_1$-bound component of the intermediate molecular weight fraction in vivo and in
vitro was obtained using anion exchange chromatography. Further optimization of extraction
procedures may be required before in vitro specificity can be demonstrated.

[1]Supported by National Science Foundation Grant
NSF-PCM80-14942 and a U. C. Davis Graduate Research Award
to C. C. Lashbrook.

INTRODUCTION

Gibberellins are an important class of plant hormones involved in the regulation of growth and development in higher plants. Much is known about the hormone's elaborate biosynthetic pathway and the marked effects of exogenous gibberellin (GA) application on such physiological processes as seed germination, stem elongation, and flowering. Evidence that gibberellins may act at the level of gene expression has recently triggered considerable interest (1). Despite such advances, however, the actual mechanism of GA action remains unclear.

The mediation of animal hormone response by proteinaceous receptors is well-characterized. The binding of hormone to these soluble or membrane-bound macromolecules has been shown to initiate a cascade of cellular events, leading to a specific physiological response. It is reasonable to speculate that the primary mechanism of plant hormone action may be the binding of hormone to a highly specific receptor. However, attempts to characterize plant hormone binding proteins in terms of known animal hormone receptor traits have often been inconclusive, particularly in in vitro assay systems (2).

The in vivo binding of [^{3}H]GA$_1$ to soluble dwarf pea proteins was first described by Stoddart et al. (3). Excised epicotyls were fed [^{3}H]GA$_1$ of high specific activity; subsequent extraction and gel filtration of the 100,000 x g cytosol revealed that two macromolecular components were labelled. These binding components were termed the high molecular weight fraction (HMW) and the intermediate molecular weight fraction (IMW).

Keith and Srivastava subsequently confirmed the specific binding of [^{3}H]GA$_1$ to the HMW and IMW fractions (4). Biologically active GAs were able to inhibit the binding of [^{3}H]GA$_1$ to a combined HMW-IMW fraction; inactive GAs or non-GAs could not. The in vivo system failed, however, to support other hormone characteristics such as exchangeability or pH sensitivity. Attempts to construct an in vitro pea binding system were unfruitful.

Here, we report on the in vitro characterization of the IMW binding fraction from dwarf pea epicotyls.

METHODS

Chemicals.

[^{3}H]GA$_1$ (41.5 Ci mmol^{-1}) was purchased from Prof. L. M. Srivastava. Radiolabelled hormone was further puri- fied using reversed phase HPLC to greater than 99% radiochemical purity. GA$_1$ was provided by Abbott Laboratories. GA$_{13}$ was obtained from Imperial Chemical Industries. GA$_4$ and GA$_{17}$ were gifts from Prof. N. Takahashi. Abscisic acid was obtained from Hoffman La Roche. Chromatographic resins were purchased from Pharmacia, Inc. All other chemicals were from Sigma Co.

Plant Material.

Imbibed dwarf pea seeds ('Progress No. 9', Lot #10750-15351; Ferry Morse Seed Co.) were sown in flats of vermiculite moistened with water and grown in the dark for 4-5 days. Etiolated seedlings were used in all studies.

In vitro Preparative Methods.

All procedures were performed at 0-4°C. The upper 1 cm of epicotyl was excised and the plumule removed. Epicotyls were extracted using a Wareing blender contain- ing 2 ml of extraction buffer per gram fresh weight of tissue. Extraction buffer was 30 mM Tris-Cl or Tris- Maleate, 0.1 g PVP/ml, 250 mM sucrose, 70 µM PMSF, 2 mM DTT, pH 7.2. Brei was pressed through four layers of cheesecloth and centrifuged at 100,000 x g for 1 hour in a Beckman L2-65B ultracentrifuge. The supernatant was reduced in volume using an Amicon stirred ultrafiltration cell with a PM-10 membrane filter (10,000 MW cutoff). Concentrated cytosol was loaded onto a bed of 2.6 x 36 cm Sephadex G-200 and eluted with 10 mM Tris-Cl or Tris- Maleate, 70 µM PMSF, 2 mM DTT, pH 7.2. Five-ml fractions were collected. A protein profile was obtained by the method of Bradford (5), using BSA as a standard. The molecular weights of the HMW and IMW peaks were estimated using BSA, blue dextran and [^{3}H]GA$_1$ as standards. Peaks of interest were pooled, concentrated on Amicon filters, and used for in vitro binding assays.

In _vitro_ Binding Studies.

Pea protein was incubated on ice in [^{3}H]GA$_1$ in the presence or absence of a 100-fold excess of unlabelled GA$_1$. [^{3}H]GA$_1$ concentrations used were in the 50 nM to 0.1 µM range. Soluble protein concentration was typically 3-8 mg ml^{-1}.

In association studies, aliquots were removed at various time points, and the bound fraction separated from the free label by G-100 gel filtration (1.0 x 29 cm). The radioactivity in each fraction was counted in a Beckman LS 8000 Scintillation Counter, and a protein profile obtained by the Bradford method.

Exchangeability was similarly assessed at various times following the addition to the [^{3}H]GA$_1$-bound sample of a 100-fold excess of GA$_1$ or GA$_{17}$.

For pH studies, pea protein was differentially pH'd using 1 M Tris or 1 M Maleate. Binding at equilibrium was assessed using G-100 gel filtration.

In specificity studies, proteinaceous fractions were incubated in [^{3}H]GA$_1$ containing a 100-fold excess of unlabelled competitor. Samples were chromatographed following the saturation of binding in the control sample.

In _vivo_ Studies.

Preparative methods were generally as described by Keith and Srivastava. Binding was evaluated using 1.0 x 29 cm beds of Sephadex G-100 eluted with 10 mM Tris-Cl, 5 mM reduced glutathione, pH 7.2. Protein determination was by the method of Bradford. All measurements were conducted on the third day of incubation.

Anion Exchange Chromatography.

One-cm diameter columns containing a 1-ml bed volume of DEAE Sepharose (Fast Flow) were used. The column was developed using a linear gradient (40 mls) of 0 to 0.175 M NaCl containing 10 mM Tris-Cl, pH 7.2.

RESULTS

Figure 1 depicts the Sephadex G-200 elution profile of 100,000 x g cytosol derived from dwarf pea epicotyls. The molecular weights of the high molecular weight fraction (HMW) and the intermediate molecular weight fraction (IMW) have been estimated to be 600 and 56 kilodaltons, respectively.

When crude cytosol was incubated in 10^{-7} $\underline{M}$ [^{3}H]GA$_1$ ± 10^{-5} $\underline{M}$ GA$_1$, subsequent G-200 gel filtration revealed low levels of specific binding in the HMW and IMW elution zones (data not shown). Specific binding of [^{3}H]GA$_1$ to the IMW fraction was greatly enhanced, however, if the IMW component was isolated from the crude cytosol by gel filtration prior to hormone addition. This enhancement of binding may result from the chromatographic removal of a low molecular weight inhibitor.

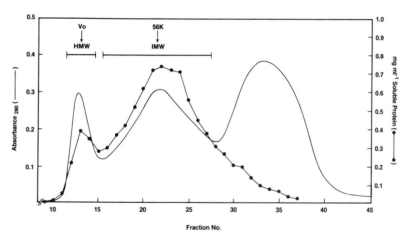

Figure 1.   Preparative G-200 gel filtration of pea cytosol.

Figure 2 depicts the binding of [^{3}H]GA$_1$ to the IMW fraction over a wide range of physiological pH values. The biphasic profile was a consistent feature. Binding optima were observed at pH intervals of pH 6.75 to 7.0 and pH 7.75 to 8.0. Further purification of the IMW fraction may be necessary to more adequately separate these pH profiles.

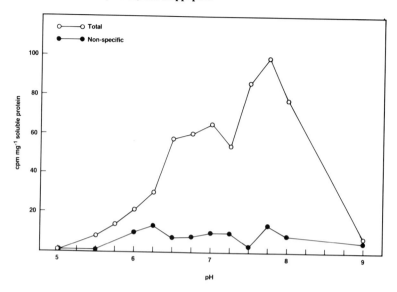

Figure 2.   pH sensitivity of $[^3H]GA_1$ binding to the IMW fraction

Hormone binding to the IMW component is rapid, saturable and exchangeable (Figure 3).   Saturation of binding protein is typically achieved by 1.5 - 2 hours; total binding levels have been observed to remain stable for periods exceeding 72 hours.   An excess of unlabelled GA_1 added at 1.5 hours rapidly displaced a significant amount of $[^3H]GA_1$ from the IMW binding site(s).   Complete displacement was evident by 26 hours. GA_{17}, a biologically inactive GA in the pea bioassay, appeared unable to exchange to any significant degree. Other specificity experiments using a variety of GAs and non-GAs were inconclusive (data not shown).

In contrast to the <u>in vitro</u>, specificity appears to be retained in an <u>in vivo</u> system.   Table 1 contains <u>in vivo</u> specificity data for a combined HMW-IMW fraction of pea. GA_1 and GA_4 significantly reduced binding of $[^3H]GA_1$ to pea protein while GA_{13}, GA_{17}, and ABA were not inhibitory.

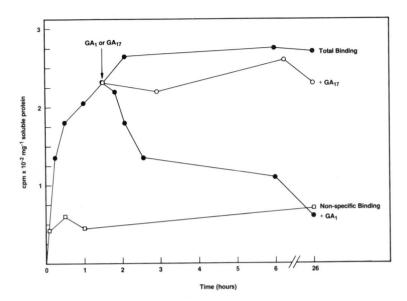

Figure 3.   [^{3}H]GA$_1$-IMW association and exchange-ability.

TABLE 1.   <u>In</u> <u>vivo</u> specificity of a combined HMW-IMW fraction after Sephadex G-100 gel filtration.

Treatment	Total Binding: % of Control
[^{3}H]GA$_1$ (control)[a]	100
+ GA$_1$[b]	29
+ GA$_4$	42
+ GA$_{13}$	112
+ GA$_{17}$	100
+ ABA	100

[a] [^{3}H]GA$_1$ concentration, 10^{-7} M.
[b] GAs and non-GAs were used at 10^{-5}M.

Evidence for a common [³H]GA$_1$-bound component of the IMW fraction <u>in vivo</u> and <u>in vitro</u> was provided by chromatographic separation of bound fractions on DEAE Sepharose columns (Figure 4). The IMW binding protein obtained under both conditions exhibited similar elution properties.

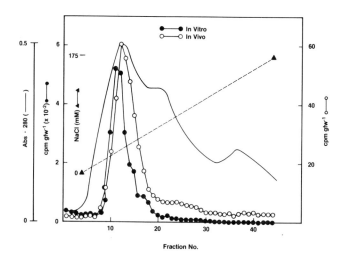

Figure 4. Separation of <u>in vivo</u> and <u>in vitro</u> binding components of the IMW fraction using DEAE Sepharose gel filtration.

DISCUSSION

Attempts to demonstrate GA binding to soluble proteins <u>in vivo</u> and <u>in vitro</u> have historically been inconclusive, primarily due to difficulties experienced in satisfying certain requirements for receptor status. It is possible that the method of extraction may be a critical parameter in determining whether receptor characteristics can be detected.

We have described an <u>in vitro</u> binding system exhibiting several hormone receptor characteristics previously undescribed in pea. The binding levels we currently observe are substantially above those obtained prior to optimization of extraction procedures. Evidence for a common $[^3H]GA_1$-bound component of the IMW fraction in vivo and in vitro has been described, yet specificity of binding is evident only in the in vivo system. It is possible that our current preparative methods do not adequately protect the IMW binding site(s). An extraction-induced alteration in protein conformation could explain our current lack of in vitro specificity.

The physiological significance of the IMW fraction is currently unclear. Further purification of the binding component(s) of this fraction will enable us to expand our current findings and may facilitate evaluation of additional undescribed receptor characteristics.

The conversion of biologically active GA_1 to inactive GA_8 metabolite by a 2β- hydroxylase has been previously described (6). We are currently evaluating procedures which might allow us to differentiate between the activities of the 2β- hydroxylase enzyme and the IMW fraction in pea. Successful resolution of these two components would resolve a major issue in gibberellin binding studies and would significantly increase the likelihood that the IMW fraction could function as a GA receptor.

ACKNOWLEDGEMENTS

We thank Mr. Robert H. Thompson for technical assistance and Ms. Nelinia Henry for preparation of the manuscript.

REFERENCES

1. Jacobsen JV, Higgins TJV (1978). In: Letham DS, Goodwin PB, Higgins TJV (eds): "Phytohormones and Related Compounds: A Comprehensive Treatise, Vol. 1," New York: Elsevier North Holland Biomedical Press.
2. Kende H, Gardner G (1976). Hormone binding in plants. Ann Rev Plant Physiol 27:267.

3.   Stoddart JL, Breidenbach W, Nadeau R, Rappaport L
     (1974). Selective binding of [^{3}H]gibberellin A$_1$ by
     protein fractions from dwarf pea epicotyls. Proc
     Nat Acad Sci 71:3255.

4.   Keith B, Srivastava LM (1980). In vivo binding of
     gibberellin A$_1$ in dwarf pea epicotyls. Plant
     Physiol 66:962.

5.   Bradford MM (1976). A rapid and sensitive method
     for the quantitation of microgram quantities of
     protein using the principle of protein-dye binding.
     Anal Biochem 72:248.

6.   Patterson R, Rappaport L, Breidenbach RW (1975).
     Characterization of an enzyme from Phaseolus
     vulgaris which hydroxylates GA$_1$ to GA$_8$. Phytochem
     14:363.

Molecular Biology of Plant Growth Control, pages 309–314
© 1987 Alan R. Liss, Inc.

PARTIAL PURIFICATION OF A GIBBERELLIN BINDING PROTEIN FROM CUCUMBER HYPOCOTYLS[1]

Nasser Yalpani and Lalit M. Srivastava

Department of Biological Sciences,
Simon Fraser University, Burnaby, British Columbia,
Canada V5A 1S6

ABSTRACT The structural specificity and kinetic data of gibberellin (GA) binding to a protein in cucumber cytosol suggest the presence of a candidate GA receptor. In this communication we report on the partial purification of this protein using $(NH_4)_2SO_4$ precipitation, ion exchange and hydroxylapatite chromatography.

INTRODUCTION

The cytosol of cucumber hypocotyls contains a protein with characteristics expected of gibberellin (GA) receptors. Results from equilibrium dialysis, gel filtration and DEAE filter paper assays have demonstrated that its binding sites are saturable (n=0.4 pmol mg^{-1} soluble protein) and bind biologically active GAs reversibly (halflife of dissociation =6 min) and with high affinity (Kd=70 nM) (1,2). Using ^3H-GA$_4$ and a series of selected GAs in an in vitro assay system we have partially characterized the GA-binding site of this protein (3). A strong correlation between affinity for the protein and in vivo growth promoting activity was observed. Only γ-lactonic C-19 GAs with a 3 β-hydroxyl and an unmethylated C-6 carboxyl group, such as GA$_4$ showed high affinity binding. Some GAs which are metabolized to GA$_4$ in cucurbits and have high in vivo activity showed little in vitro affinity for the binding site. An examination of the role of this protein in GA action requires pure preparations

[1]Supported by grant A2905 from the Natural Sciences and Engineering Research Council of Canada to L.M.S.

of this candidate GA receptor. Here we report on the partial purification of this protein using ammonium sulphate precipitation, gel filtration, ion exchange and hydroxyl-apatite chromatography.

ASSAY FOR GA-BINDING PROTEIN

Protein fractions were assayed for GA binding activity with an in vitro DEAE-filter paper assay using ^{3}H-GA$_4$ (1.6 x 10^{12} Bq mmol^{-1}, >65% purity estimated by GLC and about 83% radiopurity as estimated by TLC and scintillation counting) as described in Yalpani and Srivastava (1985). Many proteins bind to GAs resulting in apparent specific binding. In order to discriminate between this and receptor-type specific binding, each fraction was assayed with 50 nM ^{3}H-GA$_4$ in the absence and presence of a 100 fold excess of unlabelled GA$_4$ (high in vivo activity), GA$_3$ (moderate in vivo activity), and a 1000 fold excess of GA$_4$ C-7 methyl ester (GA$_4$ME) (no in vivo activity). The resulting displacement of ^{3}H-GA$_4$ from the binding sites by the competitors was used as a guide to select those fractions containing the putative receptor. In these fractions the extent of ^{3}H-GA$_4$ displacement is expected to correlate with the in vivo activity of the competitor.

In some recent experiments Amersham ^{3}H-GA$_4$ (1.4 x 10^{12} Bq mmol^{-1}, 97.6% radiopurity) was used. With this preparation background binding is reduced and apparent specific binding of GA$_4$ME to fractions containing the putative receptor is negligible.

AMMONIUM SULFATE PRECIPITATION

100,000g cytosol from the apical portions of hypocotyls of dark-grown, 1 week old Cucumis sativus cv National Pickling was prepared in 0.1 M tris-phosphate, 1 mM EDTA, 25 µM PMSF, pH 7.1 as described in Yalpani and Srivastava (1985). To this cytosol (NH$_4$)$_2$SO$_4$ was added in steps of 30, 40, 50, and 60%. The pellets obtained at each step after equilibration and centrifugation at 23,000g were assayed for specific binding as described above.

As shown in Table 1, the putative receptor precipitates with 30 to 50% (NH$_4$)$_2$SO$_4$. There is little cytosol protein precipitated with less than 30% (NH$_4$)$_2$SO$_4$. For further purification the GA binding protein was, therefore,

precipitated with 50% $(NH_4)_2SO_4$ and desalted on a Sephadex
G-25 SF column (2.5 x 12 cm) with 20 mM tris-phoshate, 1 mM
EDTA, pH 7.0. About 40% (wt/wt) of the total cytosol
proteins are removed from the preparation with this step.

TABLE 1
BINDING OF ^{3}H-GA$_4$ TO CYTOSOL CUT WITH $(NH_4)_2SO_4$

Fraction	Cytosol	$(NH_4)_2SO_4$ Pellet		
		30%	30-40%	40-50%	50-60%
^{3}H-GA$_4$	0.38^a	0.63	0.68	0.62	0.44
+ GA$_4$	0.22	0.25	0.34	0.28	0.34
+ GA$_3$	0.24	0.27	0.30	0.46	0.32
+ GA$_4$ME	0.32	0.37	0.40	0.50	0.38

^{a}Values are in pmol ^{3}H-GA$_4$ bound mg protein^{-1}

DEAE CELLULOSE CHROMATOGRAPHY

The protein fraction eluting from the desalting
column was loaded in 20 mM tris-phoshate, 1 mM EDTA, pH 7.0
onto a DE 32 ion exchange cellulose (Whatman) column (2.5 x
13.5 cm). After elution of the unbound proteins which do
not have receptor-type binding, a linear gradient of 150 mL
0.18 to 0.33 M KCl was applied. A typical elution profile
is shown in Figure 1. An assay of lyophilized column
fractions indicates that most of the GA-binding protein
elutes between 0.25 and 0.30 M KCl (fractions C and D,
Table 2). A measure of the degree of purification achieved
with this step was obtained by comparison of fresh cytosol
and G 25 eluate with the active fraction eluted from the
DE 32 column (Table 3). The active DE 32 eluate can be
concentrated on 50,000 dalton Centriflo membrane filtration
cones (Amicon) without loss of binding.

FIGURE 1. Elution profile of G-25 eluate on DEAE
cellulose column. Regions marked A to E were assayed
for GA-binding activity.

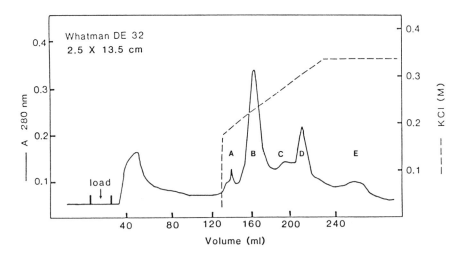

TABLE 2

BINDING OF ^{3}H-GA$_4$ TO FRACTIONS FROM DEAE CELLULOSE COLUMN
G-25 eluate was loaded on to the ion exchanger in 20mM tris-
phosphate, 1mM EDTA, pH 7.0. Fractions eluted were lyophilized
and resuspended for assay.

Fraction	A[a]	B	C	D	E
^{3}H-GA$_4$	0.12[b]	0.14	0.12	0.12	0.06
+ GA$_4$	0.08	0.09	0.02	0.04	0.02
+ GA$_3$	0.08	0.07	0.08	0.07	0.04
+ GA$_4$ME	0.09	0.07	0.13	0.11	0.07

[a]For fractions see Figure 1.

[b]Values as in Table 1.

TABLE 3
BINDING OF ^{3}H-GA$_4$ TO CYTOSOL, G-25 ELUATE AND ACTIVE
DEAE CELLULOSE FRACTION

Fraction	Cytosol	G-25 Eluate	DE 32 Eluate
^{3}H-GA$_4$	0.33[a]	0.78	0.91
+ GA$_4$	0.21	0.50	0.51
+ GA$_3$	0.23	0.54	0.61
+ GA$_4$ME	0.27	0.65	0.72

[a]Values as in Table 1.

HYDROXYLAPATITE CHROMATOGRAPHY

The GA binding protein fraction eluted from the DE 32
column was reduced in volume on ultrafiltration membrane
cones and loaded onto a hydroxylapatite column (Calbiochem)
(2.5 x 13 cm) in 10 mM potassium phosphate, 1 mM EDTA pH
7.0. The binding protein absorbed to the column matrix and
eluted with 0.4 M potassium phosphate, 1 mM EDTA, pH 7.0.
(Table 4). Nondenaturing PAGE resolves about 10 protein
bands from this fraction (data not shown).

TABLE 4
BINDING OF ^{3}H-GA$_4$ TO DE 32 ELUATE AND FRACTIONS FROM
HYDROXYLAPATITE COLUMN
The active DE 32 eluate was loaded directly on to a
hydroxylapatite column in 10 mM potassium phosphate,
1 mM EDTA, pH 7.0. Unbound fractions and those eluted with
40 mM potassium phosphate, 1 mM EDTA, pH 7.0 were assayed

Fraction	DE 32	Concentration of Eluting Phosphate Buffer	
		10 mM	0.4 M
^{3}H-GA$_4$	0.42[a]	0.24	0.37
+ GA$_4$	0.20	0.20	0.16
+ GA$_3$	0.23	0.21	0.19
+ GA$_4$ME	0.32	0.22	0.30

[a]Values as in Table 1.

DISCUSSION

The putative GA-receptor has been partially purified
by salt fractionation and column chromatography. Efforts
are currently under way to further the purity of the binding
protein. Also, nondenaturing as well as denaturing PAGE of
active and inactive column fractions are being used to allow
its identification.

REFERENCES

1. Keith B, Brown S, Srivastava LM (1982). In vitro
    binding of gibberellin A_4 to extracts of cucumber
    measured by using DEAE-cellulose filters. Proc Nat
    Acad Sci USA 79: 1515.
2. Keith B, Foster NA, Bonettemaker M, Srivastava LM
    (1981). In vitro gibberellin A_4 binding to extracts of
    cucumber hypocotyls. Plant Physiol 68: 344.
3. Yalpani N, Srivastava LM (1985). Competition for
    in vitro [^{3}H]gibberellin A_4 binding in cucumber by
    gibberellins and their derivatives. Plant Physiol 79:
    963.

Molecular Biology of Plant Growth Control, pages 315–322
© 1987 Alan R. Liss, Inc.

IN VITRO BINDING OF GIBBERELLIN A$_4$ IN EPICOTYLS OF DWARF PEA AND TALL PEA[1]

Zin-Huang Liu and Lalit M. Srivastava

Department of Biological Sciences, Simon Fraser University, Burnaby, B.C. Canada V5A 1S6

ABSTRACT In vitro gibberellin (GA) binding properties of a cytosol fraction from epicotyls of dwarf pea (Pisum sativum L. cv. Progress No. 9) and tall pea (Pisum sativum L. cv. Alaska) were investigated using [^{3}H]GA$_4$ in a DEAE filter paper assay at 0-3 C. The binding obtained is saturable, reversible, temperature labile in dwarf pea, and has a half-life of dissociation of 5-6 min. By varying the concentration of [^{3}H]GA$_4$ in the incubation medium the Kd was estimated to be 130 nM in dwarf pea and 70 nM in tall pea. The number of binding sites was estimated to be 0.66 and 0.43 pmol mg^{-1} soluble protein in dwarf pea and tall pea, respectively. Biologically active GAs, such as GA$_3$ and GA$_4$ could reduce the level of [^{3}H]GA$_4$ binding much more than the biologically inactive GA$_4$ methyl ester. Specific binding was enhanced by ammonium sulfate fractionation followed by desalting in a Sephadex G-50 column.

INTRODUCTION

Whilst it has been shown that [^{3}H]GA$_4$ binds

[1]This work was supported by grant A2905 from Natural Sciences and Engineering Research Council of Canada to L.M.S.

specifically to cytosol fractions from cucumber hypocotyls in vitro and that this binding satisfies many of the criteria associated with receptor-ligand binding (e.g., Keith et al. 1982, Yalpani and Srivastava, 1985), there has been as yet no such demonstration of in vitro binding of gibberellins to fractions from pea. Stoddart et al. (1974) reported that [^{3}H]GA$_1$ applied to dwarf pea epicotyls in vivo was bound to two protein fractions. Keith and Srivastava (1980), repeating these experiments with epicotyl sections incubated under conditions when GA metabolism was minimal, reported essentially similar results and concluded that there was evidence for high affinity, specific binding of [^{3}H]GA$_1$ to receptor protein(s).

Here we report on the kinetics of [^{3}H]GA$_4$ binding to cytosol fractions from dwarf and tall peas in vitro. [^{3}H]GA$_4$ was used instead of [^{3}H]GA$_1$ because the assay for [^{3}H]GA$_1$ binding to protein fractions using DEAE-cellulose filter discs has not yet been perfected.

MATERIALS AND METHODS

Extraction of GA-Binding Protein. Dwarf pea (Pisum sativum L. cv. Progress No. 9) and tall pea (Pisum sativum L. cv. Alaska) seeds from Western Horticultural Specialists Ltd. Calgary, Alberta, Canada, were germinated and grown for 9 days as described by Keith and Srivastava (1980). The apical 1 cm portions of epicotyls were excised and collected in ice-cold extraction buffer (100 mM potassium phosphate, 1 mM EDTA, 25 µM PMSF, pH 7.40). The pooled epicotyls were blotted, weighed, and ground with an equal volume (1:1 wt/vol) of extraction buffer, using a chilled mortar and pestle. All procedures were carried out at 0-3 C. The 100,000g cytosol and partially purified cytosol were obtained as previously described by Yalpani and Srivastava (1985).

In Vitro Binding Assay. Appropriate amounts of [^{3}H]GA$_4$ (1.6 x 10^{12} Bq mmol^{-1}, 76.5% purity as estimated by GLC, radiopurity, ca. 83% as determined by TLC and scintillation counting) in an ethanol : ethyl acetate mixture (1 : 1, v/v) were taken into 7 ml test tubes. Solutions were dried by N$_2$ gas after which tubes were cooled to 0-3 C. In the assay, defined amounts of cytosol or partially purified cytosol were added to dry [^{3}H]GA$_4$ to a final solution ranging from 5 nM to 150 nM. After incubating 1 h on ice, a 100 µl aliquot was assayed for bound hormone by DEAE-cellulose filtration as outlined by Keith et al. (1982). In order to determine nonspecifically-bound [^{3}H]GA$_4$, parallel incubations were performed in which a 100 to 1000-fold excess of nonradioactive GA$_4$ was added simultaneously with the radioactive GA$_4$. The amount of [^{3}H]GA$_4$ specifically bound was calculated by subtracting the activity bound in the presence of excess nonradioactive GA$_4$ from that bound in its absence. Other GAs and GA derivatives were substituted for nonradioactive GA$_4$ in competition experiments.

RESULTS AND DISCUSSION

The 100,000g dwarf pea cytosol was incubated with [^{3}H]GA$_4$ for 2 h and nonradioactive GA$_4$ added after 1 h. As shown in Fig. 1 the half time of dissociation at 0 C was about 5 min and, furthermore, about 60% of [^{3}H]GA$_4$ bound was exchangeable with nonradioactive GA$_4$.

The 100,000g cytosol was cut with 60% ammonium sulfate, the precipitate was washed twice with 60% ammonium sulfate, desalted on a Sephadex G-50 column, and freeze-dried. The cytosol powder was resuspended in phosphate buffer and assayed. Specific binding of [^{3}H]GA$_4$ to aliquots of resuspended ammonium sulfate precipitate (RASP) from dwarf and tall peas was studied at [^{3}H]GA$_4$ concentrations from 5 nM to 150 nM and the data plotted according to the method of Scatchard (1949). The Kd was estimated to be 130 nM in dwarf pea (Fig. 2) and 70 nM in tall pea (Fig. 3). Keith and Srivastava (1980) estimated the Kd of [^{3}H]GA$_1$ for the intermediate molecular weight proteins in dwarf pea to be 60 nM. The data presented here using [^{3}H]GA$_4$ confirm that the binding protein(s) in pea has a higher affinity for GA$_1$. The number of binding sites, n, in RASP was estimated to be 0.66 pmol mg^{-1} sol protein (Fig. 2) and 0.43 pmol mg^{-1} sol protein (Fig. 3) in dwarf and tall pea, respectively.

Binding kinetics were also studied in the presence of fixed concentrations of nonradioactive GA$_4$ and GA$_4$ methyl ester. Double-reciprocal plots as shown in Fig. 4 indicate that biologically active GA$_4$ significantly reduced [^{3}H]GA$_4$

binding.   GA₄ methyl ester also reduced the level of [³H]GA₄ binding but this may be due to the very high concentration of GA₄ methyl ester used (ca. 4000-fold excess).

The in vitro data presented here, in conjunction with the in vivo binding of [³H]GA₁ to pea epicotyl sections (Stoddart et al. 1974; Keith and Srivastava, 1981), suggest that certain protein fractions in tall and dwarf pea bind [³H]GA₄ saturably, reversibly and with high affinity. These characteristics satisfy some of the criteria for hormone-receptor binding.   Further experiments are in progress to purify the protein(s).

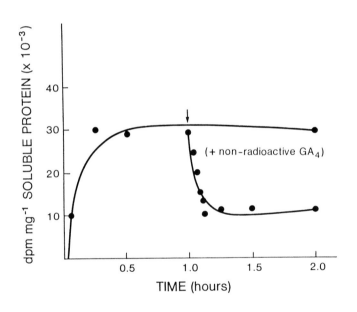

FIGURE 1. Total binding of 100,000g dwarf pea cytosol incubated in [³H]GA₄ (100 nM) and the effect of addition of nonradioactive GA₄ (100-fold) applied after 1 h (indicated by arrow) on the total binding.

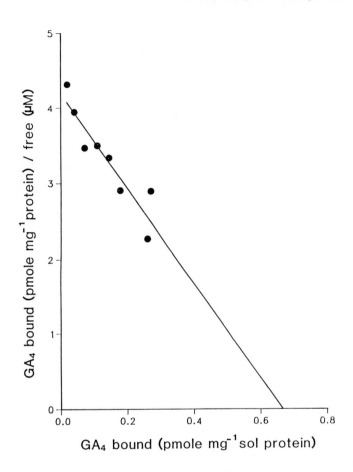

FIGURE 2.   Scatchard plot of [3H]GA4 specific binding to aliquots of resuspended ammonium sulfate precipitate (from dwarf pea cytosol) after Sephadex G-50 gel filtration.   Data were evaluated by computer regression analysis.   Kd was estimated to be 130 nM, n was estimated to be 0.66 pmol mg^{-1} sol. protein.

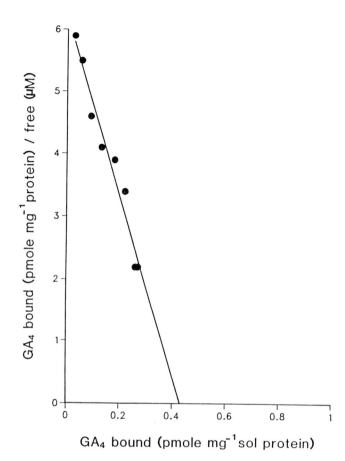

GA$_4$ bound (pmole mg^{-1}sol protein)

FIGURE 3. Scatchard plot of [^{3}H]GA$_4$ specific binding to aliquots of resuspended ammonium sulfate precipitate (from tall pea cytosol) after Sephadex G-50 gel filtration. Data were evaluated by computer regression analysis. Kd was estimated to be 70 nM, n was estimated to be 0.43 pmol mg^{-1} sol. protein.

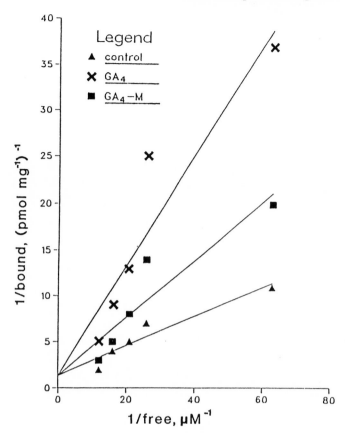

FIGURE 4. Double-reciprocal plots of [^{3}H]GA$_4$ binding to dwarf pea cytosol in the presence of nonradioactive GA$_4$ and GA$_4$ methyl ester. Samples were incubated with [^{3}H]GA$_4$ at a range of concentrations in the presence or absence of GA$_4$ (100 nM) and GA$_4$ methyl ester (0.4 mM).

REFERENCES

1. Keith B, Srivastava LM (1980). In vivo binding of gibberellin A$_1$ in dwarf pea epicotyls. Plant Physiol 66:962.

2.   Keith B, Boal R, Srivastava LM (1980). On the uptake, metabolism and retention of [^{3}H]gibberellin A_1 by barley aleurone layers at low temperature.   Plant Physiol 66:956.

3.   Keith B, Brown S, Srivastava LM (1982). In vitro binding of gibberellin A_4 to extracts of cucumber measured by using DEAE-cellulose filters.   Proc Natl Acad Sci USA 79:1515.

4.   Scatchard G (1949).   The attraction of proteins for small molecules and ions.   Ann NY Acad Sci 51:660.

5.   Stoddart JL, Breidenbach W, Nadeau R, Rappaport L (1974).   Selective binding of [^{3}H]gibberellin A_1 by protein fractions from dwarf pea epicotyls.   Proc Natl Acad Sci USA 71:3255.

6.   Yalpani N, Srivastava LM (1985).   Competition for in vitro [^{3}H]gibberellin A_4 binding in cucumber by gibberellins and their derivatives.   Plant Physiol 79:963.

Molecular Biology of Plant Growth Control, pages 323–334
© 1987 Alan R. Liss, Inc.

EFFECT OF GIBBERELLIC ACID ON LIPID DEGRADATION
IN BARLEY ALEURONE LAYERS[1]

Donna E. Fernandez and L. Andrew Staehelin

Department of Molecular, Cellular & Developmental Biology
University of Colorado, Boulder, CO 80309-0347

ABSTRACT One of the earliest effects of gibberellic
acid (GA_3) in barley aleurone layers appears to
involve the mobilization of lipid reserves stored in
lipid bodies. The biochemical data presented in this
paper suggests that lipases can be transferred from
protein body membranes to lipid bodies. The specific
activity of lipase in isolated lipid bodies increases
as much as 2-fold when the cells are exposed to GA_3.
Lipid bodies from untreated layers contain 20-25% of
the total lipase activity while those from layers
treated with GA_3 contain 75-80% of the total
activity. Transfer of lipase rather than de novo
synthesis appears to be involved since the activity in
pellets containing the membranes of protein bodies
decreases accordingly. 10 μM abscisic acid blocks
this GA_3-induced transfer of activity.

INTRODUCTION

The hormone, gibberellic acid (GA_3), induces major
changes in both the pattern of gene expression and the
organization of subcellular structures in barley aleurone
cells. The induction of a variety of hydrolytic enzymes in
isolated barley aleurone layers stimulated with GA_3 has
been extensively studied (reviews: 1, 2). Structural
changes, including the proliferation of rough endoplasmic

[1]This work was supported by a National Science
Foundation Graduate Fellowship to D.E.F. and National
Institutes of Health grant GM 18639 to L.A.S.

reticulum (3), occur as the cells become committed to the
synthesis and secretion of large quantities of secretory
proteins. These dramatic changes become apparent within
the first 4-8 hours after exposure to GA_3 and drastically
increase the demand for cellular energy and for membrane
components, including phospholipids. The principal
reserves of energy in aleurone layers are found in lipid
bodies (4) which consist of triglycerides surrounded by a
phospholipid monolayer (5). Triglycerides can be degraded
to provide cellular energy via the glyoxylate cycle or to
provide fatty acids and other components for phospholipid
synthesis. It is logical therefore that one of the
earliest effects of GA_3 in aleurone layers should involve
the mobilization of lipid reserves.

Several investigations have suggested that changes in
lipid metabolism occur soon after aleurone cells are
exposed to GA_3. The earliest documented effect of GA_3 is
an increase in the activity of two enzymes involved in
phospholipid synthesis: phosphorylcholine-cytidyl
transferase and phosphorylcholine-glyceride transferase
(6,7). These changes occur within 2 hours after adding GA_3
to isolated layers. In addition, increased incorporation
of radiolabelled choline into a microsomal fraction (8) and
radiolabelled orthophosphate into phospholipids (9) is
observed within 4 hours. However, when phospholipid
synthesis is measured using incorporation of radiolabelled
acetate or glycerol, little or no increase in synthesis is
detected (9,10,11), possible because carbon units from the
preexisting storage lipids can be used (12).

Although a great deal of attention has been focused on
phospholipid synthesis, few investigators have looked at
what must be an even earlier step: the digestion of
triglycerides. Enzymes involved in the release of energy
from fatty acids via the glyoxylate cycle appear within 1-2
days of germination in whole wheat seeds and can be induced
by GA_3 in isolated aleurone layers (13). In contrast,
neutral lipase activity does not appear until after the
first day of germination and cannot be induced by GA_3
(14). Jelsema et al. (15) have shown that uninduced
aleurone cells contain some acid lipase activity. However,
although the apparent substrates, triglycerides, are found
within lipid bodies, this acid lipase activity is located
principally in the protein bodies in cells which have not
been exposed to GA_3. Furthermore, upon exposure of the
cells to GA_3, this activity disappears from those protein
bodies.

In this paper, we propose that a neutral lipase activity associated with the protein bodies in uninduced cells is transferred to the lipid bodies when the cells are exposed to GA₃. Our data show that the transfer of activity is under GA₃ control and that it occurs early, within 1 hour after exposing the cells to GA₃.

METHODS

Materials. Aleurone layers of Hordeum vulgare L. cv. Himalaya (1979 harvest, Washington State University, Pullman, Wash., USA) were prepared as described previously (16). Layers isolated from imbibed half seeds were incubated on a rotary shaker at 26°C in 2 mM sodium-acetate buffer (pH 4.8) containing 20 mM CaCl₂, 7.5 μg/ml chloramphenicol and either 10 μM GA₃ or H₂O (17).

Isolation of lipid bodies. Lipid bodies were isolated using a modification of a procedure developed for isolating lipid bodies from corn scutellum (18). The layers were homogenized with a Virtis 45 homogenizer (Virtis Co., Gardiner, NY, USA) in grinding medium (0.6 M sucrose, 1 mM EDTA, 10 mM KCl, 1 mM MgCl₂, 2 mM dithiothreitol, 0.15 M Tricine buffer, 5 mM ε-aminocaproic acid, 1 mM benzamidine HCl, pH adjusted to 7.5 with KOH: 10 ml/100 layers). The homogenate was filtered through nylon mesh cloth with a pore size of 30 μm. The residue retained on the cloth was then ground with sand and a small volume of grinding medium in a mortar and refiltered. This residue was squeezed to remove the maximum amount of homogenate.

Lipid bodies were isolated from the filtrate using a series of sucrose step gradients (Figure 1). The filtrate was added to a centrifuge tube and overlaid with an equal volume of flotation medium (same as the grinding medium except that the sucrose concentration is lowered to 0.5 M). After centrifuging at 10,000 x g for 30 minutes in a swinging bucket rotor, the lipid bodies (LB1) float up and can be collected from the surface of the medium using a spatula. The pellet at the bottom was resuspended in a small volume of grinding medium and stored. The supernatant did not exhibit lipase activity and was discarded. The collected lipid bodies were resuspended

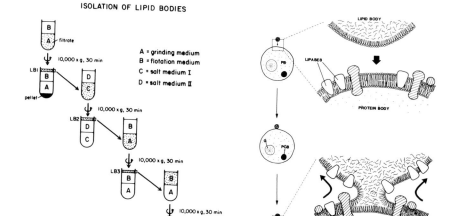

ISOLATION OF LIPID BODIES

A = grinding medium
B = flotation medium
C = salt medium I
D = salt medium II

PB-PROTEIN BODY   G-GLOBOID   PCB-PROTEIN CARBOHYDRATE BODY

FIGURE 1. Procedure for isolating lipid bodies from a homogenate of barley aleurone layers.

FIGURE 2. Model illustrating how lipases appear to be transferred from protein bodies to lipid bodies via a membrane fusion event and simple diffusion.

in 2 M NaCl, 0.4 M sucrose, and 0.15 M Tricine-NaOH buffer (pH 7.5) (salt medium I), overlaid with salt medium II (same as salt medium I except that the sucrose concentration is lowered to 0.3 M) and recentrifuged as above. There was no observable pellet and the lipid bodies (LB2) were collected and washed two more times with salt free media. The final lipid body pellicle was resuspended in a small volume of flotation medium. Fractions were stored at -20°C before being assayed for lipase activity and protein content.

   Electron microscopy. Lipid bodies and pellets were frozen with a propane jet freezer (19) and processed for freeze fracture electron microscopy as described previously (16).

Assays. Protein content was measured using the Coomassie dye binding assay of Bradford (20). Lipase activity was measured by the colorimetric method of Nixon and Chan (21) as modified by Lin and Huang (18) except that assay volumes were reduced by half. A trilinolein (Sigma Chemical Co., St. Louis, Mo., USA) emulsion stabilized with acacia powder was used as the substrate, and duplicate samples were taken every 30 minutes over a 2 hour period. Palmitic acid was used to generate the free fatty acid standard curve.

RESULTS

Our analysis of ultrarapidly-frozen barley aleurone cells by freeze-fracture electron microscopy indicates that complexes of integral membrane proteins can be transferred from protein bodies to lipid bodies during germination (16, and manuscript in preparation). Our model (Figure 2) of how this transfer occurs is as follows: (1) lipid bodies become associated with the protein bodies, (2) the monolayers of lipid bodies fuse with the outer leaflets of the bilayers of protein bodies, and (3) membrane complexes can diffuse from the protein bodies to the lipid bodies. (Transmembrane complexes with hydrophilic surfaces exposed to both the protein body matrix and the cytoplasm will remain in the protein body membrane.) In this paper, we present biochemical evidence that some of the complexes that are transferred are lipases.

Lipid bodies can be easily isolated from a homogenate of barley aleurone layers using a series of sucrose step gradients and salt washes (Figure 1). Analysis of the various fractions by freeze-fracture electron microscopy indicates that the separation of lipid bodies from other cellular components is very good. Lipid bodies and protein bodies are particularly easy to identify in freeze fracture since lipid bodies do not etch and protein body membranes have an unusually high particle density (16). Protein body membranes are found in the pellet, often as large sheets with some attached lipid bodies. The majority of the lipid bodies, however, detach from the protein body membranes during cell homogenization and float up. The lipid body fraction appears to be very pure (Figure 3) and only occasionally are lipid bodies with small, attached

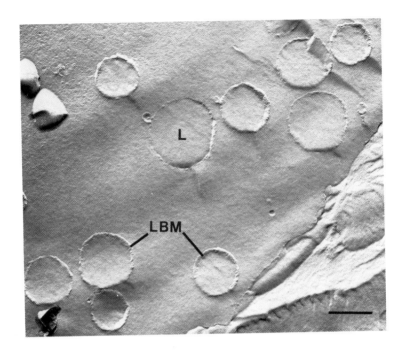

FIGURE 3. Freeze-fracture electron micrograph of the LB2 lipid body fraction. L: internal lipid, LBM: lipid body membrane. Bar: 0.3 µm, x45000.

fragments of protein body membranes found.

Lipid body fractions contain a neutral lipase which is active on trilinolein, a triglyceride of 18:2 fatty acids. Since 18:2 fatty acids are the most abundant in cereal acyl lipids (22), this lipase should also be effective in vivo. When the LB1 fraction (Figure 1) is washed with salt to remove electrostatically bound proteins, the lipase co-purifies with the lipid bodies and the specific activity increases 20-fold (Table 1).

TABLE 1
CO-PURIFICATION OF LIPASE ACTIVITY AND LIPID BODIES

Fraction	Specific activity (nmoles/min·mg)
LB1	18
LB2	62
LB4	370

The specific activity of the lipase in the isolated lipid bodies (LB2) increases as much as 2-fold when aleurone cells are incubated with 10 μM GA_3 for various lengths of time before processing (Table 2).

TABLE 2
ENHANCEMENT OF LIPASE ACTIVITY BY GA_3

Incubation time (hours)	Specific activity (nmoles/min·mg) $+GA_3$	Specific activity (nmoles/min·mg) $-GA_3$	Ratio $(+GA_3/-GA_3)$
1	104	72	1.4
4	132	62	2.1
12	62	53	1.2

This increase does not appear to be due, however, to de novo synthesis and insertion of the lipase directly into the lipid bodies. Instead, the lipase appears to be transferred from the protein bodies in a GA_3-dependent manner. Although the total lipase activity in the cells is about the same, the percent total lipase activity in the lipid body fraction increases substantially in cells which have been exposed to GA_3. At the same time, the percent total lipase activity in the pellet containing the membranes of protein bodies decreases substantially (Table 3). This increase in activity is not due to a more

efficient isolation of lipid bodies from GA₃-treated
cells. Degradation of triglycerides is apparently so
efficient that the number of lipid bodies collected from
cells which have been treated with GA₃ for 4 hours is
visibly less than the number collected from untreated
cells.

TABLE 3
DISTRIBUTION OF LIPASE ACTIVITY

	% Total lipase activity	
Treatment (4 hours)	Pellet	Lipid bodies
$-GA_3$	79%	21%
$+GA_3$	24%	76%

Many GA₃-induced changes in barley aleurone cells can
be blocked by abscisic acid (ABA), including the synthesis
of α-amylase (23), the formation of polysomes (24), and the
stimulation of phospholipid synthesis (6,8,9). The
transfer of lipase is no exception. When used in
combination with 1 μM GA₃, 10 μM ABA ((±) cis-trans isomer)
blocks the GA₃-induced transfer of lipase (Table 4). ABA
alone has no effect on the cells.

TABLE 4.
INHIBITION OF LIPASE TRANSFER BY ABA

	% Total lipase activity	
Treatment (2 hours)	Pellet	Lipid bodies
$+GA_3$	<10%	>90%
+ABA	>90%	<10%
$+GA_3/+ABA$	69%	31%

Our biochemical data also indicate that the transfer
of lipase is a very early event in the response of aleurone
cells to GA$_3$. If the cells are exposed to GA$_3$ for 45
minutes, the lipase activity remains in the pellet.
However, if the incubation is increased to 60 minutes, the
majority of the activity is found in the lipid body
fraction (data not shown).

DISCUSSION

The biochemical evidence presented in this paper
suggests that GA$_3$ controls the rate of lipid degradation in
barley aleurone cells. It is very important that lipid
metabolism be controlled in these cells. Large amounts of
hydrolytic enzymes must be synthesized and secreted using
endogenous lipid reserves as the major source of energy.
These reserves are limited: once extensive lipid
degradation starts, the storage lipids can be depleted
within 2-3 days. It would seem essential then that lipid
mobilization and enzyme synthesis be activated
simultaneously. GA$_3$ appears to be the signal that induces
both processes.

GA$_3$ controls the rate of lipid degradation by
controlling the transfer of lipase complexes from protein
bodies, where they are stored, to lipid bodies, where they
become functionally active. In this way, the level of
available energy can be raised without expending large
amounts of energy to produce lipases through de novo
synthesis. The overall energy level of the cell can also
be raised rapidly by uniting an active enzyme with its
substrate through a single membrane fusion event.

The mechanism by which GA$_3$ controls the transfer of
lipase is unclear. Based on our model, we suggest that GA$_3$
has some effect on either membrane fusion or the diffusion
of enzyme complexes in the protein body membrane. Since
its effect is not apparent for 60 minutes, it is unlikely
that GA$_3$ exerts its effect directly. GA$_3$ may achieve its
effect through a change in gene expression, by inducing a
fusion protein for example. Alternatively, it may act
through a cascade of biochemical events like those that
accompany hormone action in animal cells (25). Jelsema et
al. (26) have reported that GA$_1$ binds preferentially to
protein body membranes in wheat aleurone cells. It is
possible that this binding could bring about local changes
that mediate actual transfer of lipase activity.

In summary, our results indicate that lipase complexes can be transferred between protein body membranes and lipid bodies and that exposure to GA$_3$ is required for this transfer to occur. The transfer is complete within 1 hour after exposing the cells to GA$_3$ and is therefore one of the earliest documented effects of the hormone in barley aleurone layers.

ACKNOWLEDGMENTS

We would like to thank Dr. Anthony Huang for his advice on lipase assays, Ms. Lisa Maynard for her help in getting this project started, Trish Moore and Dr. Stuart Craig for helpful comments on the manuscript, and all the members of the Staehelin laboratory for helpful discussions.

REFERENCES

1. Chrispeels MJ (1976). Biosynthesis, intracellular transport and secretion of extracellular molecules. Ann Rev Plant Physiol 27:19.
2. Varner JE, Ho TD (1976). The role of hormones in the integration of seedling growth. In Papaconstantinou J (ed): "The Molecular Biology of Hormone Action," New York: Academic Press, p 173.
3. Jones RL (1969). Gibberellic acid and the fine structure of barley aleurone cells: I. Changes during the lag-phase of α-amylase synthesis. Planta 87:119.
4. Clarke NA, Wilkinson MC, Laidman DL (1983). Lipid metabolism in germinating cereals. In Barnes PJ (ed): "Lipids in Cereal Technology", London: Academic Press, p 57.
5. Yatsu LY, Jacks TJ (1972). Spherosome membranes: half-unit membranes. Plant Physiol 49:937.
6. Johnson KD, Kende H (1971). Hormonal control of lecithin synthesis in barley aleurone cells: regulation of the CDP-choline pathway by gibberellin. Proc Nat Acad Sci 68:2674.
7. Ben-Tal Y, Varner JE (1974). An early response to gibberellic acid not requiring protein synthesis. Plant Physiol 54:813.

8. Evins WH, Varner JE (1971). Hormone-controlled synthesis of endoplasmic reticulum in barley aleurone cells. Proc Nat Acad Sci 68:1631.
9. Koehler DE, Varner JE (1973). Hormonal control of orthophosphate incorporation into phospholipids of barley aleurone layers. Plant Physiol 52:208.
10. Firn RD, Kende H (1974). Some effects of applied gibberellic acid on the synthesis and degradation of lipids in isolated barley aleurone layers. Plant Physiol 54:911.
11. Wilkinson MC, Laidman DL, Galliard T (1984). Two sites of phosphatidylcholine synthesis in the wheat aleurone cell. Plant Sci Let 35:195.
12. Varty K, Laidman DL (1976). The pattern and control of phospholipid metabolism in wheat aleurone tissue. J Exp Bot 27:748.
13. Doig RI, Colborne AJ, Morris G, Laidman DL (1975). The induction of glyoxysomal enzyme activities in the aleurone cells of germinating wheat. J Exp Bot 26:387.
14. Tavener RJA, Laidman DL (1972). The induction of lipase activity in the germinating wheat grain. Phytochem 11:989.
15. Jelsema CL, Morré DJ, Ruddat M, Turner C (1977). Isolation and characterization of the lipid reserve bodies, spherosomes, from aleurone layers of wheat. Bot Gaz 138:138.
16. Fernandez DE, Staehelin LA (1985). Structural organization of ultrarapidly frozen barley aleurone cells actively involved in protein secretion. Planta 165:455.
17. Jones RL, Varner JE (1967). Bioassay of gibberellins. Planta 72:155.
18. Lin Y-H, Huang AHC (1984). Purification and initial characterization of lipase from the scutella of corn seedlings. Plant Physiol 76:719.
19. Bradford MM (1976). A rapid and sensitive method for the quantitation of microgram quantities of protein utilizing the principle of protein-dye binding. Anal Biochem 72:248.
20. Gilkey JC, Staehelin LA (1986). Advances in ultrarapid freezing for the preservation of cellular ultrastructure. J Electron Microsc Tech 3: in press.

21. Nixon M, Chan SHP (1979). A simple and sensitive colorimetric method for the determination of long-chain free fatty acids in subcellular organelles. Anal Biochem 97:403.

22. Morrison WR (1983). Acyl lipids in cereals. In Barner PJ (ed): "Lipids in Cereal Technology", London: Academic Press, p 11.

23. Chrispeels MJ, Varner JE (1966). Inhibition of gibberellic acid induced formation of α-amylase by abscission II. Nature 212:1066.

24. Evins WH, Varner JE (1972). Hormonal control of polyribosome formation in barley aleurone layers. Plant Physiol 49:348.

25. Marx JL (1984). A new view of receptor action. Science 224:271.

26. Jelsema C, Ruddat M, Morré DJ, Williamson FA (1977). Specific binding of gibberellin A_1 to aleurone grain fractions from wheat endosperm. Plant & Cell Physiol 18:1009.

Molecular Biology of Plant Growth Control, pages 335–344
© 1987 Alan R. Liss, Inc.

ETHYLENE BINDING PROTEINS

M.A. Hall, C.J. Howarth, D. Robertson, I.O. Sanders,
A.R. Smith, P.G. Smith, R.J. Starling, Zhao-Da Tang[1]
C.J.R. Thomas and R.A.N. Williams

Department of Botany & Microbiology, University College of
Wales, Aberystwyth, Dyfed SY23 3DA, Wales, U.K.

ABSTRACT   Recent work on ethylene binding and
metabolising systems in Phaseolus vulgaris, Pisum
sativum and Vicia faba is described and the possible
involvement of these processes in the mechanism of
action of ethylene is discussed.

INTRODUCTION

It is now well established that binding sites for
ethylene exist in various tissues and organs of a variety of
plant species (see e.g. 1), although a definite role as
functional receptors cannot as yet be assigned to them.
Equally, the existence of systems in higher plants capable
of metabolising ethylene is now unquestioned and their
possible involvement in the mechanism of action of ethylene
has been suggested (see e.g. 2).   This contribution
describes our most recent findings in these two areas and
addresses the question of the involvement of binding and
metabolism in the mechanism of action of ethylene.

MATERIALS & METHODS

With one exception, all the methods used in the work
outlined in this contribution have been fully described
elsewhere viz. assay of ethylene binding activity (3,4),
extraction and purification of the binding protein from

[1]Dept. of Biology, Nankai University, Weijin Road 94,
Tianjin, People's Republic of China.

Phaseolus vulgaris (5,6,7), extraction and assay of the
ethylene monooxygenase from Vicia faba (8) and in vivo assay
of ethylene oxidation in Pisum sativum (9).

For the simultaneous assay of binding and metabolising
functions in Pisum sativum and Vicia faba the following
technique was employed.

Tissue (usually 20 1cm epicotyl sections) or cell-free
extracts were placed in sealed flasks incorporating a
minivial containing 1 cm^3 3M NaOH. $^{14}C_2H_4$ was then injected
to give concentrations in solution of between 5.10^{-9}M and
5×10^{-7}M. In an equivalent batch, $^{12}C_2H_4$ was injected in
addition to give a concentration in solution of 5×10^{-6}M.
The flasks were incubated for various periods up to 24 h,
the minivials were removed and radioactivity in CO_2 and
ethylene oxide estimated as in reference 8. The tissue or
extracts were vented with an air flow of 160 dm^3min^{-1} for 1
min, a minivial containing 1 cm^3 0.25M $HgClO_4$ was added and
the vials were resealed and heated for 4 h at 60°C.

Radioactivity detected in the $HgClO_4$ trap was
considered as reversible C_2H_4 binding and the radioactivity
not released by heating taken as tissue-incorporation from
ethylene oxide.

RESULTS & DISCUSSION

Ethylene binding in Phaseolus vulgaris

Early work demonstrated the presence in developing
cotyledons of Phaseolus vulgaris of a binding site with a
high affinity and specificity for ethylene (3,4).
Subsequently, by means of marker enzyme studies (10) and
high resolution autoradiography (11) it was shown that the
binding site was localised on elements of the endomembrane
system, particularly endoplasmic reticulum and protein body
membranes. Despite the presence of this binding site with
all the expected properties of an ethylene receptor there is
no known role for ethylene during the ontogeny of Phaseolus
cotyledons; equally, attempts to demonstrate a link between
the binding site and the control of protein synthesis or
protein kinase activity in cell-free preparations have met
with no success.

On the other hand, given the extreme simplicity of the
structure of ethylene and the high affinity and specificity
of the binding site for the growth regulator it seems likely
that the binding domain in Phaseolus cotyledons will show

strong homologies with binding domains for ethylene, both in
other parts of the same species and even in other species.
For this reason we embarked on investigations into the
extraction and purification of the ethylene binding site
from Phaseolus cotyledons. Preliminary studies established
that unlike for example the membrane-bound auxin binding
sites in maize (12) the ethylene binding site was not
readily released from the membranes. It emerged that only
extraction with detergents did not drastically reduce
ethylene binding. Table 1 provides a comparison of the
characteristics of both membrane-bound and detergent-
solubilised binding sites. In general, the properties of
the two types of preparation do not differ significantly,
the major exceptions being marked downward shifts in the pH
sensitivity of, and rate constants of association and
dissociation of detergent solubilised sites. It is worth
noting that the solubilised protein is not only extremely
hydrophobic but also highly asymmetric. Recent work
indicates that, in common with many integral membrane
proteins, the ethylene binding site in Phaseolus is a glycoprotein. The hydrophobic nature of the binding site protein
has seriously complicated efforts to purify it since any
treatment which tends to disassociate detergent and protein
leads to precipitation of the latter. Nevertheless we have
now achieved a significant degree of purification by a
combination of pH precipitation, selective proteolysis,
detergent partitioning and most importantly FPLC.

TABLE 1

SUMMARY OF THE PROPERTIES OF THE ETHYLENE BINDING SITE FROM
PHASEOLUS VULGARIS

	Membrane bound	Triton solubilised & partially purified protein
K_D ethylene (M)	0.88×10^{-10}	5.5×10^{-10}
K_i propylene (M)	5.6×10^{-7}	0.9×10^{-7}
K_i acetylene (M)	1.03×10^{-5}	2.0×10^{-5}
K_i CO_2 (M)	1.15×10^{-3}	∞
k_1 $(M^{-1}s^{-1})$	2.97×10^3	0.042×10^3
k_{-1} (s^{-1})	1.4×10^{-6}	2.3×10^{-8}(calc)
pH optimum	$7.5 - 9.5$	$4 - 7$
Sedimentation coefficient		2.2s
Stoke's radius (nm)		6.1
Frictional ratio		2.4
Molecular weight		52-60,000

As mentioned above, the Phaseolus binding protein has
all the characteristics of a receptor but is situated in a
tissue in which there is no known developmental or
biochemical response to ethylene. Nevertheless in the same
plant the primary ethylene response in abscission zones is
well characterised in terms of induction of cell wall
degrading enzymes such as cellulase and endopolygalac-
turonase (13). We have demonstrated that an ethylene
binding site exists in the abscission zones but progress
using conventional methods is rendered impossible by the
fact that the maximum specific activity obtainable either by
^{14}C or ^{3}H labelling of ethylene is insufficient to enable
localisation or accurate measurement of binding site
concentration in a zone a few cells thick having a low
abundance of binding sites.
   In an effort to resolve this problem we have embarked
on a programme to produce polyclonal antibodies to the
binding site and hence to develop an immunoassay for use in
probing the transduction of the response. It must be
admitted however that here again the necessity to maintain
the protein in detergent is creating technical
difficulties.
   Once these problems have been resolved however it is
feasible to envisage the production of monoclonal antibodies
to the ethylene binding domain itself and, assuming that
there are strong homologies between such domains in all
species to extend our studies both to the pea systems
described below and also to systems such as tomato fruits
where again, at least some of the primary events in the
ethylene response are well characterised and hence offer
opportunities to study the transduction of the response.

Binding and Metabolism of ethylene by Vicia faba

   A system capable of metabolising ethylene rapidly at
physiological concentrations was described by Jerie & Hall
(14). Subsequent work indicated that the system had a high
affinity for ethylene and was capable of metabolising
physiologically active analogues of ethylene such as
propylene. The K_is obtained for such analogues, derived
from competition experiments, showed a close parallelism
with their relative physiological effectiveness. Other work
in this laboratory (8) and by Beyer's group (pers. comm.)
demonstrating a dependence on reducing power in the form of
NADPH and incorporation of O^{18} into the ethylene oxide
produced indicated that the enzyme responsible was an NADPH

dependent monooxygenase catalysing the reaction

$$C_2H_4 + O_2 + NADPH + H^+ \rightarrow C_2H_4O + NADP^+ + H_2O$$

Early studies indicated that the enzyme was membrane-bound but subsequent work with cell-free extracts suggests that a significant proportion of the enzyme is soluble and/or is readily released from the membranes. It is unclear whether the enzyme is cytochrome P-450 linked since the simplest test for such a connection is inhibition by carbon monoxide. However in this case carbon monoxide acts as a competitor of ethylene and 50% inhibition is achieved at 6 x 10^{-7}M CO in the aqueous phase. This represents a $[CO]/[O_2]$ ratio of about $2.45 \cdot 10^{-3}/1$ which is some two orders of magnitude less than that required to inhibit established plant P-450 enzymes where the carbon monoxide is competing with oxygen.

Because of the high rates of metabolism observed in cell-free preparations, measurement of ethylene binding in <u>Vicia</u> using the methods employed with <u>Phaseolus</u> proved impracticable. However, development of the techniques described in the 'Materials & Methods' section enabled a simultaneous assessment of binding to be made indicating the presence of a binding site of high affinity present in quantitites comparable to those observed in <u>Phaseolus</u> cotyldons. So far it has not proved possible to separate binding from metabolising activity; equally, binding activity in <u>Vicia</u> is much more labile than in <u>Phaseolus</u>.

Binding and metabolism of ethylene in <u>Pisum</u> <u>sativum</u>

Early work by Beyer (see e.g. 2) demonstrated the presence of ethylene metabolising systems in peas and other species. This metabolism differs from that in <u>Vicia</u> in that there appear to be two separate systems, one metabolising ethylene to ethylene glycol (derived from ethylene oxide) and the other metabolising ethylene directly to CO_2 (termed TI and OX respectively). Equally, the rates of metabolism by either system are very low at physiological ethylene concentrations (0.1 - 10 μl l^{-1} gas phase). On the other hand a close relationship was demonstrated in a number of systems including peas, between rates of metabolism and sensitivity to ethylene and Beyer (2) suggested that metabolism of ethylene was linked to the mechanism of action of the growth regulator.

Further work in our own laboratory demonstrated that

the systems metabolising ethylene in peas had relatively low
affinities for ethylene thus explaining the low rates of
metabolism at physiological ethylene concentrations. On the
other hand the affinities of the systems in peas for
propylene were very similar to that in Vicia faba (Table 2).

TABLE 2
PROPERTIES OF THE ETHYLENE METABOLISING SYSTEMS IN
VICIA FABA AND PISUM SATIVUM

| | Vicia faba[1] | Pisum sativum[2] | |
		OX	TI
K_m ethylene (M)[3]	2.5×10^{-8}	0.9×10^{-6}	1.6×10^{-6}
V_{max} moles $g^{-1}cm\ h^{-1}$	6.4×10^{-10}	2.4×10^{-10}	4.5×10^{-10}
K_i propylene (M)	5.0×10^{-6}	7.0×10^{-6}	3.7×10^{-7}
K_i CO (M)	6.0×10^{-7}	–	–
K_m NADPH (M)	3.7×10^{-5}	–	–
pH optimum	8.0	–	–

[1] Cell-free preparations
[2] In vivo measurements
[3] All concentrations given for aqueous phase

We have recently embarked on work designed to
characterise the ethylene metabolising systems in Pisum in
vitro. So far, it has proved impossible to prepare systems
capable of oxidising ethylene to CO_2 in vitro but some
success has attended our efforts with the system
metabolising ethylene to ethylene oxide. Although this work
is in an early stage of development it does not appear that
the TI system in peas is directly analogous to that in Vicia
since it is not NADH- or NADPH- dependent.
    Until recently, there was no evidence that metabolism
and ethylene binding were simultaneously present in the same
tissue and it was suggested that both might represent
reflections of the same system. Thus, binding sites of the
Phaseolus type were envisaged as controlling responses with
long 'on-off' times with metabolising systems reflecting
'receptors' related to developmental responses which react
rapidly to application of ethylene.
    It is now clear however that both binding and

metabolism may be present in the same tissue. The work in
Vicia faba has already been referred to but we have examined
the situation in peas more extensively.

The results of experiments in peas designed to measure
metabolism and binding simultaneously in epicotyl tips are
shown in figs 1 and 2. The two figures taken together lead
to two conclusions. Firstly, there exist in peas one
binding site with a high affinity for ethylene (0.49 x
10^{-9}M, corresponding to 0.099 µl l^{-1} gas phase ethylene)
with high rate constants of association and dissociation and
one other with a somewhat lower affinity (3 x 10^{-9}M,
corresponding to 0.63 µl l^{-1} gas phase ethylene) with rate
constants of association and dissociation comparable to

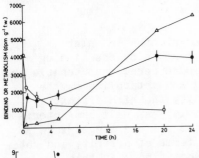

Figure 1

Association (●) and
dissociation (O) curves for
ethylene binding and for
metabolism (Δ) by 5 day old
1cm etiolated epicotyl tips
of Pisum sativum

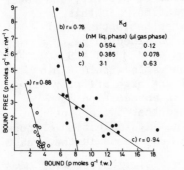

Figure 2

Scatchard plots of ethylene
binding by 5 day old 1 cm
etiolated epicotyl tips of
Pisum sativum. 1 h incub-
ation (O), 20 h incubation
(●)

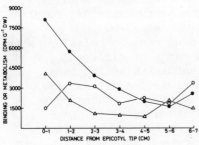

Figure 3

Binding and metabolism of
ethylene by 1 cm serial
sections of 5 day old etiolated epicotyls of Pisum
sativum. Metabolism to
CO_2 (O), metabolism to
ethylene oxide (●),
ethylene binding (Δ).

those in _Phaseolus_. The concentration of high affinity site
appears to increase during incubation over 20 h; it is not
clear whether the site with lower affinity is absent
initially or whether the technique failed to detect it in
the early stages of incubation. It is also clear that
metabolism either to ethylene oxide or CO_2 is essentially
absent from intact tissue but develops during incubation.

Fig. 3 summarises the results from a similar experiment
to those in figs 1 and 2 which measured metabolism and
binding in serial sections down 5 day old etiolated pea
epicotyls. The quantity of ethylene binding and the rate of
TI are highest in the top 1 cm portion of the epicotyl; OX
appears to occur generally throughout the tissue.

Conclusions

It is clear that the original hypothesis which
suggested that ethylene metabolism and ethylene binding
reflected two types of receptor is rendered less tenable by
the observation that both binding and metabolism may be
present in the same tissue. Equally the high affinity
binding site in peas, although it is less well characterised
than that in _Phaseolus_ does possess all the appropriate
characteristics of a receptor in terms of affinity and rate
constants of association and dissociation. In this
connection it is worth noting in parenthesis that half
maximal developmental responses are obtained in pea
epicotyls at concentrations of $4.5 \times 10^{-10}M$ and $4.5 \times 10^{-9}M$
(15) and an effect on growth is observable within 6 min
(16). In our experiments the K_D of the high affinity site
is $4.9 \times 10^{-10}M$ and the half-time for association is < 15
min. Equally, it is well established that two of the
components of the triple response, namely inhibition of hook
closure and isodiametric cell expansion, occur in the area
where binding activity is highest.

However, the fact remains that ethylene metabolism
exists and it cannot, except in one or two species, function
as a means for controlling endogenous ethylene
concentrations. Furthermore, it does show correlations with
sensitivity to ethylene in a wide range of different
ethylene-responsive systems and inhibition of ethylene
metabolism in peas by Ag^+ and CO_2 closely parallels the
inhibitory effects of these substances upon the response to
ethylene (2). A new dimension has been added to these
considerations by Beyer's observation (2) that ethylene
oxide while ineffective alone synergises strongly with

ethylene in a number of systems.

The evidence to date suggests one of two not mutually exclusive scenarios. Firstly, the systems metabolising ethylene may represent modified or 'degraded' receptor proteins; alternatively, the TI system at least may represent a totally separate system which provides an 'effector' for the full expression of an ethylene response - namely ethylene oxide.

Clearly, both of these suggestions represent only tentative hypotheses which can only be verified or rejected after much further work on the rigorous characterisation of the systems responsible.

ACKNOWLEDGEMENTS

We are grateful to the AFRC and SERC for support of this work.

REFERENCES

1. Smith AR, Hall MA (1985). Ethylene binding. In Roberts JA, Tucker GA (eds): "Ethylene and Plant Development," London: Butterworths, p. 101.
2. Beyer EM Jr (1985). Ethylene metabolism. In Roberts JA, Tucker GA (eds): "Ethylene and Plant Development," London: Butterworths, p. 125.
3. Bengochea T, Dodds JH, Evans DE, Jerie PH, Niepel B, Shaari AR, Hall MA (1980). Studies on ethylene binding by cell-free preparations from cotyledons of Phaseolus vulgaris L. I. Separation and characterisation. Planta 148:397.
4. Bengochea T, Acaster MA, Dodds JH, Evans DE, Jerie PH, Hall MA (1980). Studies on ethylene binding by cell-free preparations from cotyledons of Phaseolus vulgaris L. II. Effects of structural analogues of ethylene and of inhibitors. Planta 148:407.
5. Thomas CJR, Smith AR, Hall MA (1984). The effect of solubilisation on the character of an ethylene-binding site from Phaseolus vulgaris L. cotyledons. Planta 160:474.
6. Thomas CJR, Smith AR, Hall MA (1985). Partial purification of an ethylene-binding site from Phaseolus vulgaris L. cotyledons. Planta 164:272.

7. Hall MA (1985). Studies on the mechanism of action of ethylene. In "Proceedings of the 16th FEBS Congress, Part C," The Hague: VNU Science Press, p. 383.
8. Smith PG, Venis MA, Hall MA (1985). Oxidation of ethylene by cotyledon extracts from Vicia faba L. Planta 163:97.
9. Evans DE, Smith AR, Taylor JE, Hall MA (1984). Ethylene metabolism in Pisum sativum L.: kinetic parameters, the effects of propylene, silver and carbon dioxide and the comparison with other systems. Plant Growth Regul. 2:187.
10. Evans DE, Bengochea T, Cairns AJ, Dodds JH, Hall MA (1982). Studies on ethylene binding by cell-free preparations from cotyledons of Phaseolus vulgaris L.: subcellular localisation. Plant, Cell & Environ. 5: 101.
11. Evans DE, Dodds JH, Lloyd PC, apGwyn I, Hall MA (1982). A study of the subcellular localisation of an ethylene binding site in developing cotyledons of Phaseolus vulgaris L. by high resolution autoradiography. Planta 154: 48.
12. Venis MA (1977). Solubilization and partial purification of auxin binding sites of corn membranes. Nature 266:268.
13. Sexton R, Lewis LN, Trewavas AJ, Kell P (1985). Ethylene and abscission. In Roberts JA, Tucker GA (eds): "Ethylene and Plant Development", London: Butterworths, p. 173.
14. Jerie PH, Hall MA (1978). The identification of ethylene oxide as a major metabolite of ethylene in Vicia faba L. Proc. R. Soc. Lond. B. 200:87.
15. Burg SP, Burg EA (1967). Molecular requirements for the biological activity of ethylene. Plant Physiol. 42:144.
16. Warner LN, Leopold AC (1971). Timing of growth regulator responses in peas. Biochem. Biophys. Res.

Molecular Biology of Plant Growth Control, pages 345–359
© 1987 Alan R. Liss, Inc.

THE TURION:A BIOLOGICAL PROBE FOR THE MOLECULAR ACTION OF ABSCISIC ACID[1]

Cheryl Smart[2], Jane Longland and Anthony Trewavas

Department of Botany, Edinburgh University,
Edinburgh, U.K.

ABSTRACT The abscisic acid-induced turion is presented
as a model for studies aimed at understanding the action
of ABA. The model has been previously characterised at
the ultrastructural, biochemical and molecular level
and we report here further studies on the development
of the unusual cell wall of the turion. In the
induction of turions we have a developmental,
qualitative phenomenon (the switching of a primordium
between alternative pathways of morphogenesis). The
response can only be triggered for a limited time in the
life of the primordium and shows a precise dependence on
ABA concentration. The presence of the unique
sensitivity window suggested to us that this model would
be ideal in a search for ABA receptors, but after
extensive studies using this and other ABA-sensitive
tissues, no ABA-binding proteins could be detected worth
the term 'receptor'.

INTRODUCTION

As with all developmental processes in plants, dormancy
is believed to result in part as a consequence of the
activity(ies) of plant growth regulators. The role of one
such regulator, abscisic acid (ABA) as a possible initiator
and/or regulator of bud dormancy is the subject of this study.

[1]This work was supported by grants from the S.E.R.C.
and A.F.R.C.
[2]Present address: Department of Botany, Downing Street,
Cambridge University, Cambridge, U.K.

The aquatic plant Spirodela polyrrhiza (a member of the
Lemnaceae) normally produces daughter fronds from each of
two meristematic pockets. However, it will also undergo
dormant bud (turion) formation under a variety of conditions.
Signals which have been shown to induce turion formation
include; short photoperiod, low temperature, high carbon
dioxide, high sugar, low nitrate, low sulphate, low phosphate,
high light intensity and overcrowding (1-9). Upon reception
of an appropriate environmental signal, a new programme of
biochemical responses is initiated within the meristematic
primordium which results in the production of the morphologically distinct dormant turion. The mature turion
abscises from the mother frond, sinks and remains dormant
until conditions are favourable for 'germination'. The
ability of this plant to form turions in response to
exogenous ABA (10,11) makes it an ideal model for studies
aimed at understanding development and dormancy and the role
of ABA in the regulation of these processes. Conceivably,
there are several pathways leading to the events that are
ultimately manifested in turion development, although it is
possible that ABA may play a role in controlling turion
formation in vivo, since turion formation induced by short
days is accompanied by the release of ABA into the medium,
the time of appearance of the released ABA correlating with
the onset of turion formation (12).

Features of the Model

   Turion formation. Addition of ABA to a culture of
S.polyrrhiza results in growth inhibition at concentrations
as low as 10^{-9}M, growth being completely arrested at 10^{-5}M.
Over a single order of magnitude range around 10^{-7}M, ABA
also induces the production of turions. The range of turion
producing concentrations of ABA is very narrow, turion
production having a clearly defined threshold, optimum, and
upper limit (Fig. 1).
   ABA-induced turions germinate rapidly when transferred
to ABA-free medium. They possess no vernalisation requirement necessary for the germination of turions induced by
manipulations of temperature, photoperiod, light intensity
or sucrose concentration; although turions formed in response
to nitrate deficiency also germinate readily when supplemented with nitrate without pretreatment at low temperature (6).

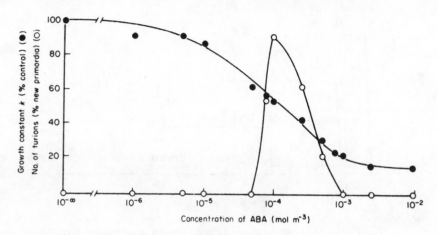

FIGURE 1. The effect of ABA on growth and turion production in <u>Spirodela</u> <u>polyrrhiza</u>. Five fronds were innoculated onto half-strength Hutner's medium supplemented with various concentrations of ABA, and their daily growth was measured by counting frond number. The growth of each culture was plotted logarithmically against time and the linear regression (k) of each treatment was calculated. The growth constant K is shown as a percentage of the control value (●). At certain concentrations of ABA, turions were formed. The number of turions produced is expressed as a percentage of the total number of new primordia formed after 8d incubation in ABA (o).

Sensitivity window. One important feature of turion formation is that only primordia below a certain developmental stage can be induced by ABA to develop into turions. The destiny of a particular primordium is determined by the time it is about 0·7mm long. All the cells of a primordium at this or a younger developmental stage are responsive to ABA in that they all develop into turion cells, and subsequently after a few days' growth a fully formed turion is formed. The production of semi-turions (with a turion-like proximal end) from primordia at later developmental stages (0·8-1·3mm) is indicative of a change in sensitivity to ABA during the development of the frond, and therefore presumably a change in cell sensitivity to ABA (Fig. 2).

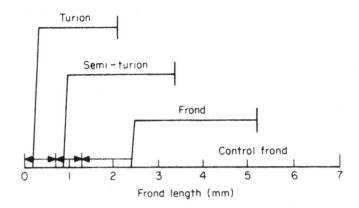

FIGURE 2. A summary of frond types produced during turion formation with 10^{-7}M ABA. Fronds of 0·7mm or less in length develop into turions which attain a length of 2mm after 7d; fronds longer than 0·7mm and shorter than 1·3mm develop into semi-turions which attain a length of 3·4mm after 7d; and fronds longer than 1·3mm develop into vegetative fronds which seldom exceed 5·2mm in length after 7d. Control fronds (no ABA) often reach a length of 7mm in the same time period.

Detailed analysis of individual frond growth curves and anatomical studies allow for the period of maximum sensitivity to ABA to be calculated on the basis that turion and vegetative primordia are morphologically and anatomically indistinguishable until they are 0·45mm long, and that the vegetative primordium is no longer able to develop into a turion once it is over 0·7mm long. In terms of turion formation there is a sensitivity window to ABA lasting some 14–20 hours in the normal developmental life of frond cells. Providing cells experience the appropriate signal during this period they initiate a new programme which eventually leads to turion formation (13) (Fig. 3).

Turion characteristics. The turion is characterised by its small size, reniform shape, and dark-brown colouration. The mesophyll is undifferentiated and totally lacking the substantial aerenchyma development found in the vegetative frond. Stereological analysis shows that the tissues differ quantitatively only in three main respects: air space formation, vacuolation, and starch and cell wall material accumulation. During development the cells of the turion,

while reaching the same final size as the vegetative frond
cells, accumulate numerous starch grains, thick cell walls,
and large deposits of anthocyanins and tannins at the expense
of the vacuolar expansion characteristic of the normal
maturity process (14).

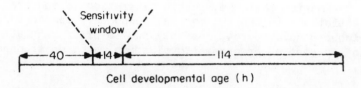

Cell developmental age (h)

FIGURE 3. The period of maximum sensitivity to ABA for
the turion-forming response. This was calculated from a
detailed analysis of both individual frond growth curves and
ultrastructural determinations. The sensitivity window is
approximately 14h wide and cells within this developmental
frame can respond to ABA by developing into turion cells.

Biochemistry of turion formation. The developmental
process leading to the formation of the ABA-induced turion is
accompanied by a repression of nucleic acid and protein
synthesis in ABA-sensitive tissue. DNA synthesis in the
developing turion is inhibited within 3h of ABA addition,
followed by a repression of protein synthesis after 24h. The
inhibitory effect of ABA on protein synthesis is general but
the synthesis of several turion-specific novel proteins is
induced within 24h of ABA addition. The rapid general
inhibition of protein synthesis at early stages of turion
formation cannot be accounted for by the change in levels of
translatable mRNA, indicating an effect of ABA at the
translational level. The specific alteration to the pattern
of in vivo labelled proteins may result, however, from control
of the level of specific mRNAs for these particular proteins.
Only after 3d in ABA, when the developing primordium is
committed to the turion developmental pathway, is there a
total inhibition of the production of mRNA leading to the
onset of the dormant state (15) (Fig. 4).
     The developmental phenomenon of ABA-induced turion
formation appears to bear little similarity to the rapid
physiological response of stomata to ABA. If the initial
action of ABA in eliciting either the turion production or
the growth inhibition response was at the membrane level,

which in turn controlled cellular processes, one might expect
rapid changes in certain ion flux.... However, while the
turion has very different ion flux and concentration
characteristics compared to the vegetative frond (Ca^{2+}, K^+,
and Cl^- efflux measured by compartmental analysis), there is
no rapid effect (up to 18h) of ABA on these fluxes through
either the plasmalemma or the tonoplast. This is consistent
with the notion that the drastic alteration in ion fluxes and
concentrations in the turion are a secondary consequence of
ABA-induced development, possibly due to prior regulation by
ABA of enzymes of processes involved in membrane transport
(16).

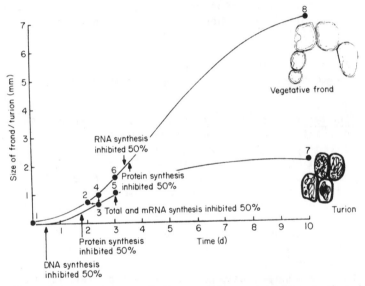

FIGURE 4. Summary of the events occurring during the
development of the vegetative frond (steps 1,2,4,6 and 8) and
the ABA-induced turion (1,3,5 and 7). Vegetative frond; step
1-2, main period of cell division; step 2-6, main period of
cell expansion; step 6-8, main period of cell separation and
differentiation. Turion: step 1-3, cell division; step 3-5,
cell expansion; step 5-7, starch, cell wall and anthocyanin
accumulation. Vegetative fronds before step 2 can switch to
the turion programme via step 3. Turions before step 3 can
switch to the vegetative frond programme via step 4. After
step 2 vegetative frond cells can no longer switch to the
turion developmental programme, and after step 5 turion cells
can no longer switch to the vegetative frond programme.

Two features of the model were the subject of further study. One was the presence of a thicker cell wall in the turion as compared with the vegetative frond, and the other was the unique sensitivity window for turion formation which we were anxious to exploit in a search for ABA receptors.

RESULTS AND DISCUSSION

Cell Wall Analysis and Plastic Extensibility Variations.

The presence of a thicker cell wall in the turion encouraged us to approach an assessment of cell wall structure and properties as a means for understanding the action of abscisic acid.

Fronds and turions were incubated for long periods on ^{14}C-glucose to reach effective isotopic equilibrium, the cell wall purified, hydrolysed to component monosaccharides and the radioactivity in each sugar determined after paper chromato-graphy. The data is shown in table 1 and they reveal some surprising differences. Turion cell walls have an almost complete absence of apiose and a very considerable abundance of glucose. We are satisfied that the glucose is not contaminating starch but we know no more about it.

The reduction in apiose content is perhaps more interest-ing. As a cell wall sugar apiose is of very limited taxonomic distribution, being largely confined to hydrophytes. Its actual role is uncertain but the pathway of synthesis has been characterised and it is incorporated into cell wall material via conversion of UDP-glucuronic acid to UDP-apiose. This same enzyme, termed UDP-xylose/UDP-apiose synthase, is also able to convert UDP-glucuronic acid to UDP-xylose (17,18). Incorporation of apiose into the cell wall is carried out by UDP apiosyl transferase, an enzyme characterised in Lemna minor (19).

We have prepared ^{3}H-glucuronic acid and used this as a precursor to assess the in vivo activity of both of these enzymes. Our estimates indicate that the activity of both declines when young fronds (putative turions) are exposed to abscisic acid but the decline is slow, commencing some 15h after treatment. Perhaps the notable feature is that the formation of UDP-xylose from UDP-glucuronic acid seems un-affected or is even slightly increased. It is known that the proportions of UDP-apiose to UDP-xylose formed by the synthase can be altered in vivo by incubation conditions (20). This in vivo change could therefore result from the ion flux alterations which we have previously characterised.

TABLE 1
% COMPOSITION OF NON-CELLULOSIC MONOSACCHARIDES[a]

	Fronds	Turions
Apiose	21•4	0•1
Arabinose	11•3	5•4
Fucose	1•3	0•6
Galactose	8•4	4•8
Galacturonic acid	25•6	15•0
Glucose	13•0	63•2
Glucuronic acid	5•8	5•3
Rhamnose	4•6	0•4
Xylose	8•6	5•2

[a]Fronds and turions were exposed to D-(U-^{14}C)-glucose (250 Ci/mol) in their mineral salts growth medium for 8d. They were homogenised in 1mM $CaCl_2$/50mM NaH_2PO_4 pH 6•5, and the cell wall material pelleted by centrifugation at 2000g for 5 min. The turion cell wall extract also contained large starch granules. These were solubilised and removed by treatment with 90% DMSO. The non-cellulosic monosaccharides were hydrolysed by autoclaving for 1h in 1M trifluoroacetic acid. A series of paper chromatographic and electrophoretic systems resolved the resulting monosaccharides qualitatively. The quantitative analysis was obtained by the scintillation counting of 1cm sections along the length of the chromatography paper, and relating these to the position of standard monosaccharides. The results show the average of 3 such analyses.

However, this drop in apiose incorporation into the cell wall is associated with two other phenomena. Firstly, we have measured the pattern of growth inhibition by ABA using time-lapse photography and image analysis. At a turion-inducing concentration, growth becomes severely impaired some 12-16h after treatment. Secondly, the physical properties of the wall show an associated change. Using the Instron stress/strain analyser we have measured plastic extensibility of the cell wall as a proportion of the total extensibility. Young fronds and turions are quite different. For fronds about 3mm in size plastic extensibility averages about 25% total extensibility and for turions it is about 5%. In the smallest fronds we could use in the Instron (1•75-2•5mm in diameter), there is a very pronounced drop in plastic

extensibility between 12 and 15 hours after ABA treatment
commensurate with the inhibition of growth and associated with
the decline in apiose incorporation.

This decrease in plastic extensibility in the developing
turion and the corresponding increase in cell wall thickness
and rigidity, may be necessary to accomodate an increased
turgor pressure within the turion cell due to the accumulation
of sugars. Treatment of fronds with a turion-inducing
concentration of ABA results in a large increase in the level
of sucrose and fructose and to a lesser extent, glucose
(Fig. 5).

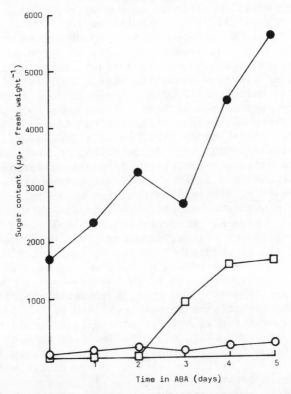

FIGURE 5. The effect of 10^{-7}M ABA on the sugar content
of fronds. Sugars were extracted with hot 80% ethanol,
evaporated to dryness and redissolved in water. Glucose was
determined by the peroxidase/glucose oxidase assay, sucrose by
the same assay after invertase treatment and fructose by the
method in (21). ● sucrose, □ free fructose, ○ free glucose

Attempts to Detect ABA-Binding Proteins.

The presence of a window of sensitivity to ABA for turion formation in Spirodela suggested to us that this might be a promising tissue with which to detect binding proteins for ABA. Since Spirodela polyrrhiza tissue can only be obtained in moderate yield and initial experiments proved negative we extended our search for binding proteins to other ABA-sensitive tissues. These were maize and wheat coleoptiles, maize mesocotyls and leaves, pea epicotyls (both Teltham First and Alaska), zucchini hypocotyls, barley aleurones, wheat and barley embryos, Spirodela turions and Lemna minor fronds. In addition we included pea apical tissue although we do not know if it is ABA-sensitive.

The techniques we used to try and detect ABA-binding proteins in homogenates were a variety of standard centrifugal procedures, pellets produced by various centrifugal forces and times, equilibrium dialysis, DEAE cellulose filter binding and Sephadex column procedures. These are all methods used to detect binding in other systems. These were accompanied by using a variety of buffers of differing strengths and pH's ranging from pH 7.5 to pH 5.5. In addition we used the Keith and Srivastava method (22) for detecting in vivo binding on sliced coleoptiles, epicotyls, hypocotyls, aleurones, Spirodela fronds and turions and other ABA-sensitive members of the Lemnaceae. This technique uses whole cells and simply treats the plasmalemma as equivalent to dialysis tubing in equilibrium dialysis. Accumulation of label above that expected for tissue volume, provided it is displaceable by unlabelled ABA is assumed to represent specific binding.

The binding experiments on homogenates used routinely a hot 10^{-9}M ^{3}H-ABA and a cold 10^{-6}M ABA treatment. We spent approximately 18 months and over 100 experiments progressively exploring each tissue in turn but with virtually all experiments negative except for some slight binding in wheat and barley aleurone (Fig. 6). The level of binding in aleurone tissue is so low that we question its significance and the degree of variation with such procedures makes curve fitting very uncertain.

Initially, experiments using the in vivo technique (22) seemed more promising. We obtained retention of higher concentrations of ^{3}H-ABA than would be anticipated on a fresh weight basis (assuming no anion trapping). We also obtained evidence that preincubation in 10^{-6}M ABA seemed sometimes to increase the level of ABA-binding, and turions often gave much higher amounts of binding than Spirodela fronds. Some

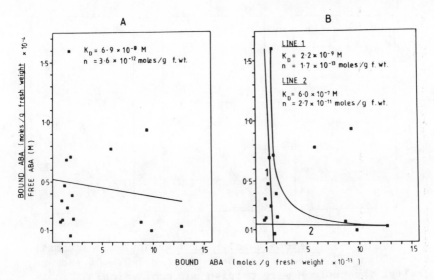

FIGURE 6. Scatchard analysis of ABA-binding to barley
aleurone particulate fraction. A barley aleurone 500-39000g
crude membrane fraction was resuspended in 0·01M Tris-HCl
(pH 7·0) and 0·003M MgCl$_2$. ^{3}H-ABA was added to a final
concentration ranging between 10^{-7}M and 10^{-9}M (=FREE ABA).
After 2h incubation at 25°C the membrane fraction was re-
pelleted and the radioactivity in the pellet extracted into
methanol. Parallel incubations containing non-radioactive
ABA at concentrations of (×1000 concentration of FREE ABA)
determined the amount of specifically bound ABA (=BOUND ABA)
by displacing it.

   K$_D$ (dissociation constant) = -(1/slope)
n (concentration of binding sites) = intercept on x axis
A shows one line representing a single binding site. B is an
interpretation of two separate binding sites, represented by
the two lines.

of this binding was displaceable. However, when we carried
out these experiments at different pHs (Fig. 7a and 7b) then
it became obvious that the apparent level of binding was very
pH dependent in a way that looked much more to us like an
effect on uptake. We were very careful in these experiments
to keep all solutions between 0 and 0·5°C (i.e. in melting
ice) but an effect on ABA metabolism cannot be discounted.
Extraction of labelled ABA and thin layer separation revealed
only a single band coincident with ABA.

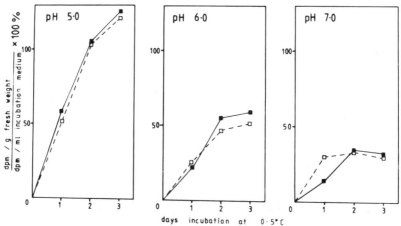

FIGURE 7a. Spirodela polyrrhiza fronds were incubated
at 0·5ºC in H/2 medium at pH 5·0, 6·0 and 7·0. At each pH,
parallel incubations were carried out containing:
A. 10^{-9}M ^{3}H-ABA ("HOT")━■━
B. 10^{-9}M ^{3}H-ABA + 10^{-6}M non-radioactive ABA ("HOT + COLD")-□--
Samples of fronds and incubation media were taken at daily
intervals, and scintillation counted.

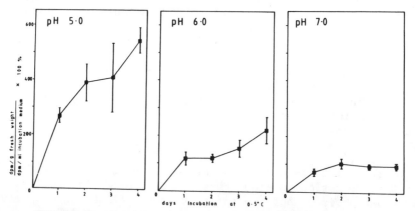

FIGURE 7b. Barley aleurone layers were incubated at
0·5ºC in 0·01M Tris-HCl, 0·0003M $MgCl_2$ containing 10^{-9}M ^{3}H-
ABA at pH 5·0, 6·0 and 7·0. Samples of aleurones and incubation medium were taken at daily intervals, and scintillation
counted for ^{3}H-ABA content.

13

Preincubation of tissue in labelled ABA for many days
to produce 'in vivo binding' and subsequent homogenisation
in labelled ABA coupled with pelleting and equilibrium
dialysis failed to reveal any binding which could account for
the excessive accumulation in the tissue. We concluded that
this method is not a reliable indicator of ABA binding.

At the end of the series of binding experiments, ABA-
binding was compared with auxin-binding using maize coleoptile tissue. Auxin binding was easily detected but none for
ABA.

We conclude that if ABA receptor proteins are present
they are either:
1) Not detectable by these methods due to structural damage
in isolation.
2) Present in such small amounts that ^{3}H-ABA is of in-
sufficiently high specific activity to detect them although
we used the highest specific activity that can be obtained.
3) We are working with the wrong concentration range of ABA;
binding perhaps might be more like that of some neuro-
transmitters operating at 10^{-4}M, thus very weak and difficult
to detect in the absence of an antagonist. We (Smart and
Hanke) are currently investigating the pattern of ABA
concentration achieved in the frond on treatment with turion-
inducing concentrations of exogenous ABA. This should
indicate (at least if the receptor is intracellular), whether
this could indeed be the major problem in ABA receptor
detection by reversible binding methods.

The evidence supporting the existence of ABA receptors
(at least in guard cells) is strong but there is nothing in
these tissues remotely similar to that recently reported by
Hornberg and Weiler (23) even though they are all sensitive
to ABA action. We believe from discussions with others that
our experience is not uncommon.

ACKNOWLEDGEMENTS

We should like to thank Mike Venis and Steve Fry for
help and advice on binding and cell wall analysis.

REFERENCES

1. Jacobs DL (1947). An ecological life-history of Spirodela
polyrrhiza (greater duckweed) with emphasis on the turion
phase. Ecol Monog 17:437.

2. Henssen A (1954). Die Dauerorgane von Spirodela polyrrhiza (L.) Schleid. in physiologischer Betrachtung. Flora 141: 523.

3. Czopek M (1959). Researches on the physiology of formation and germination of turions in Spirodela polyrrhiza (L.) Schleiden. Acta Biol Crac Ser Bot 2:76.

4. Perry TO (1968). Dormancy, turion formation and germination by different clones of Spirodela polyrrhiza. Plant Physiol 43:1866.

5. Newton RJ, Shelton DR, Disharoon S, Duffey JE (1978). Turion formation and germination in Spirodela polyrrhiza. Am J Bot 65:421.

6. Sibasaki T, Oda Y (1979). Heterogeneity of dormancy in the turion of Spirodela polyrrhiza. Plant Cell Physiol 20:563.

7. Malek L, Cossins EA (1983). Senescence, turion development, and turion germination in nitrate- and sulfate-deficient Spirodela polyrrhiza. Relationship between nutrient availability and exogenous cytokinins. Can J Bot 61:1887.

8. Augsten H, Jungnickel F (1983). Zur Steuerung der Turionenausbildung bei Spirodela polyrrhiza (L.) Schleiden durch Licht, Phosphat und Zucker. Wiss Zeit Ernst-Moritz-Arndt Univ Greif Mat Natur Reihe 3-4:64.

9. Beer S (1985). Effects of CO_2 and O_2 on the photosynthetic O_2 evolution of Spirodela polyrrhiza turions. Plant Physiol 79:199.

10. Perry TO, Byrne OR (1969). Turion induction in Spirodela polyrrhiza by abscisic acid. Plant Physiol 44:784.

11. Stewart GR (1969). Abscisic acid and morphogenesis in Lemna polyrrhiza L. Nature 221:61.

12. Saks Y, Negbi M, Ilan I (1980). Involvement of native abscisic acid in the regulation of onset of dormancy in Spirodela polyrrhiza. Aust J Plant Physiol 7:73.

13. Smart CC, Trewavas AJ (1983). Abscisic acid-induced turion formation in Spirodela polyrrhiza L. I. Production and development of the turion. Plant Cell Environ 6:507.

14. Smart CC, Trewavas AJ (1983). Abscisic acid-induced turion formation in Spirodela polyrrhiza L. II. Ultrastructure of the turion; a stereological analysis. Plant Cell Environ 6:515.

15. Smart CC, Trewavas AJ (1984). Abscisic acid-induced turion formation in Spirodela polyrrhiza L. III. Specific changes in protein synthesis and translatable RNA during turion development. Plant Cell Environ 7:121.

16. Smart CC, Trewavas AJ (1984). Abscisic acid-induced turion formation in Spirodela polyrrhiza L. IV. Comparative ion

flux characteristics of the turion and the vegetative frond and the effect of ABA during early turion development. Plant Cell Environ 7:521.

17. Roberts RM, Shah RH, Loewus F (1967). Inositol metabolism in plants IV. Biosynthesis of apiose in Lemna and Petroselinum. Plant Physiol 42:659.

18. Gustine DL, Yaan DH-S, Kindel PK (1975). Uridine diphosphate D-glucuronic acid cyclase and uridine diphosphate D-glucuronic acid carboxy-lyase I from Lemna minor. Purification, characterization, and separation from uridine diphosphate D-glucuronic acid carboxy-lyase II. Arch Biochem Biophys 170:82.

19. Pan Y-T, Kindel PK (1977) Characterization of particulate D-apiosyl- and D-xylosyltransferase from Lemna minor. Arch Biochem Biophys 183:131.

20. Matern U, Grisebach H (1977). UDP-apiose/UDP-xylose synthase. Subunit composition and binding studies. Eur J Biochem 74:303.

21. Roe JH (1934) A colorimetric method for the determination of fructose in blood and urine. J Biol Chem 107:15.

22. Keith B, Srivastava LH (1980). In vivo binding of gibberellin A_1 in dwarf pea epicotyls. Plant Physiol 66: 962.

23. Hornberg C, Weiler EW (1984). High-affinity binding sites for abscisic acid on the plasmalemma of Vicia faba guard cells. Nature 310:321.

Molecular Biology of Plant Growth Control, pages 361–370
© **1987 Alan R. Liss, Inc.**

CALCIUM AS A SECOND MESSENGER IN THE RESPONSE
OF ROOTS TO AUXIN AND GRAVITY[1]

Michael L. Evans, Karl-Heinz Hasenstein,
Charles L. Stinemetz, and James J. McFadden

Department of Botany, Ohio State University
Columbus, Ohio 43210

ABSTRACT   It has been proposed that Ca^{2+} acts as a
second messenger in the action of auxin on cell
elongation. We examined this model in the case of
auxin inhibition of root elongation in maize. We found
that Ca^{2+} mimicks the inhibitory action of auxin on
roots, that auxin sensitivity is reduced in roots with
reduced Ca^{2+}, and that auxin enhances the efflux of
radioactivity from root protoplasts preloaded with
$^{45}Ca^{2+}$. We also tested the involvement of Ca^{2+} in the
gravitropic growth response, and especially the role
that calmodulin might play in root gravitropism. We
found that calmodulin activity is concentrated in the
root tip region and that application of calmodulin
inhibitors to the cap retards gravitropism. In roots
of a cultivar requiring light for gravitropic
sensitivity, illumination causes a 4-fold increase in
the CaM activity of the cap and this precedes the
light-induced development of gravitropic competence.

INTRODUCTION

Although it is known that Ca^{2+} is important to the
growth and development of plants, the precise role played by
Ca^{2+} remains obscure. The evidence for a role for Ca^{2+} as a
second messenger in the action of animal hormones has drawn
attention to the possibility that Ca^{2+} might serve a similar

[1]This work was supported by National Science Foundation
grant PCM 8305775 and by grant NAGW-297 from the National
Aeronautics and Space Administration.

role in plant cells, perhaps through calmodulin (CaM).
Hertel (1) suggested that the movement of auxin across
membranes of intracellular compartments is coupled to a
gating mechanism allowing release of Ca^{2+} from those
compartments into the cytoplasm. He proposed that the
elevation of cytoplasmic Ca^{2+} causes the effects attributed
to the hormone. In view of the sensitivity of root growth
to auxin and Ca^{2+} as well as to environmental factors such
as light and gravity, we used primary roots to investigate
the interaction of Ca^{2+} and auxin in the control of cell
elongation and gravitropism.

MATERIALS AND METHODS

Seedlings of maize (Zea mays L., cv Merit and cv B73 x
Missouri 17) were raised as in ref. 2 and used when about 3
d old. Seedlings referred to as low Ca^{2+} (LC), intermediate
Ca^{2+} (IC), and high Ca^{2+} (HC) were raised as follows: low
Ca^{2+}, soaked and raised in distilled water followed by
treatment with 1 mM EGTA prior to experimentation;
intermediate Ca^{2+}, soaked and raised in distilled water;
high Ca^{2+}, soaked and raised in 10 mM $CaCl_2$.
Root protoplasts were prepared by a procedure modified
from the method of Lin (3). To measure the effect of auxin
on Ca^{2+} efflux, the protoplasts were incubated in 1 uM
$^{45}CaCl_2$ (specific activity about 770 MBq x mg^{-1}) for 90 min.
They were then washed and resuspended (10^6 protoplasts per
ml) in 0.6 M mannitol, 25 mM MES/TRIS buffer pH 6, 1 mM
$CaCl_2$, 0.2 mM $MgCl_2$. At appropriate times after addition of
auxin, the protoplasts were pelleted and radioactivity in
the supernatant was measured.
CaM activity in root sections was measured both by
CaM-dependent phosphodiesterase (PDE) activation (4) and NAD
kinase activation (5).

RESULTS

The Calcium-dependence of Auxin Action on Root Growth.
Figure 1 shows growth inhibition in roots of IC and HC
maize seedlings by Ca^{2+}. The severity of inhibition
increased with increasing concentrations of Ca^{2+} but, for
all concentrations in the range of 0.1 to 5 mM, inhibition
was transient.
The pattern of growth inhibition by sub-maximal doses
of IAA was similar to that induced by 1 mM Ca^{2+} (Fig. 2).
Although the lag preceding inhibition was longer with IAA,

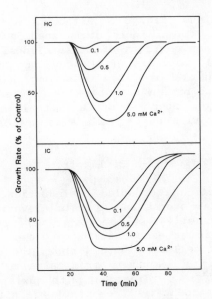

FIGURE 1. Effect of Ca^{2+} on the growth of roots of IC and HC seedlings (cv. B73 x Missouri 17).

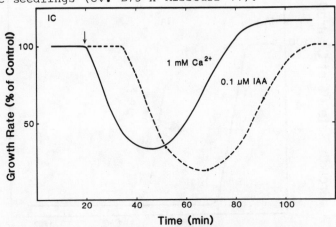

FIGURE 2. Comparative action of IAA and Ca^{2+} on the growth of roots of IC (cv. B73 x Missouri 17).

both 1 mM Ca^{2+} and 0.1 µM IAA caused strong growth inhibition followed by complete recovery.

As a further test of a possible link between Ca^{2+} and auxin action, we compared the auxin sensitivity of roots of

HC and LC seedlings (Fig. 3). Although IAA severely inhibited the growth of roots of HC seedlings, it exerted only a weak effect on roots of LC seedlings. Subsequent addition of $CaCl_2$ to roots of LC seedlings growing in the presence of 10 µM IAA resulted in strong inhibition of growth. This inhibition cannot be attributed to the growth-inhibiting properties of Ca^{2+} itself since inhibition by this concentration of Ca^{2+} alone is weaker and transient (Fig. 1).

Effect of Auxin on Calcium Efflux from Root Protoplasts.

If auxin action is coupled to induction of Ca^{2+} release from intracellular stores and if this elevation of cytoplasmic Ca^{2+} activates homeostatic Ca^{2+} pumping mechanisms, one might expect auxin to cause enhanced Ca^{2+} efflux from root cells. Figure 4 shows the effect of auxin on the efflux of radioactivity from maize root protoplasts preloaded with $^{45}Ca^{2+}$. Treatment with auxin resulted in a concentration-dependent enhancement of the efflux of previously absorbed $^{45}Ca^{2+}$.

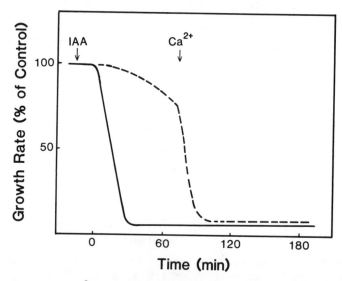

FIGURE 3. Ca^{2+}-dependence of auxin action on root elongation. HC roots (———). LC roots (----). IAA (10 µM) was added to both HC and LC roots at the first arrow. Ca^{2+} (0.5 mM $CaCl_2$) was added to the auxin-treated LC root at the second arrow (cv B73 x Missouri 17).

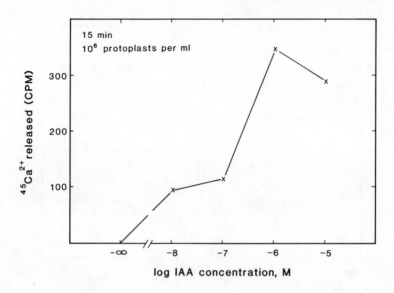

FIGURE 4. Auxin-induced release of $^{45}Ca^{2+}$ from maize root protoplasts (cv B73 x Missouri 17). Ca^{2+} efflux measured 15 min after exposure to auxin.

These results are consistent with the idea that auxin treatment of root cells may induce the release of Ca^{2+} from intracellular stores and that this may lead to enhanced Ca^{2+} efflux.

Effect of Diltiazem, a Calcium Entry Blocker, on the Kinetics of Root Growth Inhibition by Auxin.

If the reestablishment of rapid growth in seedling roots exposed to auxin (Fig. 2) is linked to Ca^{2+} homeostasis, one would expect to be able to modify the kinetics of auxin action on root growth by modifying one or more components of the homeostasis mechanism. We attempted this by pretreating roots with the Ca^{2+} channel blocker (6), diltiazem (DTZ) prior to addition of auxin (Fig. 5). Although detailed studies on the effects of DTZ on Ca^{2+} entry into plant cells remain to be done, there is evidence that plant cell membranes contain Ca^{2+} channels sensitive to Ca^{2+} entry blockers (7,8). Treatment with 25 μM DTZ

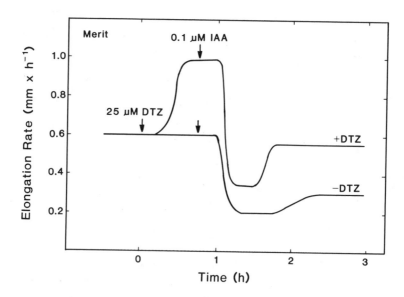

FIGURE 5. Effect of DTZ on growth and the auxin response of maize roots. Lower curve: response to 0.1 µM IAA (no DTZ pretreatment). Upper curve: DTZ added at first arrow.   IAA added at second arrow (cv Merit).

enhanced growth by about 65 %. Although the addition of 0.1 uM IAA inhibited growth strongly in both DTZ-treated and control roots, the absolute growth rate of DTZ-treated roots remained higher. Also the recovery from auxin inhibition was more rapid and the growth rate following recovery was greater in DTZ-treated roots.

The Interaction of Calcium and Auxin in the Gravitropic Response of Roots.

There is evidence that Ca^{2+} plays a role in the gravitropic response of roots (9,10). The importance of Ca^{2+} in gravitropism raises the question of the role that CaM might play in the gravitropic response. We examined this question by: 1) measuring the occurrence and distribution of CaM activity in maize roots, 2) testing the effects of localized application of CaM antagonists on root gravitropism, and 3) testing correlations between light induction of CaM activity and gravitropic sensitivity in roots of a maize cultivar known to require light for gravitropism.

We were surprised to find that strong CaM activity was detectable in intact root sections simply by adding the sections to an in vitro assay medium (Table 1).

TABLE 1

ENHANCEMENT OF NAD KINASE ACTIVITY BY BOVINE
CALMODULIN OR EXCISED MAIZE ROOT TIPS

Medium	CaM[a]	RT[b]	% Activity[c]
Complete[d]	+	−	100
+EGTA	+	−	0
Complete	−	+	90
+EGTA	−	+	15

[a] Bovine brain CaM, 2.5 units per ml.
[b] Twenty 1-mm apical root tip segments.
[c] NAD kinase activity as % of that induced by 2.5 units per ml bovine brain CaM in complete assay medium.
[d] Complete NAD kinase assay medium (5).

The CaM activity was reversible by EGTA (Table 1) and inhibited by the CaM antagonists trifluoperazine (TFP), chlorpromazine, and calmidazolium (data not shown). The activity appeared not to arise from supply by the tissue of some component of the assay medium other than CaM since root tip sections were not able to substitute for any other component of the medium. Also, the CaM activity appeared not to arise from discharging cap cells or from cut surfaces since the activity did not change with variable numbers of cut surfaces or prewashing of the segments. Although we have not yet tested the possibility that microflora contribute to the CaM activity, this seems unlikely in view of the photo-dependence of CaM activity in root tips of light-sensitive cultivars of maize (see below).

We examined the distribution of apoplastic CaM activity in maize roots and found CaM activity to be about 4-fold higher in the apical mm than in more basal 1-mm sections (data not shown). High levels of CaM were also found by Allan and Trewavas in the tip of seedling roots of pea (11). We tested the involvement of CaM in gravitropism by

examining the effect on gravitropism of application of CaM
inhibitors to the root cap. A 1 ul drop of 1 uM TFP applied
directly to the cap inhibited the gravitropic response
without inhibiting overall growth (data not shown).

A further indication of the involvement of CaM in the
graviresponse was obtained in experiments in which
photo-induced changes in CaM activity were measured in root
tips of dark-grown seedlings of the cultivar Merit, a
cultivar known to require light for root gravitropism (12).
The CaM activity in the apical mm of roots of dark-grown
seedlings (gravi-insensitive) was only about 25% as high as
the CaM activity in the apical mm of roots of light-grown
(gravi-sensitive) seedlings. Figure 6 shows the time course
of photo-enhancement of gravitropism and root tip CaM
activity in dark-grown seedlings of the Merit cultivar of
maize.

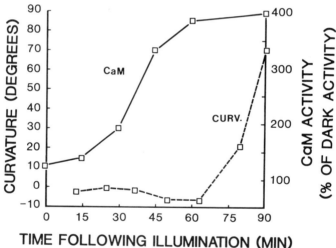

TIME FOLLOWING ILLUMINATION (MIN)

FIGURE 6. Comparative time courses of photo-enhancement
of CaM activity and gravitropic competence in roots of
dark-grown maize seedlings. CaM activity measured as
stimulation of in vitro PDE activity by excised segments and
shown as percent of activity in dark-grown roots. Similar
results were obtained using the NAD kinase assay. CaM
activity of 20 1-mm apical segments from dark-grown roots
equivalent to 0.56 units bovine brain CaM per ml.

After the onset of illumination there was a lag of about 1 h before horizontally-oriented roots began downward curvature. The increase in CaM activity was evident within 30 min. The CaM activity in the apical mm of the root was nearly maximal by the time gravitropic curvature was initiated.

DISCUSSION

The data indicate that Ca^{2+} may serve a second messenger function in the action of auxin on root cell elongation. Calcium mimicks the action of IAA on root growth, and roots raised under conditions expected to reduce endogenous Ca^{2+} exhibit little sensitivity to auxin. These findings are in line with the proposal that the initial action of auxin on root cells is to cause elevation of cytoplasmic free Ca^{2+} through release of Ca^{2+} from internal storage pools and/or through enhanced influx across the plasmalemma. There is evidence for the occurrence of Ca^{2+} transport systems which would tend to return cytoplasmic Ca^{2+} levels to normal following a step-up perturbation (13,14). A regulatory mechanism of this sort would be consistent both with the transient nature of growth inhibition by Ca^{2+} or auxin and with the finding that auxin stimulates the release of radioactivity from root protoplasts preloaded with $^{45}Ca^{2+}$. The observation that DTZ can accelerate the recovery of roots from auxin inhibition is also consistent with the this model.

Our experiments indicate that CaM may be important to the action of Ca^{2+} in the gravitropic response. CaM is concentrated in the gravisensing tip, and specific inhibitors of CaM retard the gravitropic response. In roots which require light for gravitropic sensitivity, illumination induces a 4-fold increase in the apoplastic CaM activity of the cap region, and the increase precedes the development of gravitropic competency.

Although these findings are consistent with the possibility that CaM is involved in the action of Ca^{2+} in gravitropism, we have no information on the specific role CaM might play. Experiments are planned to determine whether CaM is involved in the gravi-stimulated polar transport mechanism for Ca^{2+}, in the physiological action of Ca^{2+} asymmetry, or both.

REFERENCES

1. Hertel R (1983). The mechanism of auxin transport as a model for auxin action. Z Pflanzenphysiol 112: 53.
2. Mulkey TJ, Kuzmanoff KM, Evans ML (1981). Correlations between proton efflux patterns and growth patterns during geotropism and phototropism in maize and sunflower. Planta 152: 239.
3. Lin W (1980). Corn root protoplasts. Isolation and general characterization of ion transport. Plant Physiol 66: 550.
4. Sharma RK, Wang JH (1979). Preparation and assay of the Ca^{2+} dependent modulator protein. Adv Cyclic Nucleotide Res 10: 187.
5. Harmon AC, Jarrett HW, Cormier MJ (1984). An enzymatic assay for calmodulins based on plant NAD kinase activity. Anal Biochem 141: 168.
6. Van Breeman C, Hwang O, Meisheri KO (1981). The mechanism of inhibitory action of diltiazem on vascular smooth muscle contractility. J Pharmacol Exp Ther 218: 459.
7. Reiss H-D, Herth W (1985). Nifedipine-sensitive calcium channels are involved in polar growth of lily pollen tubes. J Cell Sci 76: 247.
8. Hetherington AM, Trewavas AJ (1984). Binding of nitrendipine, a calcium channel blocker, to pea shoot membranes. Plant Sci Lett 35: 109.
9. Lee JS, Mulkey, TJ, Evans ML (1983). Gravity-induced polar transport of calcium across root tips of maize. Plant Physiol 73: 874.
10. Lee JS, Evans ML (1985). Polar transport of auxin across gravistimulated roots of maize and its enhancement by calcium. Plant Physiol 77: 824.
11. Allan E, Trewavas A (1985). Quantitative changes in calmodulin and NAD kinase during early cell development in the root apex of Pisum sativum L. Planta 165: 493.
12. Feldman LJ, Gildow V (1984). Effects of light on protein patterns in gravitropically stimulated root caps of corn. Plant Physiol 74: 208.
13. Schumaker KS, Sze H (1985). A Ca^{2+}/H^+ antiport system driven by the proton electrochemical gradient of a tonoplast H^+-ATPase from oat roots. Plant Physiol 79: 1111.
14. Dieter P, Marme D (1980). Ca^{2+} transport in mitochondrial and microsomal fractions from higher plants. Planta 150: 1.

Molecular Biology of Plant Growth Control, pages 371–380
© 1987 Alan R. Liss, Inc.

PHYTOHORMONE METABOLISM IN PSEUDOMONAS SYRINGAE SUBSP
SAVASTANOI

by Francisco Roberto and Tsune Kosuge

Department of Plant Pathology, University of California
Davis, CA 95616

INTRODUCTION

Skoog and Miller (1) first described the effects of
auxin and cytokinin balances in promoting morphogenic
responses of normal tobacco callus tissue. Upon addition of
varying concentrations of auxin and cytokinin to culture
media, they observed that high cytokinin-auxin ratios
promoted formation of shoots from the callus tissue;
intermediate ratios maintained the tissue in a rapidly
proliferating, undifferentiated state; low ratios promoted
the formation of roots.. Mounting evidence in recent years
has unequivocally indicated that the principles for
phytohormone control of morphogenesis in plant callus
cultures elucidated by Skoog and Miller also apply to tumor
formation in plants.

A causal relationship between phytohormone production
and tumorigenicity was first demonstrated by Braun (2) who
found that bacteria-free tobacco crown gall tumor tissue
could grow in culture without addition of auxin and
cytokinins. Normal (untransformed) tobacco callus tissue
would grow only on media supplemented with the
phytohormones. In recent years, work in several
laboratories (as cited by Gelvin, 3) demonstrated that
tumorigenicity is associated with the presence of a large
plasmid called Ti in the crown gall pathogen, Agrobacterium
tumefaciens. During the infection process, a region of this
plasmid, called T-DNA, is transferred from the bacterium and
integrated into the host genome. Once integrated, genes on
the T-DNA are expressed and confer the tumorigenic condition
to host cells. By use of transposon mutagenesis, Garfinkel
and Nester (4) and others (3) constructed mutants of A.
tumefaciens which gave rise to shoot- and root-forming

tumors when inoculated into tobacco plants. Mutations that
were associated with the production of morphologically
distinct tumors occurred in specific loci in the T-DNA. The
loci were designated tms-1 and -2, and tmr; inactivation of
one of these loci conferred the unique tumor phenotypes.
When such tumors were analyzed for phytohormone content, the
cytokinin/auxin ratios were found to be highest in tissues
of shoot-forming tumors, intermediate in wild type
undifferentiated tumors and lowest in the rooty tumors (5).

INDOLEACETIC ACID PRODUCTION AND VIRULENCE IN P. SAVASTANOI

The association between IAA production and virulence is
also apparent in the interaction between the tumorigenic
bacterium, Pseudomonas syringae subsp savastanoi (P.
savastanoi), and its hosts, oleander and olive. Production
of IAA confers virulence in P. savastanoi because mutants
deficient in IAA production fail to incite the production of
tumors on oleander plants (6). When genes for IAA
production are transferred from parental strains to these
mutants, both IAA synthesis and full virulence are restored
(7, 8). In contrast to A. tumefaciens, genetic information
from P. savastanoi is not transferred to its hosts; instead
tumor formation by the plant is a response to the continued
secretion of phytohormones by the pathogen.
P. savastanoi produces IAA by the sequence of reactions
L-tryptophan----> indoleacetamide-------> indoleacetic
acid. The enzymes and their genetic determinants are:
tryptophan-2-monooxygenase (iaaM) and indoleacetamide
hydrolase (iaaH) (7, 8, 9). These genes (IAA genes) occur
on a plasmid, pIAA, in oleander strains and on the
chromosome of olive strains. Strains can be readily cured
of pIAA by selecting for resistance to alpha methyl
tryptophan. Such mutants are deficient in IAA synthesis and
show attenuated virulence. Loss of IAA production also
occurs by deletions of the IAA genes and inactivation of
iaaM by IS elements (10).

IS-ELEMENTS INACTIVATE GENES FOR INDOLEACETIC ACID SYNTHESIS

In three iaa⁻ mutants of P. savastanoi, loss of
virulence has been attributed to movable elements IS-51 and
IS-52 which insert into and inactivate the monooxygenase
gene, iaaM (10). Although the IS-elements differ

significantly in nucleotide sequence, they nevertheless
share the common property of transposition into <u>iaaM</u> causing
loss of virulence and IAA production. The site of
integration in <u>iaaM</u> is marked by the consensus sequence,
purine-A-G, in the target gene. (However, the integration
may occur at different locations in the gene.)   Both IS-51
and IS-52 have characteristics in common with those found in
other IS-elements: IS-51 has borders comprised of inverted
repeats of 26 bp whereas IS-52 has inverted repeats of 12 bp
(T. Yamada, P. Lee and T. Kosuge, unpublished results).

CONVERSION OF IAA TO INDOLEACETYL LYSINE

It was known for sometime that oleander strains of $\underline{P.}$
$\underline{savastanoi}$ converted IAA to indoleacetyl-ξ-L-lysine. The
enzyme catalyzing the synthesis of IAA-lysine from IAA
requires ATP and a divalent metal cation such as Mg^{2+} or
Mn^{2+} (11, 12). The genetic determinant (<u>iaaL</u>) for the
enzyme is borne on pIAA but is not part of the <u>iaa</u> operon
(N. L. Glass & T. Kosuge, in press, J. Bacteriol.). It is
interesting that olive isolates, unlike their counterparts
from oleander, are unable to produce IAA-lysine.

IAA-lysine synthetase may help control the pool size of
free IAA since oleander strains typically accumulate less
IAA in culture when compared with olive strains (L. Glass
and T. Kosuge, unpublished results).

IAA PRODUCTION IN OTHER TUMORIGENIC SYSTEMS

Production of phytohormones is a basis for
tumorigenicity in both crown gall tumor tissue and in
oleander knot; questions therefore were raised about the
similarities between the two systems. Gene 2 of T-DNA
encodes an amidohydrolase converting indole-3-acetamide
(IAM) into IAA (13, 14). This observation led to the
suggestion that IAA in tumors might be synthesized by the
same 2 step pathway from tryptophan previously demonstrated
in $\underline{P.}$ $\underline{savastanoi}$, and gene 1 might code for an enzyme
converting tryptophan into IAM. It has been shown that
tobacco tissue transformed with a vector containing gene 1
but without other T-DNA genes, accumulated high levels of
IAM (1500 pmol/g.fr.wt.). Normal tissue contained
comparatively little IAM (1.0 pmol/g fr.wt.) (15). Recently,
conversion of tryptophan into IAM has been successfully

demonstrated using cell-free enzyme preparations from tobacco tumor tissue (Van Onckelen et al. in press). Significant homology occurs between nucleotide sequences of iaaM and iaaH of P. savastanoi and those of genes 1 and 2 of A. tumefaciens T-DNA. Gene 1 and iaaM have about 50% perfect matches in deduced amino acid sequence; comparison of the open reading frames for these genes indicate that gene 1 encodes a polypeptide 198 amino acids longer and 22 kilodaltons larger than the proteins encoded by iaaM (16). Comparisons of the deduced amino acid sequences of iaaH and gene 2 indicate a lesser homology of about 27% for perfect amino acid matches. Moreover, strong homology has been found between gene 1 and a 24 amino acid sequence in the FAD binding domain in the FAD-linked hydroxybenzoate hydroxylase from P. fluorescens (17). The same 24 amino acid sequence of gene 1 also shows strong homology to a 24 amino acid sequence in iaaM. Since the tryptophan monooxygenase encoded by the iaaM gene is known to have an FAD cofactor it seems likely that this amino acid sequence comprises an FAD-binding site in both the gene 1 product and tryptophan monooxygenase. Since the deduced amino acid sequences of iaaM and gene 1 are highly similar, contain probable FAD-binding sites, and both catalyze the production of IAM, it seems likely that gene 1 of T-DNA encodes a protein with tryptophan monooxygenase-like activity.

ROLE OF CYTOKININS

Although IAA has been shown to be clearly necessary for the expression of virulence by P. savastanoi (6, 8), by no means has it been assumed to be the sole determinant of virulence. Direct application of IAA to the host plant produced only localized hypertrophy, without accompanying cell proliferation and failed to result in gall formation. The discovery of genes encoding auxin and cytokinin biosynthesis within the T-DNA of the Agrobacterium tumefaciens Ti-plasmid, and the role these gene products play in the establishment and morphology of crown-gall tumors lent further credence to the idea that cytokinins play an important role in gall formation by P. savastanoi (5).

Characterization by HPLC, mass spectrometry, radioimmunoassay and bioassay confirm that the cytokinins zeatin, zeatin riboside, isopentenyladenosine,

isopentenyladenine, dihydrozeatin, dihydrozeatin riboside
and a newly identified cytokinin, ribosyl-1"-methylzeatin
(19) are present in culture filtrates. When assayed during
mid-log phase growth, cultures of wild-type P. savastanoi
produce 1000 times more cytokinin than comparable cultures
of A. tumefaciens (20). This represents a level approaching
1 mg/1 (23, Roberto and Kosuge, unpublished results).
    The genetic locus encoding cytokinin biosynthetic
function in P. savastanoi has been cloned from olive isolate
EW1006Rif and is described elsewhere (20). An isopentenyl
transferase gene from P. savastanoi has recently been
sequenced and shows ca. 50% homology with tzs, the
isopentenyl transferase gene in the vir region of A.
tumefaciens T-DNA, and tmr, the corresponding gene of the T-
DNA (21).
    Initial biochemical evidence suggested a correlation
between the presence of a 42 kilobase (kb) plasmid in
oleander strain PB213 and cytokinin production (20, Roberto
and Kosuge, unpublished results). Using a cosmid clone
containing a 2.7 kb fragment of the 64 kb plasmid in
EW1006Rif, which conferred cytokinin biosynthetic capacity
on E. coli strains harboring this recombinant plasmid
(Regier, unpublished results), we have further genetic
evidence that at least the isopentenyl transferase, and
possibly an associated hydroxylase involved in zeatin/zeatin
riboside production are plasmid-encoded functions in P.
savastanoi. Further, these gene(s) appear localized to a
distinct plasmid in each of strains PB213, 2009 and EW1006,
and recent experiments suggest a common restriction fragment
found solely on the corresponding plasmids with no
chromosomal copies as yet detected (Roberto and Kosuge,
unpublished results).
    We have obtained spontaneous mutants of strain PB213
with a variety of phytohormone phenotypes, including
iaa+zeatin+, iaa-zeatin+ and iaa+zeatin-, designated PB213,
213-3 and 213-16, respectively. Analysis of phytohormone
levels and plasmid profiles are shown in Figure 1. As
expected, PB213-3, lacking a 72 kb plasmid previously
identified as pIAA2 and carrying the IAA operon produced no
detectable IAA. It is interesting that zeatin/zR levels are
below threshold limits as well. Strain PB213-16 contained
pIAA2 but lacked a 42 kb plasmid previously correlated with
cytokinin production (see above). The level of IAA detected
was slightly lower than the wild type and no zeatin/zR was
detected.
    The virulence of these strains was assessed on oleander

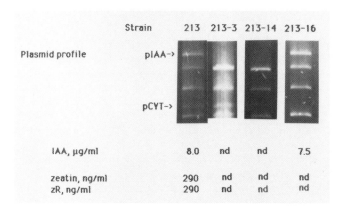

Figure 1.   Plasmid profiles of various P. savastanoi
PB213 strains and corresponding levels of phytohormones
produced by these strains.   IAA was assayed colorimetrically
after 48 h; cytokinins were quantified by HPLC after 72 h.

plants and the results are shown in Figure 2.   Inoculation
with PB213 gave rise to galls within 3 weeks, showing the
typical stem-splitting characteristic of this strain when
inoculated on oleander.   Strain 213-3 showed considerable
attenuation of symptoms since it was incapable of
synthesizing IAA (L. Glass, unpublished results).   Finally,
strain 213-16, unable to produce zeatin/zR also gave rise
to attenuated symptoms, as did 213-14, which failed to
produce either IAA or cytokinins.   It should be noted that
plants inoculated with 213-3 or 213-16 always showed a
localized region of blackened tissue, which upon closer
examination, appeared to be due to oxidized phenolic
compounds or latex.   This is most likely a reaction to the
growth of bacteria, since bacterial growth occurs even in
the absence of virulence expression (6).
     These results suggest that cytokinins are in fact,
necessary virulence factors along with IAA.   Our data
support a scenario analogous to that seen in the
transformation of plant hosts with A. tumefaciens, where
cytokinin/auxin balance controls gall formation.   To date,
however, no evidence exists that P. savastanoi transforms
its hosts, but rather, gall formation is induced and
supported by the relatively enormous amounts of
phytohormones the colonizing bacteria secrete into the host
tissues.   The hormonal responsiveness and biochemical nature

Response of oleander to infection
with P. savastanoi

8 weeks after inoculation

Strain   Control   213wt   213-3   213-14   213-16

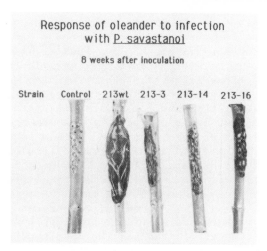

Figure 2. Virulence of various P. savastanoi PB213
strains as assayed on oleander. Plants were injured several
times with a sterile toothpick at the second terminal
internode and inoculated with 100 microliters from
exponential cultures of the indicated strains. Control was
inoculated with sterile King's medium B. Photographs were
taken eight weeks post-inoculation.

of P. savastanoi's hosts, olive, oleander and privet, may
explain why differing tumor morphologies are not seen in
these plants as have been described in tobacco plants
infected by tms (deficient in IAA biosynthesis) and tmr
(deficient in cytokinin biosynthesis) mutants of A.
tumefaciens.

SUMMARY

We have discussed the role of phytohormone production
in virulence and tumorigenicity in a plant pathogen with
indoleacetic acid (IAA) production by Pseudomonas savastanoi
as a model. The enzymes concerned and their genetic
determinants are: tryptophan-2-monnoxygenase (iaaM) and
indoleacetamide hydrolase (iaaH). The two genes, iaaM and
iaaH, occur on a plasmid, pIAA, in oleander strains and on
the chromosome of olive strains of the pathogen.

Two IS-elements, IS-51 and IS-52, resident in the bacterium have been isolated and sequenced. By integration into iaaM, they are responsible for loss of IAA production and virulence in several mutants that were isolated by selection for alpha methyltryptophan resistance. Certain strains of P. savastanoi convert IAA to its lysine conjugate; the gene for lysine conjugate synthesis occurs on pIAA but it is not part of the IAA operon. Conversion of IAA to its lysine conjugate affect the free pool size of IAA and the amount secreted into culture.

On the basis of both nucleotide and deduced amino acid sequences, the coding sequences of iaaM show significant homology with the sequences of the open reading frame of tms-1, a gene encoding IAA production in crown gall T-DNA; less, but nevertheless significant homology occurs between the coding sequences of iaaH and the open reading frame of tms-2 of crown gall T-DNA. The results suggest that the genes for IAA production in P. savastanoi and crown gall T-DNA have a common origin.

P. savastanoi strain PB213 produces zeatin, zeatin riboside, isopentenyladenosine and isopentenyladenine. Cytokinin biosynthesis appears to be directed by a 42kb extrachromosomal element, a plasmid carried within this strain. The results of virulence assays indicate that both IAA and cytokinins function as virulence factors in this plant-microbe interaction.

ACKNOWLEDGEMENTS

This material is based upon work supported by Grants PCM-8011794 and DMBO-831872 from the National Science Foundation, and Grants 81-CRCR-1-0643 and USDA-59-2063-1-1-643-0 from the Competitive Research Grants Office, United States Department of Agriculture. F. R. is supported by a fellowship from a grant made to the University of California by the McKnight Foundation.

REFERENCES

1.   Skoog F, Miller CO (1957). Chemical regulation of growth and organ formation in plant tissues cultured in vitro. Symp Soc Exp Biol 11:118-131.
2.   Braun AC (1958). A physiological basis for autonomous growth of the crown gall tumor cell. Proc Natl Acad Sci USA 44:344-349.

3. Gelvin SB (1984). Plant tumorigenesis. In: Plant-
    Microbe Interactions - Molecular and Genetic
    Perspectives. T Kosuge and EW Nester, eds.
    Mac Millan Inc. New York, pp. 243-377.

4. Garfinkle DJ, Nester EW (1980). Agrobacterium
    tumefaciens mutants affected in crown gall
    tumorigenesis and octopine catabolism. J Bacteriol
    144:732-743.

5. Akiyoshi DE, Morris RO, Hinz R, Mischke BS, Kosuge T,
    Garfinkel DJ, Gordon MP, Nester EW (1983).
    Cytokinin/auxin balance in crown gall tumors is
    regulated by specific loci in the T-DNA. Proc Natl
    Acad Sci, USA 80:407-411.

6. Smidt M, Kosuge T (1978). The role of indole-3-acetic
    acid accumulation by alpha methyl tryptophan-resistant
    mutants of Pseudomonas savastanoi in gall formation on
    oleanders. Physiol Plant Pathol 13:203-214.

7. Comai L, Kosuge T (1980). Involvement of plasmid
    deoxyribonucleic acid in indoleacetic acid synthesis in
    Pseudomonas savastanoi. J Bacteriol 143:950-957.

8. Comai L, Kosuge T (1982). Cloning and
    characterization of iaaM, a virulence determinant of
    Pseudomonas savastanoi. J Bacteriol 149:40-46.

9. Comai L, Kosuge T (1983). The genetics of
    indoleacetic acid production and virulence in
    Pseudomonas savastanoi, pp 363-366. In: A Puhler (ed.)
    Molecular Genetics of the Bacteria-Plant Interactions,
    Springer-Verlag, Berlin.

10. Comai L, Kosuge T (1983). Transposable element that
    causes mutations in a plant pathogenic Pseudomonas sp.
    J Bacteriol 154:1162-1167.

11. Hutzinger O, Kosuge T (1968). Microbial synthesis and
    degradation of indole-3-acetic acid. III. The
    isolation and characterization of indole-3-acetyl-E-L-
    lysine. Biochemistry 7:601-605.

12. Hutzinger O, Kosuge T (1968). 3-indoleacetyl-3-L-
    lysine, a new conjugate of 3-indoleacetic acid produced
    by Pseudomonas savastanoi. p. 183-194. In F. Wightman
    and G. Setterfield (ed.). Biochemistry of Plant Growth
    Substances. The Runge Press, LTD., Ottawa, Canada.

13. Schroder G, Waffenschmidt S, Weiler EW, Schroder J
    (1984). The T-region of Ti plasmids codes for an enzyme
    synthesizing indole-3-acetic acid. Eur J Biochem
    138:387-391.

14. Thomashow LS, Reeves S, Thomashow MF (1984). Crown
    gall oncogenesis: evidence that a T-DNA gene from the

Agrobacterium Ti plasmid pTiA6 encodes an enzyme that catalyzes synthesis of indoleacetic acid. Proc Natl Acad Sci USA 81:5071-5075.

15. Van Onckelen H, Rudelsheim P, Inze D, Follin A, Messens E, Hovemans S, Schell J, Van Montagu M, DeGreef J (1985). Tobacco plants transformed with the Agrobacterium tumefaciens T-DNA gene 1 contain high amounts of indoleacetamide. FEBS 181:373-376.

16. Yamada T, Palm CJ, Brooks B, Kosuge T (1985). Nucleotide sequences of Pseudomonas savastanoi indoleacetic acid genes show homology with Agrobacterium tumefaciens T-DNA. Proc Natl Acad Sci USA: 82:6522-6526.

17. Klee H, Montoya A, Horodyski F, Lichtenstein C, Garfinkel D, Fuller S, Flores C, Peschon J, Nester EW, Gordon MP (1984). Nucleotide sequence of the tms genes of the pTi A6NC octopine Ti Plasmid: Two gene products involved in plant tumorigenesis. Proc Natl Acad Sci USA 81:1728-1732.

18. Hutcheson SW, Kosuge T (1985). Regulation of 3-indoleacetic acid production in Pseudomonas syringae pv savastanoi. J Biol Chem 260: 6281-6287.

19. Surico G, Evidente A, Iacobellis MS, Randazzo G (1985). A new cytokinin from the culture filtrate of Pseudomonas syringae pv savastanoi. Phytochem. 24(7):1499-1502.

20. MacDonald EMS, Powell GK, Regier DI, Glass NL, Kosuge T, Morris RO (1986). Secretion of zeatin, ribosylzeatin and ribosyl-1"-methylzeatin by Pseudomonas savastanoi: plasmid-coded cytokinin biosynthesis. Submitted for publication.

21. Powell GK, Morris RO (1986). Nucleotide sequence and expression of a Pseudomonas savastanoi cytokinin biosynthetic gene: homology with Agrobacterium tumefaciens tmr and tzs loci. Nucl Acids Res 14(6):2555-2565.

Molecular Biology of Plant Growth Control, pages 381–389
© 1987 Alan R. Liss, Inc.

PHENOTYPE OF PLANTS DERIVED FROM HAIRY ROOTS TRANSFORMED WITH *AGROBACTERIUM RHIZOGENES*

Kitisri Sukhapinda, Anna J. Trulson,
Elias A. Shahin and Robert B. Simpson

ARCO PLant Cell Research Institute
6560 Trinity Ct., Dublin, CA 94568

ABSTRACT In a number of crop species, plants can be regenerated from hairy roots transformed by *A. rhizogenes*. These plants often display abnormal phenotype which can be characterized by changes in plant morphology, and reproductive functioning. The abnormal phenotype is associated with the integration into the plant genome of the T-DNA from the Ri-plasmid.

INTRODUCTION

Certain strains of soil bacteria, *Agrobacterium tumefaciens* and *A. rhizogenes*, are capable of infecting dicotyledonous plants, and causing tumor formation (crown gall disease; ref 1), or prolific root development (hairy root disease; ref 2), respectively, at the wound site. Virulence of the bacteria is dependent on endogenous plasmids called the Ti-plasmid (tumor- inducing; ref 3), or the Ri-plasmid (root - inducing; ref 4). The tumor or root production in the host plant is a result of transfer and integration into the plant genome of a particular DNA segment from the Ti- or Ri-plasmids (5; 6; 7). The segment from the Ti-plasmid results in the production in plant of an auxin and cytokinin (8, 9). Although it is likely that the segment from the Ri-plasmid results in the production of a plant growth regulatory substance or substances, this has not been directly demonstrated.

The portion of the Ri-plasmid DNA (T-DNA) that is transferred and integrated into plant cell DNA confers the ability to grow in the absence of exogenous phytohormones and to produce unusual amino acid and sugar derivatives (termed opines) on the transformed cells (7; 10). In addition to the T-DNA, the Ri-plasmid consists of genes which are essential for virulence and replication located outside the T-DNA. Based on the opine production, Ri-plasmids fall into three categories: the agropine type (A4, 15834, HRI,1855), the mannopine type (8196, TR7, TR101) and the newly characterized opine type (2659) (11;12). The agropine-type Ri-plasmids are very similar as a group and are quite different from the mannopine-type plasmids (4). Perhaps the most studied category of Ri-plasmid is from the A4 strain and related strains of *A. rhizogenes*, therefore, we will discuss in detail the research on this category of Ri-plasmid.

Extensive studies on the integration of the T-DNA into DNA of two tobacco tumor lines revealed that two regions of the Ri-plasmid T-DNA, designated T-left (T_L) and T-right (T_R), are integrated and stably maintained in the plant genome (13;14). Each of the T-DNA fragments spans a 15-20 kilobase region and is separated from one another by at least 15 kilobases of nonintegrated plasmid DNA. The T_R-DNA contains two genetic loci, *tms*-1 and *tms*-2, which are functionally and structurally homologous to the *tms* genes of the Ti-plasmid (14), which encode synthesis of indole-3-acetic acid, an auxin (9). Mutation analysis of the T_L-DNA defined at least four genetic loci (designated as *rol*A, *rol*B, *rol*C and *rol*D), which do not seem to be essential for the tumorigenesis of the plasmid, but affect the root growth of the transformed tumors (14). Recently, sequencing of the T_L-DNA revealed the presence of eighteen open reading frames, whose genetic function is still unknown (15). Unlike the T_R-DNA which shares extensive homology with the T-DNA of the Ti-plasmid of *A. tumefaciens,* the T_L-DNA seems to be distinct in structure and function (14).

The ability of *A. rhizogenes* to transfer its plasmid T-DNA into plant cells, and incite the formation of hairy roots can be used in mediating foreign gene transfer in dicotyledonous crop species. Whole plant can be regenerated from transformed hairy roots of many crop species such as tobacco (16; 17; 18), carrot (16), morning glory (16), potato (19), oilseed rape (20), tomato (21;22), and cucumber (23). A binary vector in *A. rhizogenes* has been used to introduce foreign genes into several plant species (21;23;24).

PHENOTYPE OF HAIRY ROOT DERIVED PLANTS

In most species, plants regenerated from hairy roots often display similar phenotypic abnormalities which involve changes in plant morphology, and changes in reproductive functioning as summarized in Table 1. These phenotypic changes in the transformed plants could be due to the effect of hormonal alteration resulting from the expression of genes located on the integrated T-DNA from the Ri-plasmid (14). In general, the transformed phenotype is a reliable indication for the presence of the Ri-plasmid T-DNA.

Leaf morphology. The change in leaf morphology of plants regenerated from hairy roots is prominent. In most species, the leaves of the transformed plants are crinkled (16;18;19;20; 22). Taylor et al.(18) reported that the leaf crinkling trait was associated with the presence of T_L-DNA. The degree of leaf crinkling may vary from species to species. Slightly crinkled leaves were observed in tobacco (16), and potato (19), whereas severely crinkled leaves were observed in certain cases of tobacco (16), oilseed rape (20), and tomato (22). Durand-Tardif et al. (25) attributed the severely crinkled leaves in the progeny of hairy root -derived tobacco plants to the change in the regulation of T-DNA expression.

In addition to leaf crinkling, the overall leaf shape is altered in many transformed plant species. Tepfer (16) showed that the mean length-to-width ratio of the leaves of transformed tobacco plants was lower than that of the non-transformed plants. The leaves of the transformed tomato plants were smaller and more pointed than those of the non-transformed plants (22). In potato, the leaves of the transformed plants were sometimes compound and more glossy than those of non-transformed plants (19).

Shoot and stem morphology. Shortening of the internodes was observed in most species including tobacco (16), oilseed rape (20), tomato (22), and cucumber (23). The transformed plants were shorter than the non-transformed plants. Ooms et al. (20), however, found that at early stage of development, the transformed potato plants had a faster shoot growth rate than the non-transformed plants, but at maturity, the transformed plants reached the same height as the non-transformed plants. Reduced apical dominance resulting in

TABLE 1
SUMMARY OF PHENOTYPES OF PLANTS DERIVED FROM HAIRY ROOTS TRANSFORMED WITH A. *RHIZOGENES*

Transformed Plant Phenotype	*Nicotiana tabacum* (16)	*Nicotiana glauca* (18)	*Convolvulus arvensis* (16)	*Daucus carota* (16)	*Solanum tuberosum* (19)	*Brassica napus* (20)	*Lycopersicon esculentum* (22, 26)	*Cucumis sativus* (23, 27)	*Medicago sativa* (28)
Leaf morphology									
Crinkled	+	+	+	+	+	+	+	--	--
Low length-to-width ratio	+	?	+	+	?	?	--	--	--
Glossy	?	?	?	?	+	?	+	--	--
Compound	?	?	?	?	some	?	some	--	--
Shoot and stem morphology									
Early vigorous growth	?	?	?	?	+	--	?	--	?
Shortened internodes	+	?	--	--	--	?	+	+	--
Adventitious roots/root knobs above ground	+	?	--	--	+	?	+	--	+
Reduced apical dominance (highly branched)	+	?	+	+	--	?	--	--	--

	C1	C2	C3	C4	C5	C6	C7	C8
Root morphology								
Vigorous growth	+	+	+	+	+	+	+	+
Abundant roots	+	?	?	+	+	+	--	?
Reduced apical dominance (highly branched)	+	?	?	+	+	+	--	?
Plagiotropic growth	+	?	--	+	+	+	--	?
Longer tuber with highly frequent and prominant eyes	na	na	na	+	na	na	na	na
Reproductive structure and functioning								
Change in flower morphology	+	?	--	?	?	+	+	--
Reduce pollen viability	some	?	--	?	?	+	+	--
Reduced seed set	some	?	+	?	+	+	?	--
Change in time and frequency of flowering	+	?	--	?	?	+	+	--
Switch from biennial to annual	na	na	+	na	na	na	na	na

(+) Observed characteristics; (--) Not observed characteristics; (?) Not described; (na) Not applicable
Not applicable

stem branching occured frequently in the transformed tobacco and carrot (16).

Root morphology. Abundant highly-branched roots were common in transformed plants. The root systems of carrot, tobacco (16), tomato (22), potato (19) and oilseed rape (20) were plagiotropic and had horizontal growth habits. Increased adventitious root production was observed in tobacco (16), and tomato (22). In potato, the transformed plants produced longer tubers with more prominent eyes (19).

Plant fertility and reproduction. Dramatic alterations of plant fertility, reproductive habits and flower morphology are frequently observed in the hairy root -derived plants. Different degrees of reduction in pollen viability were observed in transformed tobacco (16), tomato (22) and cucumbers (23). Transformed oilseed rape plants failed to produce seeds, most likely as a result of the transformation (20). In some transformed tobacco plants the stigma had a different height relative to anthers (heterostyly) and the seed capsule was smaller than in the non-transformed control plants (16). Smaller fruit and reduced seed set was also observed in transformed tomato plants (22). Change from biennial to annual flowering occurred in the transformed carrots (16).

Variation expressed during plant development. Recent observations (16; 25) indicate that the degree of variation associated with hairy root transformation may change during plant development. Tepfer (16) observed that an extreme phenotype, termed T', appeared at low frequency among progeny of transformed tobacco plants. The T' plants were characterized by extreme crinkling of the leaves and severe reduction of the internode distances. This phenotype frequently reverted to the more moderateT phenotype. Due to the similarity of the T-DNA structure in both phenotypes it was concluded that the T' phenotype was not due to large-scale deletion or insertion, nor due to transposition of the T-DNA itself. The role of the T-DNA-encoded auxins was also ruled out, since the T_R-DNA in the transformed tobacco plants did not contain the fragment responsible for auxin synthesis. The analysis of the T-DNA transcripts and their abundance in the leaves and roots of the T' and T phenotypes lead to a conclusion that the reversions of the phenotypes were due to altered regulation of T-DNA expression (25).

CONCLUDING REMARKS

A variety of morphological and reproductive abnormalities are observed in plants transformed with *A. rhizogenes*. Some of the abnormalities such as leaf crinkling, reduced apical dominance and root plagiotropism are considered typical and almost obligatory in transformed plants derived from transformed roots. However, the diversity of the observed abnormalities, and the different responses of different species suggest the complexity of the phenomenon. Although the T-DNA is now well characterized, the T-DNA genes responsible for particular morphological abnormalities are unknown. The extent of transfer of the T-DNA varies not only among different species but also among plants of the same species. Moreover, the expression of the T-DNA may be altered during plant development. The understanding of these phenomena will benefit the basic knowledge of gene expression and regulation which should subsequently lead to practical application in genetic engineering and plant breeding.

ACKNOWLEDGEMENTS

We thank Karen Long for typing the manuscript.

REFERENCES

1.   Smith EF, Townsend CO (1907). A plant tumor of bacterial origin. Science25:671-673.
2.   Riker AJ, Banfield WM, Wright WH, Keitt GW, Sagen HE (1930). Studies on infectious hairy root of nursery apple trees. J Agric Res (Washington DC) 41: 507-540.
3.   Van Larebeke N, Engler G, Holsters M, Van den Elsacker S, Zaenen I, Schilperoort RA, Schell J (1974). Large plasmid in *Agrobacterium tumefaciens* essential for crown gall-inducing ability. Nature (London)252: 169-170.
4.   White FF, Nester EW (1980). Hairy root: plasmid encodes virulence traits in *Agrobacterium rhizogenes*. J Bacteriol. 141:1134-1141..

5. Chilton M-D, Drummond MH, Merlo DJ, Sciaky D, Montoya AL, Gordon MP, Nester EW (1977). Stable incorporation of plasmid DNA into higher plant cells: The molecular basis of crown gall tumorigenesis. Cell 11:263-271.

6. Chilton M-D, Tepfer DA, Petit A, David C, Casse-Delbart F, and Tempé J (1982). *Agrobacterium rhizogenes* inserts T-DNA into the genomes of the host plant root cells. Nature (London) 295:432-434.

7. White FF, Ghidossi G, Gordon MP, Nester EW (1982). Tumor induction by *Agrobacterium rhizogenes* involves the transfer of plasmid DNA to the plant genome. Proc Natl Acad Sci USA 79: 3193-3197.

8. Binns AN, Sciaky D, Wood HN (1982). Variation in hormone autonomy and regenerative potential of cells transformed by strain A66 of *Agrobacterium tumefaciens*. Cell 31: 605-612.

9. Schroder G, Waffenschmidt S, Weiler EW, Schroeder J (1984). The T-region of Ti plasmids codes for an enzyme synthesizing indole-3-acetic acid. Eur J Biochem 138: 387-3918.

10. Byrne MC, Koplow J, David C, Tempé J, Chilton M-D (1983). Structure of T-DNA in roots transformed by *Agrobacterium rhizogenes*. J Mol Appl Gen 2:201-209.

11. Petit A, David C, Dahl GA, Ellis JG, Guyon P, Casse-Delbart F, Tempé J (1983). Further extension on the opine concept: plasmids in *Agrobacterium rhizogenes* cooperate for opine degradation. Mol Gen Genet 190:204-214.

12. Cardarelli M, Spano L, De Paolis A, Mauro ML, Vitali G, Costantino P (1985). Identification of the genetic locus responsible for non-polar root induction by *Agrobacterium rhizogenes* 1855. Plant Mol Biol 5:385-391.

13. Huffman GA, White FF, Gordon MP, Nester EW (1984). Hairy-root-inducing plasmid: physical map and homology to tumor-inducing plasmids. J Bacteriol 157: 269-276.

14. White FF, Taylor BH, Huffman GA, Gordon MP, Nester EW (1985). Molecular and genetic analysis of the transferred DNA regions of the root inducing plasmid of *Agrobacterium rhizogenes*. J Bacteriol 164:33-44.

15. Slightom J, Durant-Tardif M, Jouanin L, Tepfer D (1986). Nucleotide sequence analysis of TL-DNA of *Agrobacterium rhizogenes* agropine type plasmid: identification of open-reading frames. J Biol Chem. In press.

16. Tepfer D (1984). Transformation of several species of higher plants by *Agrobacterium rhizogenes*: sexual transmission of the transformed genotype and phenotype. Cell 37:959-967.

17. Spano L, Costantino P (1982). Regeneration of plants from callus cultures of roots induced by *Agrobacterium rhizogenes* in tobacco. Z Pflanzenphysiol Bd 106:87-92.

18. Taylor BH, Amasino RM, White FF, Nester EW, Gordon MP (1985). T-DNA analysis of plants regenerated from hairy root tumors. Mol Gen Genet 201:554-557.

19. Ooms G, Karp A, Burrell MM, Twell D, Roberts J (1985). Genetic modification of potato development using Ri T-DNA. Theor Appl Genet 70:440-446.

20. Ooms G, Bains A , Burrell M, Twell D, Wilcox E (1985). Genetic manipulation in cultivars of oilseed rape (*Brassica napus*) using *Agrobacterium*. Theor Appl Genet 71:325-329.

21. Shahin EA, Sukhapinda K , Spivey R , Simpson RB (1986). Transformation of cultivated tomato by a binary vector in *Agrobacterium rhizogenes*: Transgenic plants with normal phenotypes harbor binary vector T- DNA, but no Ri plasmid T-DNA. Theor Appl Genet (Submitted).

22. Sukhapinda K, Shahin EA, Simpson RB (1986). Genetic linkage of the Ri plasmid T-DNA and the vector T-DNA in tomato plants transformed by a binary vector in *Agrobacterium rhizogenes*. (Submitted).

23. Trulson AJ, Shahin EA, Simpson RB (1986). Transformation of cucumber (*Cucumis sativus* L.) plants with *Agrobacterium rhizogenes*. In preparation.

24. Simpson RB, Spielmann A, Margossian L, McKnight T (1986). A disarmed binary vector from *Agrobacterium tumefaciens* functions in *Agrobacterium rhizogenes*: frequent co-transformation of two distinct T-DNAs. Plant Mol Biol. In Press.

25. Durand-Tardif M, Broglie R, Slightom J, Tepfer D (1985). Structure and expression of Ri T-DNA from *Agrobacterium rhizogenes* in *Nicotiana tabacum*. Organ and phenotypic specificity. J Mol Biol 186:557-564.

26. Sukhapinda K, Shahin EA (Unpublished data).

27. Trulson A (Unpublished data).

28. Shahin EA (Unpublished data).

Molecular Biology of Plant Growth Control, pages 391–400
© 1987 Alan R. Liss, Inc.

HORMONE AND LIGHT ACTIONS IN THE DIFFERENTIATION PROGRAM OF CHLOROPLASTS

B. Parthier, J. Lehmann, S. Lerbs, W. Lerbs, R. A. Weidhase, and R. Wollgiehn

Institute of Plant Biochemistry, Acad.Sci.of the GDR, 4020 Halle (S.),German Dem. Republic

ABSTRACT   Interaction of white light and growth regulators on chloroplast development, measured as chlorophyll content and RuBPCase gene expression, has been studied in two physiological systems: Etioplast-to-chloro-plast transformation in excised pumpkin cotyl-edons, and chloroplast senescence in barley leaf segments. The former is stimulated by cytokinins due to the organ developmental program. Light increases RuBPCase gene ex-pression but is not essential. Methyl-jasmo-nate is an efficient senescence-promoting substance. Its action on chloroplast senes-cence is antagonized by cytokinin in darkness but not in light.

INTRODUCTION

Chloroplast development in angiosperm seed-lings is regarded as the result of coordinated differential gene expression of organelle consti-tuents, which are encoded in both nuclear and plastid genomes, in dependence of the external factor light. The emerging mass of information about this process has been summarized in form of five basic principles by ELLIS (1), of which only the last one is of primary concern for our topic. It states in its essence that synthesis, transport and assembly of plastid constituents can occur in both darkness and light, but illumination exerts

largely a stimulation of processes going on at
lower rates in darkness. - For the synthesis of
some constituents, however, light may likewise act
as a trigger signal.

The problem has been excellently reviewed (2)
in respect to light control of gene expression of
the major plastid polypeptides, large and small
subunits (LS and SS) of ribulose-1,5-bisphosphate
carboxylase (RuBPCase), light-harvesting chlorophyll a/b protein (LHCP), and the 32 kDa herbicide-binding thylakoid protein. According to this
review and other recent papers (3-5) there is
agreement that phytochrome mediates light control
at the transcriptional level for those plastid
polypeptides encoded in the nucleus (SS and LHCP),
however, post-transcriptional control is likewise
involved in LHCP synthesis. Light control of RuBP-
Case LS polypeptide synthesis is probably translational (6, 7).

On the other hand, little is known to which
extent endogenous factors such as phytohormones
(growth regulators) are directly or indirectly involved in chloroplast development (8). Both transcription and translation have been suggested as
processes being stimulated by cytokinins. These
compounds are well-known as anti-senescence regulators and in this respect resemble light
effects. Many open questions maintain, e.g. in
which way phytohormones modulate gene expression,
and which mode of interaction between growth regulator and light effects is involved in chloroplast
development.

Our present contribution concerns this problem under two aspects of organelle transformation
in isolated plant organs: Etioplast-to-chloroplast
formation in developing pumpkin cotyledons, and
chloroplast degradation in senescing barley leaf
segments.

RESULTS AND DISCUSSION

Effects of light and cytokinin on chloroplast formation in excised pumpkin cotyledons.

We have studied the action of cytokinins,

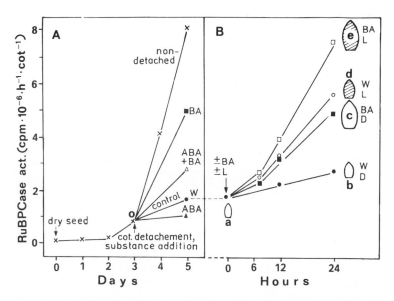

FIGURE 1.   Increase in RuBPCase activity during
pumpkin seed germination (A) and influence of 45
μM BA in light (L) or darkness (D) on detached
cotyledons (see ref. 9 for methodical details).
Letters o to e designate the different developmental stages or treatments, resp. ABA, abscisic
acid; W, water.

especially benzyladenine (BA), on plastid development in detached, etiolated pumpkin cotyledons (9),
which after illumination are transformed from storage organs to photosynthetic tissues. Changes of
two prominent chloroplast constituents, chlorophyll
and RuBPCase, were measured at various stages of
cotyledon development after treatment with 45 μM
BA and/or illumination (Fig. 1).
    Fig. 1 illustrates the increase of RuBPCase
activity during pumpkin seed germination in darkness (A) and the effect of BA and light on detached
cotyledons (B). After a lag phase the steep increase in activity corresponds with the appearance
of the primary root, which is known as the source
of endogenous cytokinins. This assumption is supported by RuBPCase activity stimulation in detached
cotyledons treated with exogenous cytokinin both

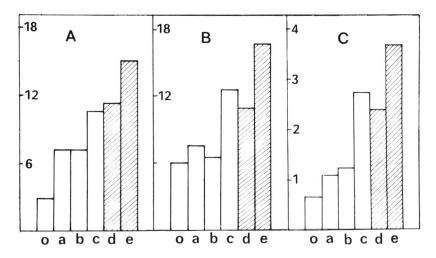

FIGURE 2.   Comparison of RuBPCase content
(A), in vivo synthesis (B) and steady-state level
of LS mRNA (C) in detached pumpkin cotyledons
treated for 24 hrs in the way outlined in Fig. 1,
B (developmental stages o to e). - A: RuBPCase
protein estimated by rocket immunoelectrophoresis
(% enzyme protein of total soluble proteins); B:
Incorporation of ^{14}C-leucine into RuBPCase protein
and quantitative determination as in A (% labeled
RuBPCase of labeled soluble proteins); C: Poly(A$^-$)
RNA dot hybridized to nick-translated LS gene from
spinach (cpm·10^{-5} per cotyledon).

in darkness and in light (Fig. 1). The kinetics of
enzyme activity is consistent with the increasing
developmental complexity of etioplast ultrastructure (not shown). Immunological methods show that
enzyme activities correspond with enzyme contents
(Fig. 2,A), and similar results can be obtained
after in vivo biosynthesis of the holoenzyme (leucine incorporation into RuBPCase protein, Fig. 2,
B), as well as for the steady-state level of translatable RNA in cell free wheat germ and E. coli
systems (9). These data correspond with our results
of quantitative dot hybridization of nick-translated RuBPCase LS gene with poly(A$^-$) RNA of pumpkin
cotyledons (Fig. 2, C). In all these cases we ob-

served two- to threefold stimulations after BA
treatment or illumination. The effects of the two
factors suggest an additive coaction in RuBPCase
gene expression at the transcription level (9),
however, interactions in other steps of gene expression, e.g. mRNA stability and translation cannot be excluded.

Thus light stimulates RuBPCase, but it is not
essential for its synthesis. Exogenous cytokinins
and probably also endogenous growth regulators are
able to stimulate the developmentally related plastid biogenesis in cotyledons (if RuBPCase regarded
as a prominent marker enzyme), but in plant cell
cultures kinetin may modulate the expression of
specific genes, e.g. LHCP, whereas the general
pattern of plastid proteins is not affected (10).

This problem can be connected with developmental changes of sensitivity ("competence") to
cytokinins in the detached cotyledons. For this
purpose radioactive leucine was incorporated in
vivo for 24 hrs into proteins of cotyledons, which
after excision (stage 0) have been pretreated for
various periods in darkness on water or BA solution. Total soluble proteins were separated on two-
dimensional gel electrophoresis/electrofocusing,
and the radioactive spots were detected by autoradiography. Fig. 3 represents the spot distribution pattern from cotyledons pretreated for 24 hrs
with BA in darkness followed by 24 hrs labeling in
darkness in the presence of BA. Those spots are
marked which are distinguished by clear changes in
radioactivity in comparison with the following
samples: (1), 24 hrs BA pretreatment, label for
24 hrs in the presence of BA in light; (2), 24 hrs
water pretreatment, label on water in light; (3),
same as (2) but label in darkness; (4), 48 hrs
water pretreatment, label on water in darkness.

The experiment shows two results. First, "BA-
specific" spots are not detectable; polypeptides
labeled under dark conditions are identical in both
BA- and water-treated cotyledons (patterns correspond with those of samples 3 and 4). However, most
of them appear earlier in development or are more
intensively labeled in BA-treated tissues. Thus BA
accelerates cotyledon development but seems not to
induce new polypeptides. Secondly, some polypep-

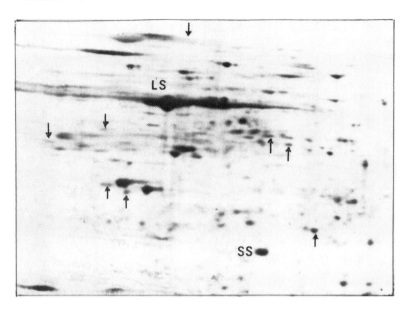

FIGURE 3. Autoradiography of the two-dimensional separation of soluble proteins of pumpkin
cotyledons after in vivo labeling with radioactive
leucine for 24 hrs. Arrows point to spots markedly
stimulated (↓) or reduced (↑) by light. See text
for details.

tides increase or decrease in label after tissue
illumination (Fig. 3). RuBPCase SS and LS polypeptides did not change markedly.

The result indicates that BA acts as stimula-
tor, not as trigger of protein synthesis in early
cotyledon development, but the hormone might exert
anti-senescence effects at later stages. Furthermore, light is able to induce or significantly enhance the synthesis of few soluble polypeptides.
Therefore, the coaction of the two factors in the
general aspect of chloroplast or organ development
can differ considerably in detail.

Effects of growth regulators and light on chloroplast senescence in excised barley leaf segments.

In a comprehensive view of leaf ontogenesis,

the development of chloroplasts is accomplished by transformation ("dedifferentiation") of these organelles into chromoplasts. It normally proceeds in the yellowing phase of the leaf. Since the chloroplast number per mesophyll cell remains essentially constant during barley leaf ageing (11), degradation of chloroplast constituents can be used as measure for organelle senescence. There is good evidence for the assumption that this process is markedly influenced by growth regulators and light (12, 13).

The following experiments demonstrate the antagonistic effects of cytokinins (benzyladenine, BA) and methyl-jasmonic acid (JA), a senescence promoting cyclopentanone derivative (14, 15), on changes in chlorophyll and RuBPCase in excised barley leaf segments floating on aqueous solutions of BA or JA, or both, in the light or in darkness. Fig. 4 shows that chlorophyll content and RuBPCase activity decrease rapidly in JA-treated tissues. Equimolar concentrations of BA are unable to reverse or prevent the senescence-promoting action of JA. However, stabilization in later periods is observed. Interestingly, JA acts more dramatic in light. This precludes the possibility of photooxidative processes in addition to the senescence response observed in darkness.

The JA-induced decrease of RuBPCase activity corresponds with the loss of RuBPCase protein content estimated by rocket immunoelectrophoresis with antibodies raised against pumpkin holoenzyme (Fig. 5). Unclear remains the obvious inactivation of the enzyme from BA-treated or water control leaves if incubated in the light (Fig. 4 vs. Fig. 5). - The almost complete disappearance of RuBPCase activity and protein after 48 hours of JA treatment in light suggest a massive degradation of this enzyme. This has been shown to take place in senescing leaves in darkness (16) due to the action of de novo synthesized acid proteases localized within vacuoles (17-19).

The senescence response caused by JA seems to differ from these findings. If total soluble protein is separated by non-denaturating gel electrophoresis, RuBPCase holoenzyme is drastically reduced under the influence of JA, but some prominent

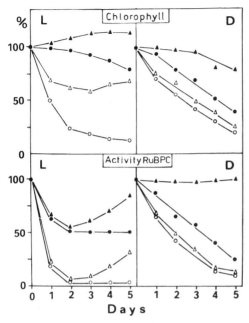

FIGURE 4. Changes in chlorophyll contents and RuBPCase activities in barley leaf segments influenced by 45 μM JA (O), 45 μM BA (▲) or both substances in equimolar concentrations (Δ); water control (●). Substance addition after excision (day O). L, light of 3000 lx; D, darkness.

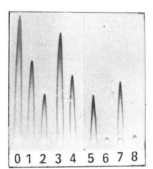

FIGURE 5. RuBPCase estimated by rocket immunoelectrophoresis in the soluble fractions of 48 hrs treated barley leaf segments in darkness (1-4) or light (5-8). O, zero time; 1 and 5, water controls; 2 and 6, 45 μM JA; 3 and 7, 45 μM BA, 4 and 8, JA + BA. Same amounts of soluble protein per rocket.

new bands appear (Fig. 6, lane 5). After SDS electrophoresis two major bands were observed with relative molecular masses of 23 and 38 kDa (not shown). Preliminary experiments suggest that the JA-induced polypeptides are probably no degradation products of the RuBPCase holoenzyme but represent newly synthesized cytoplasmic proteins, because: they fail to react with anti-RuBPCase while other bands do react; they do not occur after incubation of leaf segments with JA plus cyclohexi-

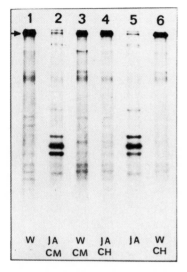

FIGURE 6. Non-denaturating polyacrylamide gel electrophoresis of soluble fractions of 48 hrs treated barley leaf segments in light. CM, chloramphenicol (300 μM); CH, cycloheximide (14 μM). Arrow points to RuBPCase holoenzyme.

mide (Fig. 6, lane 4) but are present after chloramphenicol plus JA treatment (lane 2); they are labeled after feeding leaf segments with radioactive amino acids even 18 hrs after JA pretreatment (not shown). These bands do not prominently occur in later stages of senescing leaves (water controls). - RuBPCase is found to respond most sensitive in JA-treated barley leaf tissues. Other enzymes (PEP carboxylase, plastid- and cytosol-specific aminoacyl-tRNA synthetases) have not changed activity after 28 hrs JA treatment, whereas peroxidase activity is enhanced.

Senescence of chloroplasts is vastly dependent on processes in the cytoplasmic sourrounding, since constituents of isolated chloroplasts remain much more stable than those in the leaf tissues. We were able to confirm Thimann's observations (20) in the presence of JA. Neither in light of 500 or 3000 lx nor in darkness the addition of JA and/or BA to isolated barley chloroplasts can alter the continuous loss of chlorophyll (not shown), suggesting that JA primarily affects cytoplasmic events.

REFERENCES

1. Ellis RJ (1983). In Roodyn DB (ed): "Subcellular Biochemistry." Vol. 9. New York & London: Plenum Press, p. 237.
2. Tobin EM, Silverthorne J (1985). Ann Rev Plant Physiol 36:593.
3. Harpster M, Apel K (1985). Physiol Plant 64:147.
4. Bennett J, Jenkins GJ, Hartley MR (1984). J Cell Biochem 25:1.
5. Kaufman LS, Thompson WF, Briggs WR (1985). Science 226:1447.
6. Berry JO, Nikolau BJ, Carr JP, Klessig DF (1985). Mol Cell Biol 5:2238.
7. Inamine G, Nash B, Weissbach H, Brot N (1985). Proc Natl Acad Sci USA 82:5690.
8. Parthier B (1979). Biochem Physiol Pflanzen 174:173.
9. Lerbs S, Lerbs W, Klyachko NL, Romanko EG, Kulaeva ON, Wollgiehn R, Parthier B (1984). Planta 162:289.
10. Teyssendier de la Serve B, Axelos M, Péaud-Lenoël C (1985). Plant Mol Biol 5:155.
11. Martinola E, Heck U, Dalling MJ, Matile Ph (1983). Biochem Physiol Pflanzen 178:147.
12. Thimann KV (ed) "Senescence in Plants." Boca Raton: CRC Press (1980).
13. Stoddart JL, Thomas H (1982). In Boulter D, Parthier B (eds): Encyclop Plant Physiol. Vol. 14A. Berlin, Heidelberg, New York: Springer. p. 592.
14. Ueda J, Kato J (1980). Plant Physiol 66:246.
15. Ueda J, Kato J, Jamane H, Takahashi N (1981). Physiol Plant 52:305.
16. Wittenbach VA (1978). Plant Physiol 64:604.
17. Wittenbach VA, Lin W, Hebert RR (1982). Plant Physiol 69:98.
18. Thayer SS, Huffaker RC (1984). Plant Physiol 75:70.
19. Miller BL, Huffaker RC (1985). Plant Physiol 78:442.
20. Choe HT, Thimann KV (1975). Plant Physiol 55:828.

Molecular Biology of Plant Growth Control, pages 401–411
© 1987 Alan R. Liss, Inc.

REGULATION OF THE SYNTHESIS OF TWO CHLOROPLAST PROTEINS ENCODED BY NUCLEAR GENES [1]

E.M. Tobin, J. Silverthorne, S. Flores,
L.S. Leutwiler, and G.A. Karlin-Neumann

Department of Biology, University of California,
Los Angeles, California 90024

ABSTRACT   Genes encoding the small subunit (SSU) of
ribulose 1,5-bisphosphate carboxylase and the major
light-harvesting chlorophyll a/b-proteins (LHCP)
associated with photosystem II can be regulated at the
level of transcription by the action of phytochrome.
In Lemna gibba, gene specific probes for six of the
approximately 12-14 genes in the SSU family have been
used to demonstrate that phytochrome affects the expression of these genes to differing extents.   In addition,
the RNA for two of these six does not accumulate in the
green roots.   A plant hormone, cytokinin, can also
affect the expression of Lemna genes encoding SSU and
LHCP (rbcS and cab genes, respectively).   Unlike phytochrome action, this effect apparently occurs post-
transcriptionally.
    Studies with Arabidopsis thaliana have demon-
strated that the levels of SSU and LHCP RNAs can also
be regulated by phytochrome in this species.   The conditions of germination and growth of the seedlings can
affect the details of the expression of this response.
We have isolated and sequenced three cab genes from
Arabidopsis, and found that they all encode identical
mature polypeptides.   However, there is some indication
that there may be a fourth cab gene which may in part
account for the multiple forms of LHCP found in the
thylakoid membranes.   Additionally, post-translational
modifications may also be involved in generating the
multiple forms of LHCP found in this and other plants.

[1] This work was supported by grants from N.I.H. (GM-
23167), N.S.F. (DCB-84-17606) and U.S.D.A. (83-CRCR-1-1319).

INTRODUCTION

The expression of a number of genes has been demonstrated to be affected by the action of phytochrome (1). In
many instances these genes are nuclear genes which encode
chloroplast proteins. We have studied the expression of
genes for two such proteins--the small subunit (SSU) of
ribulose 1,5-bisphosphate carboxylase (RuBisCO), and the
major light-harvesting chlorophyll a/b-protein associated
with photosystem II (LHCP). In both cases, the proteins are
synthesized in the cytoplasm as larger precursor forms,
imported into chloroplasts, cleaved to their final size, and
assembled with other chloroplast components into their
functional form. In all species studied so far, these two
proteins are encoded by gene families.

In the aquatic monocot Lemna gibba we have used cloned
hybridization probes to demonstrate that the levels of RNAs
for both kinds of proteins increase in response to
phytochrome action. In further work designed to distinguish
whether this increase is primarily a result of increased
transcription or decreased degradation of these RNAs, we
used an in vitro "run-off" transcription assay to show that
the transcription of mRNAs for these proteins can also be
increased by the action of phytochrome (2). Although the
transcription decreases in nuclei isolated from plants
placed into complete darkness, it remains at a detectable
level; thus phytochrome action has a quantitative effect in
this system. We have also begun to investigate the action
of phytochrome on the genes encoding LHCP in the dicotyledonous plant Arabidopsis thaliana.

L. gibba can be grown heterotrophically in the dark on
sucrose if a cytokinin is included in the medium. In addition, since this species of duckweed is dependent on phytochrome action for its growth in the dark, heterotrophic
growth also requires brief intermittent illumination with
red light (2 min/8h). Without these supplements, the plants
do not continue to grow at an appreciable rate; however,
they do not become senescent over a period of at least
several weeks. Presumably, in the dark the amount of endogenous cytokinins is decreased over the level in light grown
plants.

RESULTS AND DISCUSSION

Regulation of the Expression of Individual Genes Encoding SSU.

The results obtained using the cloned coding regions for rbcS and cab genes do not answer the question of whether all the genes of a family are similarly affected by phytochrome action. This is because in Lemna, as well as several other species, the coding regions of the rbcS and cab genes are conserved within each family. Since there might well be differences in the responsiveness of individual family members, we took advantage of the fact that there is divergence in the 3'untranslated regions of the six different rbcS clones we have isolated from Lemna to make specific probes for these six different genes (3,4). Genomic Southern blots demonstrate that five of these subclones hybridize to a single band while the sixth (pSSU1 3'UT) may be homologous to two additional rbcS genes. None of the subclones cross-hybridize with each other (42° C, 0.1 X SSC). These probes have been used to demonstrate that the six different sequences are expressed to different extents in the light. Those genes homologous to SSU1 are expressed at the highest level, while SSU5A, SSU5B, and SSU26 are expressed at an intermediate level, and SSU40A and SSU40B are expressed at an extremely low level. The amount of each of the sequences decreased when plants were placed into the dark and increased in response to phytochrome action. In the case of the two genes expressed at the lowest level, 40A and 40B, the increase could only be seen after growing the plants with multiple red irradiations (2 min/8hr). The response of the other RNAs to a single 1' red illumination was variable, ranging from approximately 2-fold to a 6-fold increase over the dark level 2 hours after the red illumination (3,4).

We have also used these rbcS gene specific probes to provide evidence of organ-specific expression (5). The roots of Lemna plants grown in the light are green and contain RNA coding for SSU. When Northern blots of RNA isolated from roots and fronds are compared after hybridization with rbcS gene-specific probes (Fig.1), it can be seen that two of the sequences (5B and 40A) are not detectable in the roots. Experiments are currently being performed to determine whether this difference in expression is the result of transcriptional or post-transcriptional events.

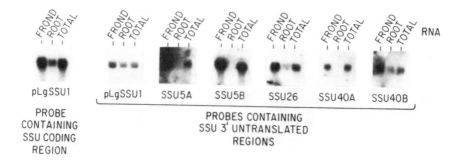

FIGURE 1. Differential expression of individual rbcS genes in roots and fronds of Lemna gibba. Samples of RNA (5 µg/lane) from FRONDS, ROOTS, or TOTAL plants were subjected to Northern blot analysis (6) with the indicated probes.

Cytokinin Interaction with the Phytochrome Response.

Earlier work (7) had demonstrated that when light-grown Lemna fronds are placed in the dark, the levels of translatable mRNAs for the SSU and LHCP decrease, and the extent and rate of this decline are less if a cytokinin (kinetin) is added to the growth medium when the plants are placed in the dark. We suspected that this effect of added cytokinin was apparent because endogenous cytokinin levels decline during the dark treatment. Recently, in a preliminary experiment, it was found that the level of the cytokinin dihydrozeatin does dramatically decline when Lemna is placed into the dark (J.P. Slovin and R.O. Morris, personal communication). Furthermore, when these dark-treated plants were assayed for changes in translatable mRNAs and protein synthesis after a red illumination, it was found that the phytochrome response apparently depended on the presence of cytokinin in the growth medium. This observation is illustrated for SSU synthesis in Fig. 2. Plants placed in darkness for four days with or without the addition of 3 X 10⁻⁶ M kinetin on the third day were given 2 min red illumination 1.5 h. before labeling and harvesting. This experiment demonstrates that the SSU synthesis is high in the light-grown plants, but greatly decreases in the dark treated plants (DARK and DARK + KINETIN). In the absence of kinetin, red illumination does not appear to result in increased synthesis of SSU (DARK + 2'RED), but if kinetin is present (DARK + KINETIN + 2'RED), the increase in SSU synthesis is readily apparent.

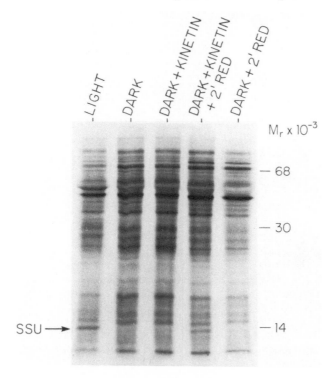

FIGURE 2. Cytokinin enhancement of the effect of red light on SSU synthesis. Fluorogram of soluble proteins from plants labelled in vivo for 1 h with ^{35}S-methionine after electrophoresis on a 12.5% polyacrylamide-SDS gel. Equal counts were loaded on each lane. LIGHT: light-grown plants; DARK: plants placed into the dark for 4 days; + KINETIN: 3 X 10^{-6} M kinetin added after 3 days in the dark; + 2' RED: 2 min red light given 1.5 h before labeling and harvest.

We extended our observations on this apparent interaction of a cytokinin with the phytochrome response by using cloned probes to determine effects of a cytokinin with and without red illumination on the expression of cab and rbcS genes. In order to demonstrate that the cytokinin effect has a certain degree of specificity, probes for two chloroplast genes (encoding the large subunit of RuBisCO and the 32-kDa herbicide binding protein), and two other nuclear genes, β-actin and a sequence known to be more abundant in the dark than in light-grown plants (a Lemna cDNA clone designated p106, S. Flores, unpublished data), were used. Cytokinin, alone or in addition to red illumination, can be shown to increase the amounts of the SSU and LHCP RNAs, but not the other four RNAs (8). However, there is evidence that unlike the response of these RNAs to phytochrome action, the increase in response to 0.5 uM benzyladenine is not due to increased transcription of the genes, but rather to an effect on a post-transcriptional process (8).

In in vitro "run-off" experiments designed to assay transcriptional activity, it can be demonstrated that the increase in the RNA levels seen after benzyladenine and light treatment cannot be attributed solely to an increase in the transcription of the rbcS and cab genes (9). The stimulation of transcription by the action of phytochrome does occur in the absence of benzyladenine, and the benzyladenine alone has little or no effect on transcription. The effect of the benzyladenine addition on the RNA levels is not very rapid: little effect is seen unless it has been present for at least 8 hours (9). The experimental evidence is consistent with the hypothesis that in the dark in the presence of an exogenous cytokinin, the degradation of the SSU and LHCP RNAs is decreased, perhaps because the RNAs are protected by a greater degree of association with ribosomes than in the absence of the added cytokinin.

Light-harvesting Chlorophyll a/b-Proteins from Lemna gibba and Arabidopsis thaliana.

Genes. The LHCPs associated with Photosystem II have been found to be encoded by gene families in all species examined to date. We have sequenced two of the genes from Lemna, AB19 and AB30, and have found that although they encode proteins which are highly homologous, the transit peptides and the amino-terminal ends of the mature proteins are divergent (10,11). This finding raises the questions of whether there is any functional difference between the two types of LHCP and which of the genes give rise to the polypeptides of slightly different rates of electrophoretic migration found in the thylakoids in the LHCII complex.

We have also isolated and sequenced three genes encoding LHCPs from Arabidopsis thaliana (12). This dicotyledonous plant has a very small genome and appears to contain fewer cab genes than other species. The three genes we have sequenced are closely linked in the Arabidopsis genome. Comparison of the deduced amino acid sequences reveals that two of the genes are identical at the amino acid level, and the third contains a single amino acid change in the transit peptide. Genomic Southern blot analysis has been used to estimate the number of cab genes in the Arabidopsis genome (12). An Arabidopsis cab gene probe (At165) hybridizes the three bands corresponding to the fragments containing the sequenced genes and an additional fainter band which may represent a fourth cab gene.

Proteins. The function of the different types of LHCP sequence is yet to be understood. Since there is more than one type of LHCP sequence in Lemna (AB19 vs. AB30) and since multiple LHCPs are observed in LHCII complexes, the question arises as to how these two phenomena might be related. Both Lemna and Arabidopsis LHCII complexes contain multiple LHCPs (Fig. 3). If LHCII complexes are prepared by electrophoresis on a non-denaturing polyacrylamide gel (cf.11), cut out from the gel and then electrophoresed on a fully denaturing SDS-polyacrylamide gel, the Lemna complex can be shown to contain at least three bands and the Arabidopsis complex at least two bands which react with polyclonal antibodies (4) raised against LHCII polypeptides from Lemna. Immunoprecipitation of in vitro translation products of Lemna poly(A) RNA prepared in a wheat germ extract using anti-LHCP antibodies results in a single major band of immunoprecipitated precursor at about 32,000 (4, 13). In addition, RNA that can be translated into this same sized band is hybrid-selected by clones of either AB19 (14) or AB30 (J. Silverthorne, C.F. Wimpee, unpublished data). However, when the same antiserum (against Lemna LHCP) is used against Arabidopsis poly(A) RNA translation products, two precursor bands are observed. This discrepancy in the numbers of precursor and mature forms of the LHCPs may reflect post-translational modifications which are made in vivo but not in the wheat germ extract. Multiple forms of in vitro synthesized LHCP have also been observed after processing and uptake of a single precursor type into isolated Lemna chloroplasts (11). This observation may result from the activity of an LHCP-specific protease or modification enzyme during the uptake reaction and could also occur in vivo. However, these results may also indicate that the multiple LHCPs can arise from different genes. These hypotheses are currently under investigation.

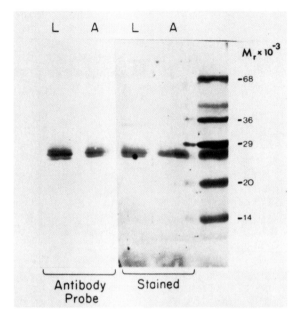

FIGURE 3. Comparison of Lemna and Arabidopsis LHCII proteins. Blots of isolated LHCII proteins after denaturation with SDS and electrophoresis on a 12.5% polyacrylamide-SDS gel. The stained blot is shown on the right and the bands which reacted with the Lemna LHCP antiserum are shown on the left. Alkaline phosphatase coupled to goat anti-rabbit antiserum was used for detection of the primary antibodies. L: Lemna; A: Arabidopsis.

Phytochrome Regulation of Cab Gene Expression in Arabidopsis.

We are interested in performing a genetic and molecular analysis of the phytochrome transduction pathway whereby red light stimulates the transcription of certain nuclear genes. To this end, we have initiated an investigation of phytochrome-regulated gene expression in two plant species which are amenable to transformation by Agrobacterium -- Arabidopsis thaliana and Nicotiana tabacum. Seeds were germinated for 18-24 h in constant white light (WL), in 2 min red light every 8 h for 24 h (RL) or in 1 mM gibberellic acid (GA) in total darkness. After these treatments were administered, the seeds were placed in darkness for 4-6 d (Arabidopsis) or 8 d (tobacco). The resultant seedlings were given a single minute of red light, placed in total darkness for two hours and then used to prepare total RNA for Northern blot analysis. A single minute of red light is sufficient to induce LHCP mRNA in WL, RL and GA-germinated Arabidopsis seedlings and in GA-germinated tobacco seedlings. Thus, the response to red light is independent of the method used to germinate the etiolated seedlings. In Arabidopsis, the effect of red light on RL-germinated seedlings can be reversed by immediate treatment with 10 min far-red light, indicating that the response is under phytochrome regulation. However, preliminary experiments have suggested that full reversal cannot be achieved when the seeds were germinated in the dark in the presence of GA. This observation raises the question of whether the GA or the lack of previous light treatment results in the mRNA induction exhibiting a very low fluence response to red light (cf. 15).

CONCLUSION

It is clear that the regulation of the synthesis and assembly of nuclear-coded chloroplast polypeptides involves many factors, including the interaction of hormonal and phytochrome pathways. By use of a combination of in vivo, in vitro and genetic studies it will be possible to elucidate some of these mechanisms in the near future.

ACKNOWLEDGMENTS

We thank Ruth Butalla for her technical assistance. Leslie Leutwiler and Susan Flores were supported in part by N.R.S.A. fellowships. George Karlin-Neumann was supported by a fellowship from the ARCO Plant Cell Research Institute.

REFERENCES

1. Tobin EM, Silverthorne J (1985). Light regulation of gene expression in higher plants. Annu Rev Plant Physiol 36:569.
2. Silverthorne J, Tobin EM (1984). Demonstration of transcriptional regulation of specific genes by phytochrome

action. Proc Natl Acad Sci (USA) 81:1112.

3. Tobin EM, Wimpee CF, Karlin-Neumann GA, Silverthorne J, Kohorn BD (1985). Phytochrome regulation of nuclear gene expression. In Arntzen CJ, Bogorad L, Bonitz S, Steinback KE (eds): "The Molecular Biology of the Photosynthetic Apparatus," Cold Spring Harbor: Cold Spring Harbor Laboratory, p 373.

4. Wimpee CF, Silverthorne J, Yamada T, Tobin EM. Differential expression of individual genes encoding the small subunit of ribulose 1,5-bisphosphate carboxylase in Lemna gibba. Manuscript in preparation.

5. Silverthorne J, Tobin EM. Differential expression of individual rbcS genes in green Lemna roots and fronds. Manuscript in preparation.

6. Thomas, PS (1980). Hybridization of denatured RNA and small DNA fragments transferred to nitrocellulose. Proc Natl Acad Sci (USA) 77:5201.

7. Tobin EM, Turkaly E (1982). Kinetin affects rates of degradation of mRNAs encoding two major chloroplast proteins in Lemna gibba L. G-3. J Plant Growth Regul 1:3.

8. Flores S, Tobin EM (1986). Benzyladenine modulation of the expression of two genes for nuclear encoded chloroplast proteins in Lemna gibba. Planta, in press.

9. Flores S, Tobin EM (1986). Benzyladenine regulation of the expression of two nuclear genes for chloroplast proteins. In Fox JE, Jacobs M (eds): "Molecular Biology of Plant Growth Control," New York: Alan R. Liss, in press.

10. Karlin-Neumann GA, Kohorn BD, Thornber JP, Tobin EM (1985). Sequence and structure of a chlorophyll a/b-protein gene containing an intron with the structure of a transposable element. J Mol Appl Genet 3:45.

11. Kohorn BD, Harel E, Chitnis PR, Thornber JP, Tobin EM (1986). Functional and mutational analysis of the light-harvesting chlorophyll a/b-protein of thylakoid membranes. J Cell Biol 102:972.

12. Leutwiler LS, Meyerowitz EM, Tobin EM (1986). Structure and expression of three light-harvesting chlorophyll a/b-binding protein genes of Arabidopsis thaliana. Nulc Acids Res 14:4051.

13. Tobin EM (1981). White light effects on the mRNA for the light-harvesting chlorophyll a/b-protein in Lemna gibba L. G-3. Plant Physiol 67:1078.

14. Stiekema WJ, Wimpee CF, Silverthorne J, Tobin EM (1983). Phytochrome control of the expression of two nuclear genes encoding chloroplast proteins in Lemna gibba L. G-3. Plant Physiol 72:717.

15. Shinkle JR, Briggs WR (1985) Physiological mechanism of the auxin-induced increase in light sensitivity of phytochrome-mediated growth responses in <u>Avena</u> coleoptile sections. Plant Physiol 79:349.

Molecular Biology of Plant Growth Control, pages 413–423
© 1987 Alan R. Liss, Inc.

MOLECULAR EVENTS IN PHOTOREGULATED GREENING IN
BARLEY LEAVES

Winslow R. Briggs, Egon Mösinger,
Alfred Batschauer, Klaus Apel, and
Eberhard Schäfer

Department of Plant Biology, Carnegie Institution
of Washington, 290 Panama St., Stanford, California 94305
(W.R.B.)

Biologisches Institut II der Albert-Ludwigs-Universität
Freiburg i. Br., Federal Republic of Germany
(E.M. and E.S.)

and

Botanisches Institut der Christian-Albrechts-Universität
Kiel, Federal Republic of Germany (A.B. and K.A.)

ABSTRACT A brief pulse of red light reduces the lag
phase for chlorophyll accumulation in subsequent white
light in dark-grown barley leaves. It also induces an
increase in the relative abundance of the mRNA for the
light-harvesting chlorophyll a/b-binding protein
(LHCP) and a decrease in that of the mRNA for the
NADPH-dependant protochlorophyllide oxidoreductase
(reductase). Finally, it induces an increase in transcription for the LHCP mRNA in vitro and a corresponding decrease in transcription for the reductase mRNA.
Fluence-response studies at the levels of chlorophyll
accumulation, mRNA abundance, and transcriptional
activity of nuclei from irradiated plants indicate
that all three responses show both very low fluence
and low fluence components, spanning eight orders of
magnitude of light dose. The low fluence response
persists in all three cases for many hours, as does
the very low fluence component for mRNA abundance and
chlorophyll accumulation. The very low fluence
response for transcriptional activity is transient.

The very low fluence responses where measured do not show far-red reversibility, while the low fluence responses do. Kinetic and quantitative considerations indicate that while phytochrome regulation of mRNA abundance - and hence eventually of greening - may be partly at the transcriptional level, phytochrome regulation must occur at other levels as well.

INTRODUCTION

We have known for almost thirty years that a brief pulse of red given to dark-grown seedlings will eliminate most or all of the lag normally found in chlorophyll accumulation when these seedlings are placed in continuous white light (see 1 for review). This red-light effect has been assumed to be the consequence of phytochrome phototransformation both on the basis of its action spectrum (2, 3) and its far-red reversibility (2, 4). However, as Virgin (2) noted some 25 years ago, the reversal mechanism is "poorly developed" in dark-grown wheat, a conclusion subsequently reached for dark-grown pea seedlings (4, 5).

More recently, a number of laboratories have shown that a pulse of red light brings about dramatic changes in the abundance of several different cytoplasmic mRNAs coding for proteins that are intimately involved in the greening process (see 6, 7, 8, 9, for reviews). The involvement of phytochrome was established by the demonstration of far-red reversibility. When far red light was administered immediately subsequent to red treatment, it either cancelled or very much reduced the ultimate red-induced change. In the case of Lemna (10, 11, 12) and barley (13, 14, 15, 16, 17), measured changes in mRNA abundance are preceded by changes in transcription as measured by in vitro transcriptional activity of isolated nuclei - assayed by determining radioactivity incorporated into RNA hybridizable to appropriate cloned cDNAs. Hence in these cases it seems reasonable to assume that phytochrome regulation of the abundance of particular mRNA transcripts is in part at the level of transcription. Regulation at other levels in addition is not excluded. Possible candidates for regulatory steps include the following: processing, transport, or degradation of mRNA; synthesis, processing, transport, or degradation of the corresponding protein; or in the case of greening, the synthesis of chlorophyll itself.

It is now well established that there are two classes of responses to red light that are readily separable on their fluence requirements (18, 19, 20). One of these, the so-called low-fluence response (LF, see 20, 21), is potentiated by fluences between roughly 10^1 and 10^4 μmol/m^2 while the other, the so-called very low-fluence response (VLF, see 20, 21), has its entire effective range shifted down four orders of magnitude or more, to fluences between roughly 10^{-4} and 10^{-1} μmol/m^2. LF responses are those showing the normal phytochrome property of far-red photoreversibility, and presumably require a major fraction of the phytochrome in the Pfr form for saturation. VLF responses, on the other hand, require only minute amounts of Pfr for saturation - as little as 1% is often sufficient. Since far red light alone will produce up to 3% Pfr (depending upon the far red source) at photostationary equilibrium, it is obvious that LF responses are *not* far-red reversible, and should in fact be inducible by far-red light itself. Indeed, the only real evidence that VLF responses are phytochrome-mediated is that their action spectrum (18) resembles the absorption spectrum of Pr. Although in the remaining part of this paper we shall treat VLF responses as though they were mediated by phytochrome, the reader should keep this caution in mind.

The present study was designed to study some of the events between phytochrome phototransformation and acceleration of greening in white light. The hypothesis being tested is that there is a linear sequence of events between the photoact and final expression, including in order (among other things) the following: changes in transcription of genes coding for proteins involved one way or another in the greening process; an increase (or decrease) in mRNA abundance as a consequence; and an increase (or decrease) in the corresponding proteins. Three questions are being addressed. First, does phytochrome control occur at the transcriptional level? Second, if it does not, at which step(s) does it occur? Third, if it does, can it also occur at other steps, and if so which ones? In this brief paper, we address these questions with respect to two proteins involved in the greening process in the primary leaves of barley (and in other higher plants): the light-harvesting chlorophyll *a/b*-binding protein (LHCP) which shows positive regulation by Pfr (13, 15, 16) and the

NADPH: protochlorophyllide oxidoreductase (reductase) which shows negative regulation by Pfr (14, 15). We present preliminary fluence-response and kinetic data on mRNA abundance and relative *in vitro* transcription. We also present comparable data for the effect of Pfr on chlorophyll accumulation in white light, considered an accurate reflection of the accumulation of the LHCP (see 22).

MATERIALS AND METHODS

Barley seeds (*Hordeum vulgare* L., var. Carina) were sown in moist vermiculate and grown in absolute darkness for 6 days at 25 C at a humidity near saturation. At the time of harvest, the primary leaves extended roughly 6-9 cm above the coleoptile tips. For chlorophyll measurements, involving various white light treatments, seedlings were selected from which the apical 6.5 cm of primary leaves could be harvested without any coleoptile tissue, and the harvest was carried out under white light. For isolation of RNA or nuclei, requiring far larger numbers of seedlings, precise measurement of leaf length was not possible, but efforts were made to minimize contamination by coleoptile tissue, and the leaves harvested were roughly 7 cm long. These latter harvests were carried out in 5 min or less under dim green light.

Red and far red light sources were those described by Heim and Schäfer (23). For greening experiments, white light was obtained in a 25 C growth room from mixed fluorescent and incandescent light at 4 W m^{-2}. Chlorophyll measurements were made according to the method of Moran and Porath (24). Nuclei were isolated by the technique of Willmitzer and Wagner (25), *in vitro* transcription measurements made as described by Mösinger *et al.* (17), and the abundance of specific mRNAs was determined as described by Batschauer and Apel (15).

RESULTS AND DISCUSSION

Preliminary studies confirmed that, as is the case with other dark-grown seedlings (1, 2, 3, 4), the lag period for chlorophyll accumulation normally encountered in white light can be completely eliminated in barley when Pfr is present at an appropriate time prior to the onset of white light treatment. A four-hour dark period yielded the optimal response and a 4 h white light treatment was sufficient to assay the magnitude of the red light effect with good resolution (results not shown). Hence this protocol was used for subsequent experiments on greening *per se*. Red light treatments in the VLF range were effective in reducing the lag by about 50 %, treatments between the top of the VLF range and the bottom of the LF range were without further effect, and treatments in the LF range were once again effective, eliminating the remainder of the lag. Thus the effect of red light on greening of the primary leaf of barley shows almost exactly the same fluence-response relationships as the effect of red light on suppression of mesocotyl elongation and promotion of coleoptile elongation in oat seedlings (19). As predicted by the earlier studies, the very low fluence component was not reversible by far-red light whereas the LF component was.

TABLE 1

EFFECTS OF RED LIGHT AS MEASURED BY *IN VITRO* TRANSCRIPTION, mRNA ABUNDANCE, AND CHLOROPHYLL ACCUMULATION IN BARLEY PRIMARY LEAVES MEASURED AFTER 3 H DARKNESS

	Dark Control	Percent dark control	
		VLF	LF
Total transcription	100	132	137
rRNA transcription	100	94	81
LHCP transcription	100	99	324
LHCP mRNA abundance	100	210	452
Reductase transcription	100	66	39
Reductase mRNA abundance	100	75	18
Chlorophyll (after 4 h WL)	100	152	185

Tables 1 and 2 show the results of experiments testing the effects of fluences of red light selected to be either in the VLF and LF ranges (10^{-1} and 10^3 $\mu mol/m^2$, respectively) on some of the components of interest in greening. Table 1 shows changes after a 3 h dark period, and Table 2 after 4.5 h of darkness following the red light

pulse.
    Inspection of these two tables shows several clear
trends. First, there is a small effect of red light on
overall transcription and rRNA transcription, as measured
in vitro, mostly evident as an increase near 4.5 h and

TABLE 2
EFFECTS OF RED LIGHT AS MEASURED BY *IN VITRO*
TRANSCRIPTION AND CHLOROPHYLL ACCUMULATION
IN BARLEY PRIMARY LEAVES MEASURED
AFTER 4.5 H DARKNESS

	Dark Control	Percent dark control	
		VLF	LF
Total transcription	100	113	140
rRNA transcription	100	89	152
LHCP transcription	100	101	205
Reductase transcription	100	75	41
Chlorophyll (after 4 h WL)	100	156	190

then primarily in the low fluence range. The changes
observed in the transcripts for specific genes are
considerably more dramatic. The level of LHCP transcripts
shows no change whatsoever either at 3 or 4.5 hours
following VLF pulses. LF pulses, by contrast, produce a
strong increase, and one which is considerably higher at 3
than at 4.5 h following red light treatment. These results
confirm both the effect on LHCP transcription *in vitro* and
the transient nature of this response reported previously
(17, 26; see 6) for 5-day-old seedlings (at 5 days, the
response was considerably stronger than that reported
here). The results for the reductase, shown here confirm
earlier experiments (17) that red light reduces the level
of *in vitro* transcription of the mRNA, and show in addition
that the effect occurs following both VLF and LF fluences,
but somewhat more in the latter case.
    At the level of mRNA abundance, the situation is
somewhat different. Whereas VLF red light had no apparent
effect on LHCP transcription *in vitro* (Table 1), it
strongly increased the relative abundance of LHCP mRNA as
measured at 3 h. This result is perhaps not surprising

since previous work (17, 26) showed the effect of red light on LHCP transcription to be transient, and of shorter duration following far-red light than red (see 26). As mentioned above, far-red light would be expected to induce VLF responses only (18, 19, 20). Indeed, preliminary experiments with nuclei isolated 1.5 h after VLF fluences of red light consistently yielded *in vitro* transcription rates above those for the dark controls (though owing to the small size of the effect considerable variability was encountered; results not shown). It is thus reasonable to hypothesize that the increase in LHCP mRNA abundance noted 3 hours after VLF red light treatment reflects a transient burst of transcriptional activity following the treatment, a burst which has subsided at 3 h.

Examination of results obtained with the reductase lead to a different conclusion. Here decreases are seen both in *in vitro* transcription and in mRNA abundance at both fluences. However, the effect on transcription is somewhat larger in the VLF range than that on mRNA abundance, whereas in the LF range the reverse is the case. Complete fluence-response curves to be published elsewhere confirm that whereas VLF irradiation markedly decreases reductase transcription *in vitro*, it has a minimal effect on mRNA abundance, as measured at 3 h. By contrast, LF treatment has only a minor additional effect on transcription, but a dramatic effect on mRNA abundance. Hence, although transcription of reductase mRNA *in vitro* shows some control by irradiation *in vivo*, the changes observed are insufficient to account for observed decreases in reductase mRNA abundance, and regulation at the level of mRNA stability must also be occurring.

Clearly phytochrome is affecting more than one step in the sequence of events from transcription on. Colbert *et al.* (27, Quail PH, personal communication) report a similar phenomenon with another mRNA showing down-regulation in the presence of Pfr: that for phytochrome itself. Though both transcription of the mRNA (measured as above in isolated nuclei) and mRNA abundance are decreased rapidly by red irradiation, the decrease in transcription is inadequate to account for the decrease in mRNA abundance.

The changes in chlorophyll accumulation as assayed after 4 h in white light (Tables 1 and 2) are consistant with the changes in LHCP mRNA abundance noted above. There is, however, a possible discrepency which should be noted. Kinetic studies to be published elsewhere indicate that the time courses in darkness for the effect of VLF or LF red light in eliminating the lag phase in chlorophyll synthesis in subsequent white light are different: both the rate of change and the maximum extent of change show a clear dependence on Pfr concentration. The difference in rate is detectable within a few minutes following irradiation. Mosinger *et al.* (26, see 6) found Pfr dependence for the maximum extent of increase in *in vitro* transcription, but little difference in initial rate whether they used saturating red light (LF) or far-red light (equivalent to VLF) to induce the responses.

There are at least two possible explanations for the discrepency. If, as proposed by Bennett (22), extractable chlorophyll is indeed a reliable indication of the amount of LHCP present *in vivo*, then another light-regulated step for accumulation of chlorophyll must be hypothesized. First, since changes in mRNA abundance for the LHCP are consistant with transcriptional changes measured *in vitro*, it is not necessary that such a second regulatory step be at the level mRNA stability; it could equally well appear later in the transduction chain - e. g. on chlorophyll synthesis itself. Chlorophyll synthesis is well known to be limited by δ-aminolevulinic acid (ALA) concentration in the tissue (28), and ALA levels are reported to be under phytochrome regulation both in etiolated *Sinapis* (29) and in etiolated maize seedlings (30). On the other hand, it could be that after an LF dose of red light, LHCP mRNA transcript stability is higher during the white light treatment than after a VLF dose. These and perhaps, other possibilities require further exploration.

Since Mösinger *et al.* used plants one day younger than those used for the present work, these experimental results need confirmation before a firm conclusion can be reached. An added caution is that there is no certainty that the *in vitro* transcription assay used in this, or for that matter in many other studies is indeed a completely accurate reflection of the transcriptional status of the nuclei at the moment of extraction. The potential limitations and problems with the assay are discussed in detail elsewhere (6).

In summary, and to address the questions posed in the Introduction, phytochrome as Pfr probably does regulate the reductase and LHCP genes at the transcriptional level. However, such regulation is not sufficient to account for the kinetic and fluence-response behavior of the various componants of the system. In the case of the reductase (as in the case of phytochrome itself, see 27), an effect of Pfr in decreasing mRNA stability seems likely as well. With the LHCP, the mRNA changes are at least consistent with the measured transcriptional changes. It is the accumulation of chlorophyll itself which is not consistent. In this case there could be a direct effect of Pfr on chlorophyll accumulation itself through regulation of ALA synthesis, an effect which limits greening. Limitation on chlorophyll accumulation would then strictly limit abundance of the LHCP (see 22), regardless of the abundance of mRNA. On the other hand, a differential effect of VLF and LF red light treatment on subsequent mRNA stability in white light can not be ruled out.

˒ACKNOWLEDGEMENTS

The work described in this paper was largely carried while the senior author was on sabbatical leave at the University of Freiburg, Federal Republic of Germany as a Senior U. S. Scientist Awardee of the Alexander von Humboldt-Foundation. This paper is CIW-DPB publication #934.

REFERENCES

1.   Virgin HI (1972). Chlorophyll biosynthesis and phyto-chrome action. In Mitrakos K, Shropshire W Jr (eds): "Phytochrome," New York: Academic Press, p. 371.
2.   Virgin HI (1961). Action spectrum for the elimination of the lag phase in chlorophyll formation in previously dark grown leaves of wheat. Physiol. Plantarum 14:439.
3.   Raven CW, Spruit CJP (1972). Induction of rapid chlorophyll accumulation in dark grown seedlings. I. Action spectrum for pea. Acta Botan. Neerl. 21:219.
4.   Raven CW, Spruit CJP (1972). Induction of rapid chlorophyll accumulation in dark grown seedlings. II. Photoreversibility.
5.   Raven CW, Shropshire, W Jr (1975). Photoregulation of logarithmic fluence response curves for phytochrome control of chlorophyll formation in *Pisum sativum* L. Photochem. Photobiol. 21:423.
6.   Schäfer E, Briggs WR (1986). Photomorphogenesis from signal perception to gene expression. Photobiochem.

Photobiophys. In press.

7. Sharma R (1985). Phytochrome regulation of enzyme activity in higher plants. Photochem. Photobiol. 41:747.

8. Thompson WF, Kaufman LK, Watson, JC (1985). Induction of plant gene expression by light. BioEssays 3:153.

9. Tobin EM, Silverthorne, J (1985). Light regulation of gene expression in higher plants. Annu. Rev. Plant Physiol. 36:569.

10. Silverthorne J, Tobin, EM (1984). Demonstration of transcriptional regulation of specific genes by phytochrome action. Proc. Nat. Acad. Sci. U. S. 81:1112.

11. Stiekema WJ, Wimpee CF, Silverthorne J, Tobin, EM (1983), Phytochrome control of the expression of two nuclear genes encoding chloroplast proteins in *Lemna gibba* L. G-3. Plant Physiol. 72:717.

12. Tobin EM (1981). Phytochrome mediated regulation of messenger RNAs for the small subunit of ribulose-1,5-bisphosphate carboxylase and the light-harvesting chlorophyll *a/b* protein in *Lemna gibba*. Plant Mol. Biol. 1:35.

13. Apel K (1979). Phytochrome-induced appearance of mRNA activity for the apoprotein of the light-harvesting chlorophyll *a/b* protein of barley *(Hordeum vulgare)*. Eur. J. Biochem. 97:183.

14. Apel K (1981). The protochlorophyllide holochrome of barley *(Hordeum vulgare* L.). Phytochrome-induced decrease of translatable mRNA coding for the NADPH: protochlorophyllide oxidoreductase. Eur. J. Biochem. 120:89.

15. Batschauer A, Apel K (1984). An inverse control by phytochrome of the expression of two nuclear genes in barley *(Hordeum vulgare* L.). Eur. J. Biochem. 143:593.

16. Gollmer I, Apel K (1983). The phytochrome-controlled accumulation of mRNA sequences encoding the light-harvesting chlorophyll *a/b*-protein of barley *(Hordeum vulgare* L.). Eur. J. Biochem. 133:309.

17. Mösinger E, Batschauer A, Schäfer E, Apel K (1985). Phytochrome control of *in vitro* transcription of specific genes in isolated nuclei from barley *(Hordeum vulgare)*. Eur. J. Biochem. 147:137.

18. Blaauw OH, Blaauw-Jansen G, van Leeuwen WJ (1968). An irreversible red light-induced growth response in *Avena*. Planta 82:87.

19. Mandoli DF, Briggs WR (1981). Phytochrome control of

two low-irradience responses in etiolated oat seedlings. Plant Physiol. 67:733.218.

20. Mandoli DF and Briggs WR (1982). The photoperceptive
    sites and the   function of tissue light piping in
    photomorphogenesis of etiolated oat seedlings. Plant
    Cell Environ. 5:137.

21. Briggs WR, Mandoli DF, Shinkle JR, Kaufman LS, Watson
    JC, Thompson WF (1985). Phytochrome regulation at the
    whole plant, physiological, and molecular levels. In
    Columbetti G, Lenci F, Song P-S (eds.): "Sensory
    Perception and Transduction in Aneural Organisms," New
    York, London: Plenum, p. 265.

22. Bennett J (1981). Biosynthesis of the light-harvesting
    chlorophyll *a/b* protein. Polypeptide turnover in
    darkness. Eur. J. Biochem. 118:61.

23. Heim B, Schäfer E (1984). The effect of red and
    far-red light in the high irradiance reaction of
    phytochrome (hypocotyl growth in dark-grown *Sinapis
    alba* L.). Plant Cell Environ. 7:39.

24. Moran R, Porath D (1980). Chlorophyll determination in
    intact tissue using *N,N*-dimethylformamide. Plant
    Physiol. 65:478.

25. Willmitzer L, Wagner KG (1981). The isolation of
    nuclei from tissue-cultured plant cells. Exp. Cell
    Res. 135:69.

26. Mösinger E, Batschauer A, Schäfer E, Apel K (1986).
    Comparison of the effects of exogenous native
    phytochrome and in vivo irradiation on in vitro tran-
    scription in isolated nuclei from barley (*Hordeum
    vulgare*). Submitted to Planta.

27. Colbert JT, Christensen AH, Peters WK, Quail PH
    (1985). Regulation of phytochrome gene expression at
    the transcriptional level. Proc. First Int. Congress
    Plant Mol. Biol. (Abstract) p. 23.

28. Beale SI (1978). δ-Aminolevulinic acid in plants: its
    biosynthesis, regulation, and role in plastid develop-
    ment. Annu. Rev. Plant Physiol. 29:95.

29. Masoner M, Kasimer H (1985). Control of chlorophyll
    synthesis by phytochrome I. The effect of phytochrome
    on the synthesis of 5-aminolevulinic acid in mustard
    seedlings. Planta 126:111.

30. Klein S, Katz E, Neeman E (1977). Induction of δ-
    aminolevulinic acid formation in etiolated maize
    leaves controlled by two light reactions. Plant
    Physiol. 60:335.

Molecular Biology of Plant Growth Control, pages 425–439
© 1987 Alan R. Liss, Inc.

STRUCTURAL FEATURES OF THE PHYTOCHROME MOLECULE AND FEEDBACK REGULATION OF THE EXPRESSION OF ITS GENES IN <u>AVENA</u>[1]

Peter H. Quail[+*], Richard F. Barker[‡2], James T. Colbert[+3]
Susan M. Daniels[+4], Howard P. Hershey[+5], Kenneth B. Idler[‡]
Alan M. Jones[+], James L. Lissemore[+]

Department of Botany[+] and Genetics[*]
University of Wisconsin
Madison, WI 53706
and
Agrigenetics Advanced Research Laboratory[‡]
Madison, WI 53716

ABSTRACT Structural studies indicate that the phytochrome molecule is a nonglobular dimer of subunits that consist of two major domains: a larger, globular NH_2-terminal domain containing the single chromophore, possibly housed in a hydrophobic cavity; and a smaller, more open COOH-terminal domain apparently bearing the dimerization site(s). Binding studies with previously isolated monoclonal antibodies indicate that one class, specific for the NH_2-terminal 6 kDa of the phytochrome polypeptide (Type 1), can be subdivided into two groups that recognize at least two spatially separate

[1]This work was supported by NSF grant nos. PCM 8003792 and DMB8302206; USDA SEA grants 59-2551-1-1-774-0, 81-CRCR-1-744-0 and 85-CRCR-1-1578; and Agrigenetics Research Associates; JLL is the recipient of a NSF Graduate Fellowship; JTC supported by NIH Training Grant 5T32GM07215.
[2]Present address: Plant Breeding Institute, Trumpington, Cambridge, CB2 2LQ England.
[3]Present address: Botany Department, Colorado State University, Fort Collins, CO 80523.
[4]Present address: Department of Plant Pathology, University of Wisconsin, Madison, WI 53706.
[5]Present address: Central Research and Development, E. I. du Pont de Nemours & Co., Wilmington, Delaware 19898.

epitopes. One group, designated Type 1A, exhibits 4-
to 5-fold higher affinity for Pr than for Pfr, shifts
the peak of Pfr absorbance from 730 nm to 724 nm, and
induces nonphotochemical reversion of Pfr to Pr.
Another antibody, designated Type 1B has equal affinity
for Pr and Pfr and does not perturb the photoreceptor's
spectral properties. Thus these antibodies identify
two spatially separate subdomains within the
NH_2-terminal 6 kDa of phytochrome: one containing the
Type 1A epitope that undergoes a photoconversion-
induced conformational change and may interact with the
Pfr chromophore; and another containing the Type 1B
epitope with no evidence for involvement in either
process. The rate of decline in phytochrome mRNA
levels is the same in etiolated Avena seedlings exposed
either to a 5 s red light pulse or to continuous white
light, whether assayed as translational activity or as
hybridizable sequence using cDNA probes. Together with
previous data, these results indicate that the response
represents a decrease in phytochrome mRNA abundance as
opposed to altered translatability, and that
phytochrome is the principal, if not the exclusive,
photoreceptor mediating this effect.

INTRODUCTION

Phytochrome controls plant development via altered
expression of selected genes. Recent studies provide direct
evidence that regulation occurs at the transcriptional level
for those genes examined (15, 18, 20). Our approach to
investigating the molecular basis for this regulation
involves defining the structural properties of the
photoreceptor molecule (13) and characterizing the feedback
control that phytochrome exerts over the expression of its
own gene (3-5).

PRIMARY AND SECONDARY STRUCTURE

The complete amino acid sequence of Avena phytochrome
has been deduced recently from cloned nucleotide sequences
(9). The polypeptide is 1128 amino acids long with a single
chromophore-attachment site on the NH_2-terminal side at
Cys-321. Secondary structure analysis according to Chou and
Fasman (1) predicts that the chromophore is attached at a

ß-turn in a mildly hydrophilic region. Hydropathy analysis (12) reveals extensive hydrophobic regions adjacent to the chromophore-attachment site (9), consistent with a possible role in forming the hydrophobic cavity postulated from chemical probe data to house the tetrapyrrole (7, 19).

TERTIARY AND QUATERNARY STRUCTURE

Controlled proteolytic digestion indicates that the native phytochrome polypeptide is folded into two major operational domains: A larger, globular, NH_2-terminal domain of ~ 74 kdaltons (kDa) that carries the chromophore, and a smaller, possibly more extended, COOH-terminal domain of ~ 55 kDa that appears to carry the contact site(s) for dimerization of the photoreceptor (Fig. 1) (10, 11, 14, 21-24).

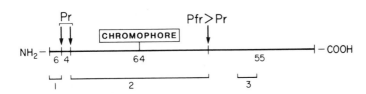

Fig. 1. Peptide map of 124-kDa <u>Avena</u> phytochrome. Positions of the major proteolytic cleavage sites are indicated by the arrows with the molecular masses (kDa) of the derived peptides indicated below. The NH_2-terminal sites are cleaved almost exclusively in the Pr form; the central site is preferentially cleaved as Pfr. Polypeptide segments bearing epitopes for Type 1, 2 and 3 monoclonal antibodies are indicated by square brackets.

Part of the evidence supporting this notion is presented in Fig. 2. Purified 124-kDa <u>Avena</u> phytochrome has been subjected to partial tryptic digestion, size-fractionated by size exclusion chromatography (SEC) and the individual fractions subjected to SDS-PAGE, electroblotted and probed with polyclonal antibodies. The undigested 124-kDa species migrates, under the non-denaturing conditions of the SEC column, substantially ahead of the position expected of a globular 248-kDa molecule. These and other data indicate that the native photoreceptor is a nonglobular dimer with a

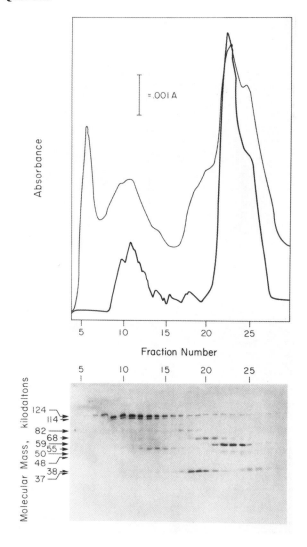

Fig. 2.   SEC of trypsin-digested 124-kDa phytochrome
purified from Avena.   Top: Absorbance profile from
TSK3000SW column under non-denaturing conditions. A_{656}
(———); A_{280} (————).   Bottom: Immunoblot analysis
of column fractions following SDS-PAGE.   Blot probed with
polyclonal antibodies against 124-kDa Avena phytochrome.

frictional ratio of 1.3 (10). The 82-,68- and 59-kDa digestion products migrate on the SEC column as globular monomers of the size indicated by SDS-PAGE. That these species are released from the NH_2-terminal end of the polypeptide is indicated by the associated chromophore, detected spectroscopically (top panel, Fig. 2), and reactivity toward our Type 2 monoclonal antibodies that bind to an epitope located in the NH_2-terminal domain (6, 10). The migration of the 55- and 37-kDa products ahead of their expected monomeric size on the SEC column is consistent with these fragments having remained dimerized after digestion. Their reactivity with our Type 3 monoclonal antibodies confirms that they are derived from the COOH-terminal domain (6), and indicates that the dimerization site is within 40 kDa of the COOH-terminus (10).

PROBING PHYTOCHROME STRUCTURE WITH MONOCLONAL ANTIBODIES

A complete understanding of the molecular mechanism of phytochrome action will require characterization of the structural changes that occur upon Pr ⇄ Pfr transformation. One approach to identifying the regions of the molecule undergoing conformational changes is to test for phototransformation-induced alterations in binding of site-specific probes to the photoreceptor.

We have generated three classes of monoclonal antibodies, designated Types 1, 2 and 3, which are directed against epitopes located, respectively, within 6 kDa, between 10 and 74 kDa, and between 88 and 97 kDa from the NH_2-terminus (Fig. 1) (6). Competitive binding assays indicate that the epitopes for each class of antibody are topologically separate in the native molecule as well as within the primary sequence (Fig. 3). The complete cross-competition of the two Type 3 antibodies with one another indicates recognition of the same or spatially very close epitopes. On the other hand, subclasses of only partially cross-competing antibodies within each of the Type 1 and 2 classes, suggest recognition of different epitopes within the relevant polypeptide segment. For example, Type 1 clone 1.3G7F reduces the binding to phytochrome of Type 1 clone 1.9B5A by only ~ 30%, suggesting sufficient spatial separation of the epitopes to result in only partial steric hindrance (Fig. 3). Similar data have been obtained in other laboratories (5, 16).

The majority of our Type 1 monoclonal antibodies (now

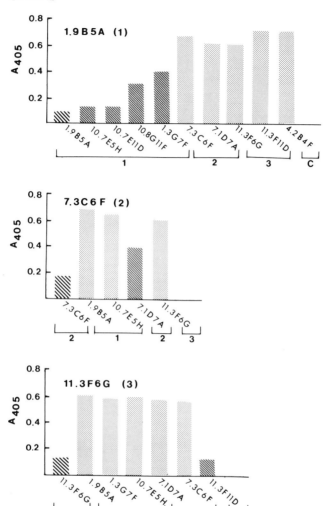

Fig. 3. Competitive ELISA binding assay for discrimination
between epitopes recognized by different monoclonal
antibodies according to Cordonnier et al. (5).   The test
antibodies are indicated in bold type at top left of each
panel: Top, 1.9B5A (Type 1); Middle, 7.3C6F (Type 2);
Bottom, 11.3F6G (Type 3).   The competing antibody in each
case is indicated below each histogram and designated Type
1, 2 or 3 below square brackets.   Low A_{405} readings signify
effective competition for the phytochrome epitope; High A_{405}
readings signify ineffective competition.

designated Type 1A) bind to Pr with a 4- to 5-fold higher affinity than to Pfr (Fig. 4, top panel), confirming the observations of Cordonnier et al. (5). This result indicates that the epitope for these antibodies, located within the 6-kDa NH_2-terminal segment, is in a section of the polypeptide that undergoes a change in conformation or accessibility upon phytochrome phototransformation. This conclusion is consistent with previous demonstrations that proteolytically vulnerable sites that are exposed in this segment in the Pr form become inaccessible in the Pfr form (14, 21-24). Thus the polypeptide segments containing the proteolytically-sensitive site and the Type 1A epitope are candidates for involvement in the molecular action of phytochrome.

In contrast, the Type 1 monoclonal antibody 1.3G7F (now designated Type 1B), which recognizes an epitope spatially separate from that for most of our other Type 1A antibodies (Fig. 3), binds equally well to Pr and Pfr (data not shown). Thus the epitope for antibody 1.3G7F encompasses amino acid residues within the 6-kDa NH_2-terminal region that apparently are not directly involved in a conformational rearrangement. Likewise, all of our Type 2 and 3 monoclonal antibodies bind with equal affinity to Pr and Pfr (Fig. 4) indicating that these epitopes are also not involved in structural changes. Cordonnier et al. (5) reported similar findings for their antibodies equivalent to our Type 2.

The Type 1A monoclonal antibodies which bind differentially to Pr and Pfr also induce a shift in the absorbance peak of Pfr (Fig. 5) and accelerate dark reversion (Fig. 5, 6), confirming the data of Cordonnier et al. (5). These observations indicate that the binding of these antibodies distorts the local polypeptide structure in a way that interferes with normal Pfr-chromophore-protein interaction, a result consistent with previous studies that have established the importance of the 6-kDa NH_2-terminal segment in this interaction (11, 23, 24). The Type 1B antibodies, in contrast, have minimal effects on the spectral properties of the photoreceptor (unpublished data; Fig. 6). Together the data support the notion that the localized segments within the NH_2-terminal 6-kDa of the polypeptide that undergo phototransformation-induced conformational changes are involved in interactions with the Pfr-chromophore, whereas other segments within the 6 kDa that do not exhibit detectable structural changes may not be involved in these interactions. A similar conclusion might be drawn for the domains bearing the Type 2 and 3 epitopes

MONOCLONAL ANTIBODY AFFINITY FOR **Pr** AND **Pfr**

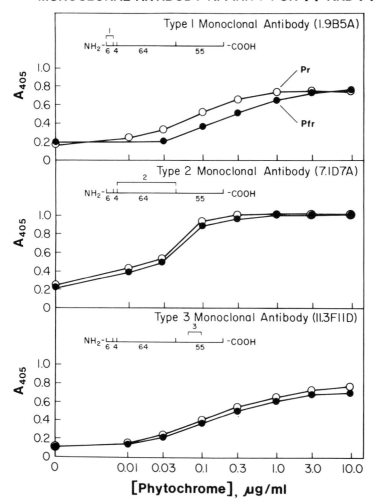

Fig. 4. Relative affinities of monoclonal antibodies for Pr
(o) and Pfr (●) assayed by ELISA according to Cordonnier
et al. (5). Phytochrome was added as Pr or Pfr at the
indicated concentrations to microtiter plate wells
containing immobilized Type 1A (top), Type 2 (middle) or
Type 3 (bottom) monoclonal antibodies. The A_{405} reading is
proportional to the amount of phytochrome bound by the
monoclonal antibody.

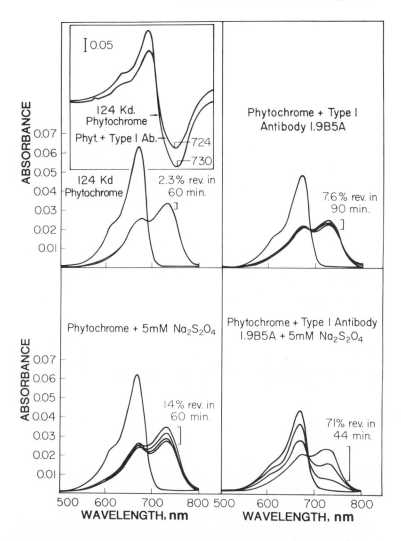

Fig. 5. Phytochrome absorbance spectra and reversion
behavior as influenced by Type 1A antibody 1.9B5A. Inset:
Phytochrome difference spectra in the absence (upper curve)
and presence (lower curve) of 1.9B5A. Main panels:
Absorbance spectra of Pr and of Pfr after increasing
intervals in the dark (see Fig. 6) in the absence (left
panels) or presence (right panels) of 1.9B5A, and in the
absence (top panels) or presence (bottom panels) of 5mM
$Na_2S_2O_4$, an accelerant of dark reversion. Experiments
performed according to Cordonnier et al. (5).

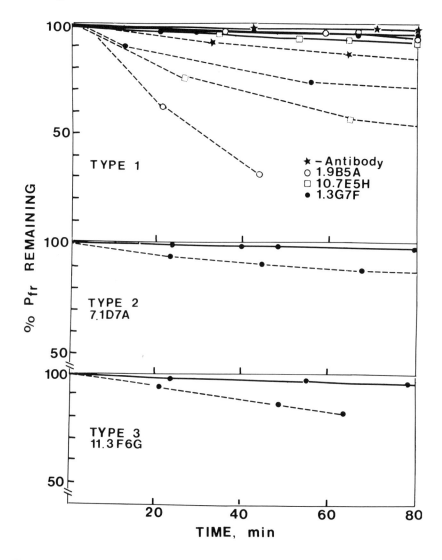

Fig. 6.   Time-course of dark-reversion of Pfr to Pr at 4°C
as influenced by Type 1 (top), Type 2 (middle) and Type 3
(bottom) monoclonal antibodies in the absence (solid curves)
or presence (dashed curves) of 5mM $Na_2S_2O_4$.

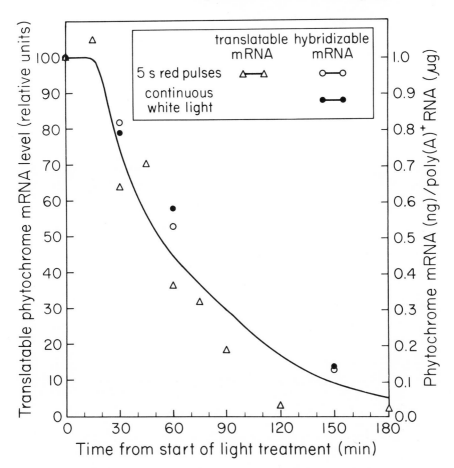

Fig. 7. Time-course of the change in <u>Avena</u> phytochrome mRNA
levels following irradiation of etiolated seedlings with
either a saturating pulse of red light at time zero (Δ, o)
or continuous white light (•). Data for translatable mRNA
(Δ) from Colbert et al. (3) and for hybridizable mRNA (o, •)
from Colbert et al. (4).

as these antibodies neither bind differentially to Pr and
Pfr nor perturb the spectral properties examined
(unpublished data; Fig. 6).

FEEDBACK DOWN-REGULATION OF PHYTOCHROME-GENE EXPRESSION

Figure 7 shows that the level of phytochrome mRNA declines rapidly after an apparent lag of about 15 min following exposure of etiolated _Avena_ seedlings to light. The kinetics of this response are the same, within the measurement error, whether the level of phytochrome mRNA that is translatable or hybridizable to a cloned phytochrome cDNA sequence is followed. This result establishes that it is the steady-state abundance, rather than the translatability of the phytochrome mRNA that is altered. Red/far-red reversibility experiments indicate that phytochrome itself induces this decrease in abundance (3, 4). The rate of decline in the phytochrome mRNA level is the same following a 5 s red light pulse and in continuous white light (Fig. 7). This result indicates that phytochrome is the predominant, if not the exclusive, photoreceptor involved in the response. Evidence from _in vitro_ run-off transcription experiments with isolated nuclei, indicates that regulation is exerted at least partially at the transcriptional level (2).

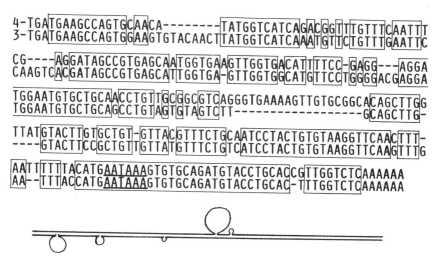

Fig. 8. 3'-untranslated nucleotide sequences from Types 3 and 4 _Avena_ mRNA (9) realigned to maximize homology (boxes). Missing nucleotides indicated by dashes; putative poly(A)+ addition signals underlined. Below is a schematic diagram indicating regions of sequence homology.

Restriction enzyme map polymorphism among phytochrome cDNA clones (9) and genomic Southern blot analysis (8) indicate that at least four phytochrome genes are transcribed in etiolated Avena. Comparison of two of these sequences indicates ~ 98% homology in the coding and 5'-untranslated regions. Colinear comparison of the 3'-untranslated regions of these sequences indicates only 34% homology (9). Realignment to account for apparent deletions or substitutions (Fig. 8) results in an overall homology of 72%. These differences in the 3'-untranslated sequences offer the potential for preparing gene-specific probes for examining the relative levels of expression of the individual phytochrome genes.

ACKNOWLEDGEMENTS

We thank Hiram Sanchez for technical assistance in producing and analyzing monoclonal antibodies; Michael Murray, Alan Christensen, James Tokuhisa and Richard Vierstra for helpful discussions; and Michael Gribskof, Fred Blattner and John Devereaux for assistance in performing computer analysis of amino acid and nucleotide sequences.

REFERENCES

1. Chou, PY, Fasman, GD (1978). Empirical predictions of protein conformation. Ann Rev Biochem 47:251.
2. Colbert, JT (1985). Regulation of phytochrome gene expression in Avena sativa L. seedlings. Ph.D. Thesis, University of Wisconsin.
3. Colbert, JT, Hershey, HP, Quail, PH (1983). Autoregulatory control of translatable phytochrome mRNA levels. Proc Natl Acad Sci USA 80:2248.
4. Colbert, JT, Hershey, HP, Quail, PH (1985) Phytochrome regulation of phytochrome mRNA abundance. Plant Mol Biol 5:91.
5. Cordonnier, MM, Greppin, H, Pratt, LH (1985). Monoclonal antibodies with differing affinities to the red-absorbing and far-red-absorbing forms of phytochrome. Biochemistry 24:3246.
6. Daniels, SM, Quail, PH (1984). Monoclonal antibodies to three separate domains on 124-kilodalton phytochrome from Avena. Plant Physiol. 76:622.
7. Hahn, TR, Song, PS, Quail, PH, Vierstra, RD (1984).

Tetranitromethane oxidation of phytochrome chromophore as a function of spectral form and molecular weight. Plant Physiol 74:755.

8. Hershey, HP, Barker, RT, Colbert, JT, Lissemore, JL and Quail, PH 1985. Phytochrome: Molecular properties, feedback regulation of mRNA levels and genomic clone isolation. In van Vloten-Doting L, Groot, G S P, Hall T (eds): "Molecular Form and Function of the Plant Genome", Plenum Press, p 101.

9. Hershey, HP, Barker, RF, Idler, KB, Lissemore, JL, Quail, PH (1985). Analysis of cloned cDNA and genomic sequences for phytochrome: Compelete amino acid sequence for two gene products expressed in etiolated Avena. Nuc Acids Res 13:8543.

10. Jones, AM, Quail, PH (1986). Quaternary structure of 124-kDa phytochrome from Avena sativa L. Biochemistry in press.

11. Jones, AM, Vierstra, RD, Daniels, SM, Quail, PH (1985). The role of separate molecular domains in the structure of phytochrome from Avena sativa. Planta 164:501.

12. Kyte, J, Doolittle, RF (1982). A simple method for displaying the hydrophobic character of a protein. J Mol Biol 157:105.

13. Lagarias, JC (1985). Progress in the molecular analysis of phytochrome. Photochem Photobiol 42:811.

14. Lagarias, JC, Mercurio, FM (1985). Structure function studies on phytochrome. Identification of light-induced conformational changes in 124-kDa Avena phytochrome in vitro. J Biol Chem 260:2415.

15. Mösinger, E, Batschauer, A, Schäfer, E, Apel, K (1985). Phytochrome control of in vitro transcription of specific genes in isolated nuclei from barley (Hordeum vulgare). Eur J Biochem 147:137.

16. Nagatani, A, Yamamoto, KT, Furuya, M, Fukumoto, T, Yamashita, A (1984). Production and characterization of monoclonal antibodies which distinguish different surface structures of pea (Pisum sativum co. Alaska) phytochrome. Plant Cell Physiol 25:1059.

17. Quail, PH, Colbert, JT, Hershey, HP, Vierstra, RD (1983). Phytochrome: molecular properties and biogenesis. Phil Trans R Soc Lond B303:387.

18. Silverthorne, J, Tobin, EM (1984). Demonstration of transcriptional regulation of specific genes by phytochrome action. Proc Natl Acad Sci USA 81:1112.

19. Song, PS (1985). The molecular model of phytochrome deduced from optical probes. In Blauer, G, Sund, H

(eds) "Optical Properties and Structure of
Tetrapyrroles", Berlin: Walter D. Gruyter & Co., p 331.

20. Tobin, EM, Silverthorne, J (1985). Light regulation of
expression in higher plants. Ann Rev Plant Physiol 36:
569.

21. Vierstra, RD, Cordonnier, MM, Pratt, LH, Quail, PH
(1984). Native phytochrome: immunoblot analysis of
relative molecular mass and in-vitro proteolytic
degradation for several plant species. Planta 160:521.

22. Vierstra, RD, Quail, PH (1982). Native phytochrome:
inhibition of proteolysis yields a homogeneous monomer
of 124 kilodaltons from Avena. Proc Natl Acad Sci USA
79:5272.

23. Vierstra, RD, Quail, PH (1983). Proteolysis alters the
spectral properties of 124 kilodalton phytochrome from
Avena. Planta 156:158.

24. Vierstra, RD, Quail, PH (1985). Spectral
characterization and proteolytic mapping in native
120-kilodalton phytochrome from Cucurbita pepo L.
Plant Physiol 77:990.

Molecular Biology of Plant Growth Control, pages 441–447
© 1987 Alan R. Liss, Inc.

THE HYDROXYPROLINE-RICH GLYCOPROTEINS OF PLANTS[1]

J. E. Varner

Department of Biology, Washington University
St. Louis, Missouri 63130

ABSTRACT Higher plants have at least three kinds of
hydroxyproline-rich glycoproteins(1). Two of these,
the extensins and the arabinogalactan proteins, are
localized in the extracellular matrix. Both the
extensins and the arabinogalactan proteins are coded
by multigene families and for both gene expression
appears to be tissue specific. Wounding (2,3),
infection (4), oligosaccharide elicitors (5), and
ethylene (5) increase gene expression and/or
accumulation of the extensins. One extensin molecule
from aerated carrot slices (3) is well characterized
as is one extensin gene from carrots (6). The most
abundant tryptic peptides of two different extensins
from tomato contain the sequence $-SerHyp_4-$ (7). Base
sequences encoding the $-SerPro_4-$sequence occur
repetitively in the carrot genomic sequence, in a
tomato partial genomic sequence and in tomato and
petunia cDNA sequences. The hydroxylation and the
glycosylations that convert the proline-rich peptide
product of the extensin gene into the hydroxylated and
glycosylated extensin monomer are only partially
characterized. The details of how the secreted
extensin monomers are associated with each other and
with other cell wall components to comprise the wall
architecture remain to be discovered. The amino acids
(lysine, tyrosine, histidine and threonine) which
occur frequently in the extensin sequences between the
$-SerPro_4-(SerHyp_4-)$ repeats have sidechains rich in
functional groups that could form electrovalent,
reversible covalent, and irreversible covalent links
with other wall components.

[1]This work was supported by a grant from NSF,
PCM8104516, a contract from DOE, FG02-84ER13255, and
an unrestricted grant from Monsanto.

INTRODUCTION

Extensins are hydroxyproline-rich glycoproteins found most abundantly in the cell walls of dicotyledonous plants (1). These glycoproteins are assumed to play a role in the structure of plant cell walls and may therefore be important in controlling growth, development, and disease resistance. Extensins accumulate in carrot root cell walls upon wounding (2) and become insolubilized in cell walls with time (8).

The soluble extensin produced in wounded carrot root tissue has been well characterized both biologically and chemically (2). It has a molecular size of ~86 kDa and consists of one-third protein and two-thirds carbohydrate; the estimated size of the polypeptide is 30 kDa. Six amino acids (hydroxyproline, serine, histidine, tyrosine, lysine, and valine) comprise 95% of the polypeptide backbone (2). The carbohydrate is composed largely of arabinose and a small amount of galactose (3). The carrot extensin is 100% in the polyproline II helical conformation, as determined by circular dichroism spectra, while the deglycosylated molecule is only 50% in the polyproline II helix (3). It appears that the oligosaccharides help in forming and stabilizing the helical structure.

The extensin of carrot roots is insolubilized in the wall with time after it is secreted (9). This insolubilization is paralleled by the formation of isodityrosine (8,9,10). Although it seems reasonable to suppose (9) that the formation of this diphenylether link between adjacent extensin molecules is responsible for the insolubilization, the only isodityrosime link so far identified is intramolecular (11). The work from Lamport's laboratory (12) made it clear that extensin has the repetitive sequence -Ser Hyp Hyp Hyp Hyp-, and that some of the serines are glycosylated with a single galactose and that many of the hydroxyprolines are glycosylated with one, two, three or four arabinose residues. The determination of the complete amino acid sequence of extensin is difficult because of the high imino acid content, the repetitive sequences and the many post-translational modifications.

Through the use of cDNA clones (6), isolated from a cDNA library constructed from wounded carrot root mRNA, as probes extensin genomic clones were isolated (13). One of these clones was characterized and found to contain an open reading frame coding for a proline-rich peptide of 306 amino acids containing an apparent signal peptide

sequence (13). In this sequence -Ser Pro Pro Pro Pro-
occurs 25 times and -Ser Pro Pro Pro Pro Thr Pro Val Tyr
Lys- occurs 8 times. Because -Ser Hyp Hyp Hyp Hyp- (12)
and -Ser Hyp Hyp Hyp Hyp Thr Hyp Val Tyr Lys- (7) are
abundant peptides in tomato cell walls it would appear
that the carrot genomic sequences encode carrot extensin.
It is therefore of interest to examine this derived amino
acid sequence in some detail.

DEDUCTIONS ABOUT THE ROLE(S) OF EXTENSIN

The repetitive structure of the carrot extensin
(Figure 1, (13)), the high degree of glycosylation of
extensin as isolated from the wall (3) its extended
structure as indicated by CD spectra (3) and by electron
microscopy (3,14) and the large number of lysine residues
interspersed between the -Ser Hyp Hyp Hyp Hyp- regions
that could interact electrovalently with the uronic acid
residues of the cell wall pectins indicate an important
role for extensin in determining cell wall architecture.
Precise elucidation of these in muro interactions would
appear to require ingenious application of biophysical
methods. In addition to the ε-amino groups of the lysine
residues, other regularly interspersed side chain
functional groups include 1) the hydroxyl group of
tyrosine which could form covalent intermolecular cross-
links (14) as well as intramolecular links (11), 2) the
sulfur of methionine, 3) the carboxyl group of glutamate,
4) the hydroxyl groups of threonine and of the
unglycosylated serines and hydroxyprolines, and 5) the
nitrogen of the imidazole ring of histidine. Mild
oxidation of the imidazole ring can open the ring to form
and aldehyde group (15) which conceivably could form a
Schiff base with the ε-amino grop of a lysine residue.
Aldehyde groups might also be formed from the action of
glycosidases and from the action of ascorbic acid in the
presence of oxygen. How many of the possible interactions
of these functional groups with other wall components are
physiologically important remains to be established.
Nonetheless these possible interactions are of interest
because of the increased accumulation of hydroxyproline-
rich proteins in response to infection (4), ethylene (5)
and oligosaccharide elicitors (5,16).
Extensin dimers isolated from aerated carrot slices
and visualized by electron microscopy most frequently have
the form of two monomers joined near the ends (14).
Higher polymers of extensin similarly formed could

MGRIARGSKMSSLIVSLLVVLVSLNLASGTTA

```
KYTYS SPPPPEH
      SPPPPEH
      SPPPPYHYE
      SPPPPKH
      SPPPPTPVYKYK
      SPPPPMH
      SPPPPYHFE
      SPPPPKH
      SPPPPTPVYKYK
      SPPPPKH
      SPAPVHHYKYKY
      SPPPPTPVYKYK
      SPPPPKH
      SPAPGHHYKYK
      SPPPPKH
      FPAPEHHYKYKYK
      SPPPPTPVYKYK
      SPPPPTPVYKYK
      SPPPPKH
      SPAPVHHYKYK
      SPPPPTPVYK
      SPPPPEH
      SPPPPTPVYKYK
      SPPPPMH
      SPPPPTPVYKYK
      SPPPPMH
      SPPPPVY
      SPPPPKHHYSYT
      SPPPPHHY
```

FIGURE 1.   Derived sequence of carrot extensin. The
entire derived sequence is shown in an arrangement
that emphasizes the repetitive character and the large
number of amino acid residues with side chain
functional groups that could interact with other wall
components.

organize a large volume with minimal extensin.

From the facts and speculations mentioned so far it should be clear that we need to know more about 1) the structure of the extensins, 2) the control of their synthesis and secretion, and 3) their in muro interactions before we can assign them a precise role in growth and development.

Further I wish to mention that in at least some cell walls there are significant accumulations of other proteins that may have structural roles. Aerated carrot slices accumulate in the cell walls a protein (Mary Tierney, private communication) corresponding to the cDNA clone that encodes a peptide having many repeats of -YTPPV- and -HKPPV- (5). Whether any of these prolines is hydroxylated is not yet known. In addition, a fraction has been extracted from pumpkin seed coat cell walls (Gladys Cassab, private communication) that is 47% glycine. And, in addition to the hydroxyproline-rich glycoprotein already reported (17) to be present in soybean seedcoats, there is also a glycine-rich protein (Gladys Cassab, private communication). In the epidermal cell walls of oat coleoptiles 25% of the amino acid content is glycine (David Pope, private communication). Finally, one of the richest sources of cell wall hydroxyproline yet found is the cortex of the soybean root nodule (Gladys Cassab, private communication).

REFERENCES

1. McNeil M, Darvill AG, Fry SC, Albersheim P (1984). Structure and function of the primary cell walls of plants. Ann Rev Biochem 53:625.
2. Stuart DA, Varner JE (1980). Purification and characterization of a salt-extractable hydroxyproline-rich glycoprotein from aerated carrot discs. Plant Physiol 66:787.
3. van Holst GJ, Varner JE (1984). Reinforced polyproline conformation in a hydroxproline-rich cell wall glycoprotein from carrot root. Plant Physiol 74:247.
4. Esquerre-Tugaye MT, Lafitte C, Mazau D, Toppan A, Touze A (1979). Cell surfaces in plant-microorganism interactions. II. Evidence for the accumulation of hydroxyproline-rich glycoproteins in the cell wall of diseased plants as a defense mechanism. Plant Physiol 64:320.

5. Esguerré-Tugaye MT, Mazau D, Pélissier B, Roby D, Rumeau D, Toppan A (1985). Induction by elicitors and ethylene of proteins associated to the defense of plants. In Key JS, Kosuge TS (eds): "Cellular and Molecular Biology of Plant Stress," New York: Alan R. Liss, p 459.

6. Chen J, Varner JE (1985). Isolation and characterization of cDNA clones for carrot extensin and a proline-rich 33 kd protein. Proc Natl Acad Sci USA 82:4399.

7. Smith JJ, Muldoon EP, Willard JJ, Lamport DTA (1986). Tomato extensin precursors P1 and P2 are highly periodic structures. Plant Physiol In Press.

8. Cooper JB, Varner JE (1984). Cross-linking of soluble extensin in isolated cell walls. Plant Physiol 76:414.

9. Cooper JB, Varner, JE (1983). Insolubilization of hydroxproline-rich cell wall glycoprotein in aerated carrot root slices. Biochem Biophys Res Commun 112:161.

10. Fry SC (1982). Isodityrosine, a new cross-linking amino acid from plant cell wall glycoprotein. Biochem J 204:449.

11. Epstein L, Lamport DTA (1984). An intramolecular linkage involving isodityrosine in extensin. Phytochem 23:1241.

12. Lamport DTA (1980). Structure and function of plant glycoproteins. In Preiss J (ed): "The Biochemistry of Plants". New York: Academic Press 3:501.

13. Chen J, Varner JE (1985). An extracellular matrix protein in plants: characterization of a genomic clone for carrot extensin. Embo J 4:2145.

14. Stafstrom JP, Staehelin LA (1986). Cross-linking patterns in salt-extractable extensin from carrot cell walls. Plant Physiol. In Press.

15. Levine RL (1985). Covalent modification of proteins by mixed function oxidation. In "Current Topics in Cellular Regulation." New York: Academic Press 27:305.

16. Showalter AM, Bell JN, Cramer CL, Bailey JA, Varner JE, Lamb CJ (1985). Accumulation of hydroxyproline-rich glycoprotein mRNAs in response to fungal elicitor and infection. Proc Natl Acad Sci USA 82:6551.

17. Cassab GI, Nieto-Sotelo J, Cooper JB, van Holst
     GJ, Varner JE (1985). A developmentally regulated
     hydroxproline-rich glycoprotein from the cell
     walls of soybean seed coats. Plant Physiol
     77:532.

Index

Date Due

			UML 735